# The Breakthrough HABITAT AND EXHIBIT MANUAL

*by*
*the editors of*
*Breakthrough Magazine:*

PUBLISHER and EDITOR
*Bob Williamson*

ART DIRECTOR and MANAGING EDITOR
*Ken Edwards*

ASSOCIATE EDITOR
*Jim Hall*

ASSOCIATE EDITOR
*Dan Blair*

# DEDICATION

God has blessed my life enormously in so many ways that I cannot begin to tell you. His benevolence is "impossible" for me to understand; for no one knows better than I, the flaws, weaknesses, and frailties of my human nature. God's loving kindness and generosity have never been more evident in my life than with His gift of allowing me to know and love Jennie and Shelby Teague. I cannot find adequate words to describe the happiness that I have enjoyed because of them. Without them, this book, *Breakthrough* Publications, and Wildlife Artist Supply Company would have never been. Their steadfast optimism, encouragement, and love have helped guide me through both the good, happy times and the painful, dark storms of life. Although I cannot begin to comprehend why God has allowed me such a wonderful blessing, I will be eternally grateful.

—Bob Williamson

This manual is compiled from information from many sources. The recommendations, procedures and precautions are presented in good faith as the results of experimentation and information from reputable wildlife artists and manuafacturers. Due to the fact that we have no control over the actual application of the information provided, the publisher and the authors disclaim any responsibility for the results, including damage by injury, whether or not caused by using the products, techniques, recommendations or suggestions referred to herein. Nor can the publisher or the authors be held responsible for the value of work alleged to be spoiled by the use of a product, technique, recommendation, or suggestion contained herein. It is the user's responsibility to make sure that products and techniques are suitable for their particular requirements.



First Edition published November 1986.
Printed in the United States of America.
Published by *Breakthrough* Publications, P. O. Box 1330, Loganville, GA 30249.

ISBN 0-925245-07-0
BP1007
The *Breakthrough* Habitat and Exhibit Manual.
Written by Bob Williamson, Jim Hall and Dan Blair.
Edited and designed by Ken Edwards.
Principal photography by Jim Hall and Ken Edwards.
Additional photography and illustrations by Wayland Adams, Sue Baker, Mark Belk, Bob Berry, Dan Blair, Terri Chidester, Bob Elzner, Alan Gaston, Jim Hass, Ralph Lehrman, Kelly Seibels, Tom Sexton and Erica Smoker.
Typesetting/formatting by Jean Coleman and Terri Chidester.
Halftones by Erica Smoker.

Rendering on cover by Bob Elzner.
Cover photo by Ken Edwards.

# CONTENTS

CHAPTER 1

# The Most Important Chapter: Safety!

*In a split second, a moment of inattention or carelessness can cause a tragedy that will be mourned for decades. Bob Williamson discusses the various harmful effects of the chemicals and tools used in habitat construction.*

***by Bob Williamson***

Your loved ones' health as well as your own health and/or life can be destroyed by being careless just "one" time. Habitat and exhibit building requires that the artist use various chemicals and tools that are potentially harmful to one's health IF USED INCORRECTLY! There are many common chemicals used in our daily lives that can be just as dangerous if used incorrectly, i.e., gasoline, charcoal lighter, etc. It has been accurately said that firearms are not any more dangerous than an empty water pipe if used in a safe, proper manner. Accidents occur because often people will not pay attention to safety rules and handle potentially dangerous items carelessly.

During my many years in the coatings industry, I have witnessed several very bad accidents. Without a single exception, all could have been avoided by observing some simple safety procedures. Accidents are not very forgiving—all too often there is no warning and no second chance. Accidents strike as fast as lightning and sometimes can be very cruel!

I will list three such accidents that could have been *"very"* easily avoided but weren't. One employee lost his eyesight while using a bench grinder. For some reason, as he was using it to sharpen a screwdriver, the stone grinding wheel literally exploded. A piece of it hit him in "one" eye. The damage and shock was so severe that he became totally and permanently blind in "both" eyes. He had a pair of safety glasses in his back pocket at the time of the accident. The bench grinder was originally equipped with a safety shield; however, the shield had been removed *because the workers didn't like it "in the way."*

On another occasion an employee was heating and melting some wax on a hot plate in an area that a five gallon bucket of flammable solvent (lacquer thinner) was carelessly stored. The bucket of solvent was accidentally tipped over and it immediately caught fire when it came in contact with the hot plate. Instantly a raging wall of fire trapped the employee in the corner of the building where he was working. A quick-thinking foreman grabbed a fire extinguisher and ran to the scene and was able to put the fire out in about five minutes time. The quick action of the foreman saved his life, but in this short period of time, the employee was changed for the rest of his life. He was screaming from the pain. Layers of burned flesh were literally hanging in stringy globs off of his arms and legs and huge blisters covered his face and body. His hair and clothes had been burned off of his body.

This employee stayed in the hospital 9 months. He endured several painful operations and skin grafts, yet is still so badly scarred that he doesn't even look like the same person. All this because of a careless oversight that could have been easily avoided.

A final accident that occurred sounds almost comical but was serious to the employees that endured it and it could have been much worse. Two employees were filling 55 gallon drums of paint. There was not enough paint to completely fill the last drum that they were working on. They needed to estimate how many gallons were in the drum so they looked through the bung opening in the drum and it was dark in there. One of them had the "bright" idea to use his lighter to look inside the drum. Guess what happened? You're right! It exploded and both employees were knocked back about five feet. The drum lid went flying up to the ceiling like a flying saucer.

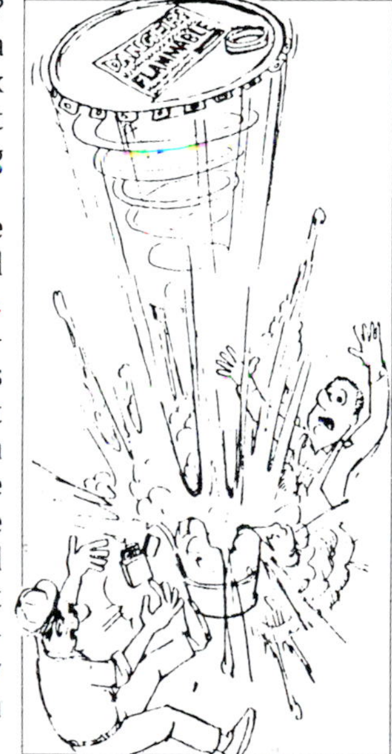


Fortunately, the employees were not critically injured but they did suffer burns on their arms and face.

The fire inspector that investigated the accident shook his head and looked incredulously at the lid of the drum that had blown up. There, less than 1" from the opening of the drum that the lighter was held over, was a red flammable warning sticker that read "Danger, This material is extremely flammable. Keep away from all heat, sparks, or open flames, etc . . ."

Don't let it happen to you! Use chemicals, tools, and equipment safely. Virtually all accidents such as fires, inhalation of dangerous fumes, allergic reactions, explosions, and the like can be avoided if the proper safety precautions and equipment are used. Don't think that you, somehow innately, know how to use every product or tool on the market. "No one" is "born" knowing such information. Don't "naively" think that a serious accident cannot happen to you. A foolish lack of regard for safety can, in an "instant," blind you or your loved ones for life, burn your studio "to the ground," ruin your health, or actually cause death.

Keep in mind also that in addition to yourself, "every" employee in your studio needs to observe these same precautions and know exactly how to use every chemical and tool that they come in contact with. All necessary safety equipment should be provided and used or the project should be delayed until it can be provided.

# Guide to Handling Chemicals Safely

Columbus Adhesive and Chemical Company recently published a Guide to handling chemicals safely. We were granted permission to reprint this excellent information in this chapter. We sincerely appreciate their making this valuable information available.

## Product Identification and Knowledge Safety Rules

1. Read and understand manufacturer's directions.
2. Know health and safety hazards of chemicals handled.
3. Handle and store chemicals properly.
4. Take all safety precautions necessary.
5. Know the first aid procedures for each chemical handled.
6. Know the location of all safety equipment and how to use it.
7. Take special precautions during reproductive years.

*Chemicals can be dangerous if improperly handled or used. Exposure to some chemicals may affect men and women differently or at different rates. Special consideration must be given to pregnancy and transfer of toxic chemicals through mother's milk. Some chemicals can cause deformities in offspring even at low concentrations.*

- **Clearly identify a chemical product before use.**
- **Read and understand the supplier's directions for use and handling.**
- **Follow recommended safety procedures in use and handling.**
- **Use proper labeling if chemicals are transferred to new containers.**

## Handling and Mixing Safety Rules

1. Know the hazards of the chemicals used.
2. Wear appropriate protective clothing.
3. Read label directions and follow them.
4. Work in well ventilated areas.
5. Store and mix chemicals in proper containers.
6. Clean up spills immediately.
7. Label all containers.
8. Know first aid procedures for each chemical used.

*When working with chemicals, proper protective clothing is necessary to avoid contact with skin, eyes, and lungs.*

- **Glasses, face shields, aprons, and gloves are commonly worn. Protective clothing must be selected for the specific chemicals being handled.**
- **Contact with spilled chemicals or their vapor can cause serious injury. Long exposure to low levels of spilled chemicals can be dangerous.**
- **Store, mix, and carry in containers that will withstand any heat generated or corrosive effects of the chemicals.**
- **Use grounded, explosion-proof equipment when mixing flammable materials.**
- **Mix chemicals in well ventilated areas.**
- **Do not mix different chemicals unless the label clearly says it is safe to do so.**
- **Wastes must be disposed of in a safe manner to prevent poisoning of the environment and later human contact.**

## Skin Contact Safety Rules

1. Be aware of the hazards of the chemicals you handle.
2. Know the first aid procedures to be followed for each chemical handled. Wear protective clothing to avoid skin contact.
3. Take prompt action if contact occurs.
4. Do not use organic solvents to clean skin.

*Skin contact by chemicals can result from splashing, immersion, or saturation of clothing. Such contact can cause dermatitis and chemical burns including blistering and tissue death. Some chemical burns can be severely disfiguring and life threatening.*

*In case of skin contact flushing the area with water for at least 15 minutes is usually advised. A soap or detergent can be used to aid in the removal of some chemicals but do not neutralize. Know your first aid procedures.*

*Know the symptoms of skin contact:*

- *Irritation*
- *Inflammation*
- *Blisters*
- *Tissue damage*

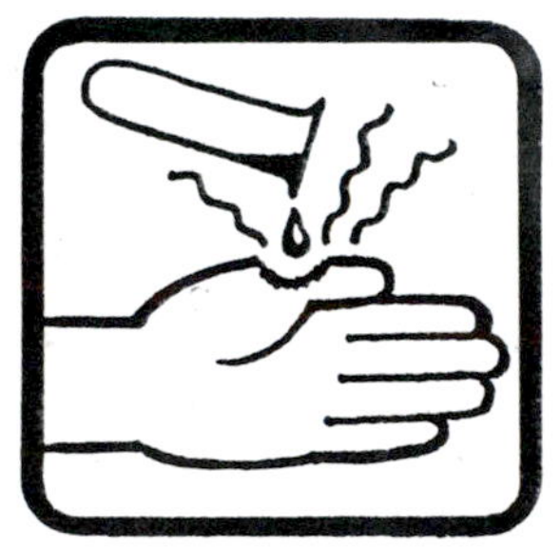

- **Chemical burns may not be immediately apparent.**
- **Know the recommended first aid for each chemical handled. Know the dangers of the chemicals you are handling and follow the proper safety procedure.**
- **Chemicals may enter the body by absorption through the skin. Organic solvents may be absorbed through the skin or may increase the rate of absorption of other chemicals.**
- **Gloves and other protective clothing may not give adequate protection against chemicals due to chemical breakdown of the material or permeation through it. Therefore:**

  **A: Select protective clothing and gloves for the specific chemicals being handled.**

  **B: Clean and inspect protective clothing and gloves after each use.**

  **C: Replace protective gloves and clothing when needed.**

- **When removing protective clothing such as gloves, wash them before removing to prevent transfer of chemical from protective clothing to skin.**
- **Contaminated or saturated clothing holds the chemical in contact with the skin.**
- **Remove contaminated clothing immediately without concern for modesty.**
- **Clothing can be removed while under the shower.**
- **Do not reuse clothing until decontaminated.**

# Eye Contact Safety Rules

**1.** Know the hazards of all chemicals handled.
**2.** Wear proper eye protection.
**3.** Do not touch or rub eyes with contaminated fingers or hands.
**4.** Know the location of eye wash stations and how to use them.
**5.** Treat injuries promptly.
**6.** Get medical attention.
**7.** Do not wear contact lenses in work areas.

*Liquid, solid, or gaseous chemicals entering the eye can cause permanent injury or blindness. Solid particles thrown into the eye can be equally dangerous. Many laboratories and plants require eye protection in all work areas. Eye protection should include safety lenses that will not shatter when struck.*

*Contact lenses cause special problems. Gases can concentrate under contact lenses causing eye damage or be absorbed by soft lenses causing them to stick in the eye. Small particles can get trapped under lenses. Soft contacts can dry out in hot, low humidity conditions. They may be hard to remove in an emergency due to eyelid spasm and if left in place will prevent proper irrigation of the cornea.*

**• Contact lenses should not be worn in the laboratory or other chemical or dusty areas.**

**• Do not wear contact lenses with a respirator.**

**• Know the symptoms of eye injury such as:**
**A: Painful burning sensation**
**B: Watering of eye**
**C: Inflammation**
**D: Sensitivity to bright light.**

**• When chemicals enter the eye, immediate washing should be done with a soft flow of water. (A strong jet of water may cause more damage.) Wash eye for 15 minutes in a soft flow of water. Hold the lids open and roll the eyeball around to wash off all of the eye.**

**• After washing, seek medical help without delay.**

**• Treat all eye injuries promptly even if pain is not present.**

# Inhalation Safety Rules

**1.** Know the hazards of the chemicals you are using.
**2.** Handle chemicals in well ventilated areas
**3.** Use a breathing apparatus when entering areas with toxic concentrations.
**4.** Do not smoke in areas where chemicals are handled.
**5.** Do not operate machinery when senses are impaired by chemicals.

*Toxic chemicals may enter the body through the lungs. These materials may be in the form of vapors, gases, dusts, or liquid drops. Exposure can result in permanent injury to the lungs, nervous system, or other body organs and may result in death.*

*Safety is everyone's responsibility! Symptoms of severe poisoning may not appear at once. As with nitrogen dioxide the damage caused may not be known until hours later and require medical help.*

**• Dangerous concentrations of fumes can build up in closed spaces or these spaces may have low oxygen levels that can cause unconsciousness and death.**

**• Proper ventilation should be provided for chemical reactions or procedures that involve hazardous or irritating fumes.**

**• Wear approved respiratory protective equipment as instructed.**

**• The temperature of a cigarette can convert some chemicals to highly toxic substances.**

**• Do not smoke in areas where chemicals are handled.**

**• Do not smell chemicals of unknown toxicity.**

**• Know symptoms of severe poisoning, such as:**
**Irritation of skin, eyes, or respiratory system.**
**Difficulty in breathing.**
**Headache/Nausea.**
**Sleepiness/Unconsciousness.**
**Poor coordination/Staggering.**

**• If dizziness, sleepiness, or nausea occurs:**
**Leave the contaminated area. Get fresh air.**
**Seek medical attention.**

# Ingestion Safety Rules

**1.** Do not use your mouth for pipetting chemicals or starting siphons.
**2.** Do not use chemical containers for food or drink.
**3.** Wash hands before eating.
**4.** Do not eat or smoke where chemicals are stored or handled.
**5.** Do not store food in refrigerators used for chemicals.

*Toxic chemicals can be ingested by using contaminated containers for food or eating with contaminated hands.*

**• Do not use laboratory glassware for food.**
**• Do not eat where chemicals are stored or handled.**
**• Do not store food in refrigerators used for chemicals.**
**• Wash hands before eating.**

*The following are general recommendations, however, first aid treatment depends on the chemical contacted.*

*Know the symptoms that appear in chemical poisoning, such as:*
- *Irritation and burning sensation of lips, mouth, and throat.*
- *Nausea and Vomiting.*
- *Diarrhea.*
- *Convulsions.*
- *Salivation.*

**• Vomiting may cause more harm than leaving the chemical in the digestive tract.**

**• DO thoroughly rinse the mouth.**

**• DO drink 3 to 4 glasses of water to dilute the chemical (unless advised otherwise).**

**• DO get medical help.**

**• DO NOT induce vomiting, except if the label or first aid recommendation says to do so.**

## Fire Prevention Safety Rules

1. Know the fire hazards of the chemicals you handle.
2. Store and handle flammable materials properly.
3. Do not smoke or create an ignition source where chemicals are stored and handled.
4. Have proper fire extinguishing and safety equipment available in case of fire.

*Fires can be started by open flames, cigarettes, hot surfaces, sparks, or other ignition sources. Heavier-than-air vapors can travel to distant ignition sources.*

- **Prevent contact of oxidizing agents with flammable materials.**
- **Burning materials may release toxic gases.**
- **The first few minutes after a fire starts are important in limiting its destructiveness.**
- **Be alert for the presence of open flames or sparking electrical equipment.**
- **Ground all electrical equipment.**
- **Fire may be prevented or extinguished by removing one of the three elements necessary for burning. Burning requires a fuel, oxygen, and temperature (heat, spark, arc, etc.). Remove one of these elements to prevent or stop burning.**
- **Fires and extinguishers are rated by class:**

**Class A Fires—Ordinary combustibles (wood, paper)**
**Class B Fires—Flammable liquids**
**Class C Fires—Electrical**
**Class D Fires—Combustible or reactive metals, metal hydrides, and organometalics.**

- **Know how to use a fire extinguisher.**

## Storage Safety Rules

1. Store chemicals in their original containers.
2. Make sure all containers are properly labeled.
3. Store incompatible chemicals in separate storage areas.
4. Store chemicals in properly constructed areas only.
5. Give special attention to hazardous materials.
6. Store chemical wastes in properly labeled containers and special storage areas.

*Chemicals should be stored in a cool, dry area away from ignition sources and in their original containers.*

- **Return chemicals to their proper location after each use with lids secured.**
- **Inspect stored chemicals to their proper location after each use with lids secured.**
- **Do not store food in refrigerators used for chemicals.**
- **Food can become contaminated if stored near dangerous chemicals.**
- **PROVIDE**
  - **Cool and dry areas**
  - **Storage temperature of 67 to 94 degrees F unless told otherwise**
  - **Good ventilation**
  - **Sturdy shelving securely fixed to floor or wall**
- **AVOID**
  - **Direct sunlight**
  - **High heat and humidity**
  - **Sources of heat and ignition**
- **Incompatible chemicals should not be stored in the same area.**

# Epoxies—What They Are; What They Can Do To You

If you are working with "EPOXIES," that means you're working with special chemical formulations needed and used by industry, commerce, and the home.

Those "EPOXIES" and their related chemicals are:

- very adhesive
- very chemical resistant
- very heat resistant
- very dimensionally stable
- very unaffected by moisture
- very unaffected by heat or cold
- very dangerous to you, unless you follow easy, simple, but very important procedures in their handling and use.

By practicing the following guidelines, these valuable chemicals can be used safely.

## What is EPOXY?

To help you work safely with these chemicals, we must agree on exactly what products are "EPOXY."

"EPOXY" isn't a thing. It's a chemist's short-hand way of describing a certain way that molecules are linked together. "EPOXY RESINS" can be a whole lot of different molecules of different materials. The "Epoxy" resins wildlife artists work with are in several different forms, each with many variations.

a. There are liquid "EPOXY" resins.
b. There are solid "EPOXY" resins.
c. There are solution "EPOXY" resins.

The last mentioned, the solution "EPOXY" resins, are usually the solid resins supplied in one or a combination of different solvents or diluents. There are different "do's" and "don'ts" for each of the types and forms of "EPOXY."

Epoxy resins alone are useless! Those good properties listed don't happen until the resin is "cured" or "hardened" by other types of chemicals called, naturally enough, curing agents and/or hardeners.

## What Are Curing Agents/Hardeners?

Think of it this way. You don't have fudge if you only have cocoa. And you don't have a useful epoxy product if you only have an epoxy resin.

The curing agent/hardener molecules react with the epoxy resin molecules and cause, for example, a liquid resin and liquid hardener to become a commercially useful solid epoxy casting, laminate, coating film, or adhesive.

Not considered curing agents, but working in the same way, are many kinds of esters, solvents, diluents, and resin modifiers. They either put the resin into a solution or react with it, resulting in a useful product for further processing.

There are different "do's" and "don'ts" for each of the curing agents, solvents, and diluents.

When you are **working** with "EPOXY" resins, and curing agents/hardeners, and the other related materials, you can't spend your time studying them. Know about them ahead of time.

You can work with these materials in safety and continued

good health if you treat them all with respect, if you don't try to short cut, and if you give yourself a margin for natural error.

Thousands of people have worked with these materials for over twenty years with very few problems, because they've always used proper equipment and have adequate ventilation, and have followed simple, careful procedures. However, ignore those procedures and "EPOXY" and related materials can be hazardous.

What *can* happen if you don't work with care? Sensitization, skin rash, dizziness, unconsciousness, poisoning, eye damage, fire or explosion burn, sickness, pain, or even death!

## The DO's and DON'T's of Epoxies:

**DO** treat every product or raw material with respect, and know how to protect yourself.
**DON'T** be careless.

**DO** be clean and stay clean in your person and your clothes.
**DON'T** let any of the products get or stay on your clothes.

**DO** protect your eyes. Wear safety glasses with side shields or chemical goggles. Contact lenses can be a hazard when working with chemicals.
**DON'T** take a chance on any material or spray getting into your eyes.

**DO** use the ventilation provided. Use approved respirators or other approved masks as provided.
**DON'T** inhale vapors of any solvents, resins, hardeners, diluents. Beware of dusts and powders from grinding or polishing operations.

**DO** Keep all possible sources of ignition (cigarettes, lighters, open electrical heaters or motors, bunsen burners . . . all sources) away from the area of open solvents, diluents, resins, curing agents.
**DON'T** gamble with fire or explosion hazards.

**DO** keep contaminated hands, gloves, cloths away from eyes and mouths.
**DON'T** assume anything is clean enough.

**DO** use disposable dippers, stirrers, bench covers; discard them regularly or when contaminated. Keep clean!
**DON'T** Let your tools, etc., turn on you and your health.

**DO** check your co-workers as well as yourself on these safety procedures. Your mistakes can hurt them; their carelessness can hurt you.,
**DON'T** guess about safety. Know!

**DO** read and follow safety-warning-hazard statements on product package labels.
**DON'T** assume this chapter says it all.

**DO** keep all containers of flammable liquids covered.
**DON'T** take a chance on fire.

## First Aid for Epoxies

KNOW where the nearest eye bath, shower, sink with water, and fire extinguishing equipment are located. NOW is the time to find out, not when there is an accident.

**Eye Contact:** Wash/flush the eyes immediately with low pressure flowing water for at least 15 minutes. Be sure the entire eye surface is flushed. GET MEDICAL ATTENTION AT ONCE.

**Skin Contact:** Remove contaminated clothing at once. Use disposable wipes to take syrupy or thick liquids off the skin before cleaning. Clean skin *immediately* with soap and water or waterless hand cleaner. DO NOT USE SHOP SOLVENTS TO CLEAN THE SKIN. If there is any sign of redness, itching, or a burn, GET MEDICAL ATTENTION PROMPTLY. Discard, do not reuse, contaminated leather articles (such as belts, shoes, watch straps, etc.).

**Dizziness, nausea, blackout:** When any of these symptoms occur as a result of breating vapors, get yourself or the affected person into fresh air immediately. Know how and be ready to use artificial respiration if breathing should stop. GET MEDICAL ATTENTION IMMEDIATELY.

**Illness or nausea from swallowing:** GET MEDICAL ATTENTION IMMEDIATELY.

NOW is the time for you and your co-workers to KNOW how to help each other, should the need arise.

JUST BE CLEAN. That means don't get chemicals on you—don't get them in you—and don't breathe them, either. Do those simple things, be careful, and there should be no problems with epoxies.

BE AWARE if you are doing these kinds of things with "EPOXIES": formulating, grinding, cutting, stirring, or applying adhesives—you can be vulnerable!

As we have said: treat *all* the "EPOXY" kinds of products like they are the worst ones around. That will give you a margin of safety.

It's also true that some things do require special care and caution. A few:

- Applying room temperature cure epoxy systems over hot pipes or metal parts vaporizes solvents or catalysts. And it is much more critical for you to have good ventilation.
- The thinner and more liquid an epoxy system is, the more chance of it's being dangerous to you from splashes, skin absorption, and breathing.
- If the epoxy system you're using cures at room temperature (instead of needing oven or other heat cure), chances are that the resin, hardener, and solvent-diluent used are more hazardous.
- Solvents, diluents, resins, all can burn. Some can explode. You can be the cleanest angel in heaven if you're not especially careful of *all* ignition sources from cigarettes and matches to electric motors and bunsen burners.
- Clean-up time can be a hazard! Don't use solvents carelessly. Do take proper care of the waste clean-up. Remember, the janitor, the guy in the next department, or the fellow from city pick-up isn't as aware as you and your co-workers about what to do and what not to do with these "EPOXY" materials.

## Use Tools Safely

Always read instructions and safety precautions provided with power tools and observe them.

- Don't use power tools carelessly.
- Do wear eye protection where called for.
- Always use properly grounded electrical plugs.
- Don't remove protective guards. (They are there to protect you.)
- Store power tools (and chemicals) in a safe place alway from children.
- Don't overload circuits.
- Always replace faulty electrical cords immediately.
- Never use power tools (or chemicals) when under the influence of alcohol or drugs.
- Always keep work areas clean.

# CHAPTER 2

*Many readers will find this chapter boring and perhaps even elementary. Those of you who are thoroughly familiar with many of these products should perhaps skim through it. Beginners are encouraged to read it in its entirety as all of these products, tools and equipment are mentioned in various chapters throughout this book. A brief description and reference as to a supplier of the item at the time of this writing is furnished for your convenience.*

# Chemicals, Tools, Equipment & Supplies

***by Bob Williamson***

**SUPPLIERS LIST**

The following suppliers distribute some of the various products mentioned in this chapter. Refer to the Supplier Number at the end of each item's discription to determine where that item can be obtained.

**10**—Wildlife Artist Supply Company (WASCO)
P. O. Box 967, 1306 West Spring, Monroe, GA 30655
Phone Toll-Free 1-800-334-8012

**20**—Riberglass, Inc.
P. O. Box 460790, Garland, Texas 75046

**30**—New Wave Taxidermy
3101 S.E. Slater St., Stuart, FL 34997 (407) 283-7270

**40**—Your local hardware or home center store.

**50**—Ohaus Scale Corporation
29 Hanover Rd., Florham Park, NJ 07932

**60**—A large art supply or hobby supply store

## Molding Materials

**POLYESTER RESIN**—There are several different kinds of polyester resins available. We recommend isophthalic resin, which is a flexible polyester resin, to be used for making molds and/or laminating (lay-up work) purposes (*Polytranspar*™ Lay-Up Resin, LR101). To create water scenes, we recommend a casting resin (*Polytranspar* Artificial Water) which hardens to a "water" clear finish. (Lay-Up Resin cures to a violet color.)

Polyester resins must be mixed with a catalyst in order for them to harden and set up. The catalyst is Methyl Ethyl Ketone Peroxide or MEKP. It is important to follow the directions for mixing "precisely" as recommended on the label of the container. Adding too little catalyst (cold mix) or improperly stirring can result in the polyester resin not setting up and remaining liquid or tacky. Too much catalyst ("hot" mix) can build excessive heat to the point that it will actually start smoking and/or self-ignite. It can also cause warping, discoloration, and cracking. Polyester resin, like plaster, heats up as it "cures" and hardens and it is normal for some heat to be given off, even when properly mixed. (See the table in Chapter 10 for catalyst ratios.)

It is imperative to observe all safety precautions exactly as outlined on the label of the container. Never mix the chemicals near a heat source and always mix them in a well-ventilated area. Use all the recommended safety equipment such as a respirator, disposable rubber gloves, safety goggles, rubber aprons, etc., and have an approved fire extinguisher on hand at all times. "Remember," the resin will emit a powerful odor as it cures necessitating a good ventilation system. Adding water to leftover resin in mixing cups after using will help reduce this odor and the risk of it catching on fire.

Dispose of excess and leftover resin, paper towels, mixing cups, and sticks, etc., by storing them outside your building. Always store these materials in a tightly sealed 5 gallon **"metal"** bucket or similar container and pour cool water over the top of the materials to cool them down and reduce the risk of a fire.

Always replace the lids to the containers of resin and catalyst immediately after use. MEKP is a strong oxidizer, which means that it can spontaneously ignite if it comes in contact with various different materials. Keep these and other chemicals out of the reach of children, and in clearly marked containers in a safe place.

The shelf life of most polyester resins is approximately six months to one year. The shelf life can be prolonged and extended by keeping the lids tightly closed and storing in a cool, dry location, away from direct sunlight. Storing these type resins near a heat source or exposing them to moisture (such as leaving the lids off) will dramatically reduce the shelf life and the resin will become stringy, lumpy, and "gelled" in the can.

Polyester resins will not cure properly (if at all) if they are exposed to moisture. Keep this in mind when making molds, Artificial Water pours, or doing laminating work. Always apply the polyester resin to dry surfaces only. (Supplier No. 10.)

***POLYTRANSPAR*™ ARTIFICIAL WATER**—This product is a crystal clear casting resin that is used primarily for recreating realistic water and ice scenes. It can be used to create tiny water droplets or large waterfalls complete with splashes. It is also great for making paperweights with imbedded objects, ice cubes, etc. The two component kit includes MEKP catalyst which is mixed together as other polyester resins are mixed and has similar properties. Follow the directions and SAFETY INSTRUCTIONS for best results. (Suplier No. 10.)

**GELCOAT**—Gelcoat is used in conjunction with lay-up resin. It is used wherever an extra smooth surface and good detail is required and is applied first, with lay-up resin applied later. Gelcoat, like polyester resins, must be mixed with MEKP in order for it to harden. It is generally sprayed using a paint spray gun or special gelcoat gun to spray the surface intended to be molded. Two fairly heavy coats are generally sufficient to build the coat up to approximately 18 mils thick. It is best to allow the gelcoat to cure overnight before applying the lay-up resin. (Supplier No. 10, Code No. GC32.)

**FIBERGLASS**—Polyester resin is brittle and has little tensile strength by itself. However, when fiberglass is added, it gains tremendous tensile strength and becomes very strong. Fiberglass generally is sold in three different mediums (pictured below, left to right): mat, chopped rope and cloth.

**Chopped fiberglass** (above center) can be purchased already chopped, or the artist can obtain a chopper gun and a roll of fiberglass rope and chop their own. A chopper gun is much like a paint sprayer and is used to spray layers of finely chopped fiberglass onto the laminating resin. If using "pre" chopped fiberglass, simply sprinkle it onto the resin by hand. Once the chop is applied, simply tamp it into the resin with a paint brush until thoroughly saturated. This is a fast way to lay-up or laminate a project.

**Fiberglass mat** (left) comes in various weights and is made of short strands of fiberglass "matted" into a sheet. A 1½ oz. mat (pictured) works very well for most projects. Fiberglass mat is applied by tearing pieces and tamping them into the wet resin with a paint brush until saturated.

**Fiberglass cloth** (right) is very heavy and resembles a woven rug. It also comes in various different weights. A 6 oz. cloth (pictured) is ideal for molds that will see repeated use and where tremendous strength is desired.

**SURFACE CURING AGENT**—Add this chemical to polyester or casting resin (in addition to catalyst) when pouring the final layer. It yields a hard "tack free" surface. *Polytranspar* Lay-Up Resin and Artificial Water, as a rule, do not require a surface curing agent as they "cure" to a hard enough tack free surface anyway. Any tacky feeling remaining can be eliminated by sprinkling with talcum powder, provided the powder won't ruin the effect (as might be the case with some Artificial Water scenes).

**COMPETITION WET LOOK GLOSS**—This multi-use high gloss resin covers imperfections such as fingerprints, scratches, and smudges on artificial water castings. This will help the artist to achieve a smoother finish than brushing catalyzed resin on the surface to achieve the same effect. Additionally, it can be used as a high gloss spray to finish coat a variety of substances. (Supplier No. 10, Code No. FP241.)

**SURFACE COAT SPRAY**—This high gloss finish can also be used to cover imperfections such as fingerprints, etc., in castings. It comes in an aerosol spray can which is very handy and inexpensive. (Supplier No. 60.)

**GEL PROMOTER**—This chemical speeds up the "cure" time of polyster and casting resins (artificial water). Again, *Polytranspar* products don't generally need a gel promoter. Preheating the resin will also speed up the "cure" time. (Supplier No. 60.)

**CEARA MOLD RELEASE WAX**—This product is used to keep polyesters from sticking to the subject that is being molded. It is commonly referred to as a separator or mold release agent. Generally all that is needed is one medium coat that is slightly "hand" buffed. (Never use power buffers.) This brand of wax is 100% yellow carnauba and it contains no silicones, acetates, or plastics. (Supplier No. 10.)

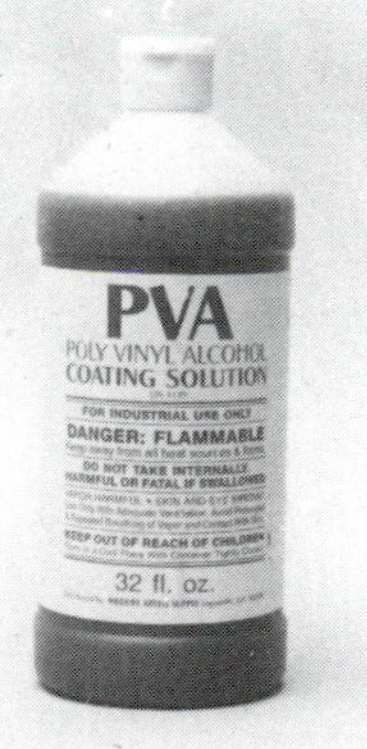


**POLY VINYL ALCOHOL (PVA)**—This product is used as a separator or mold release agent for polyesters. It is generally a transparent green color. PVA can be sprayed or applied with a brush. PVA is generally applied after waxing the subject to be molded; however, it can be used without waxing. Generally 2 to 5 coats are applied to bring it to 5 mils, allowing each to dry before applying the next. (Supplier No. 10, Code No. PVA-32.)

**TRANSPARENT DYE KIT**—The wildlife artist will find it beneficial to add color to artificial water to create a variety of effects. Although artist oils will work to color the resin, these transparent dyes are specially formulated for tinting resins and are more natural looking. Dyes don't absorb light which makes them more transparent as in actual water. The colors may be blended together with each other to create almost any color imaginable. Standard colors available in a kit are red, blue, green, yellow, amber, and pearl. These dyes are very intense and only small amounts are necessary to tint the resin. (Supplier No. 60.)

**AEROSIL, CAB-O-SIL**—These products are used to thicken polyester resin. They are microfine white pastes that when added to polyester resin will thicken the polyester to a like consistency. It is desirable to use either of these two products to avoid sagging or developing a "run." "Always" use a respirator when using these products as they can be dangerous to your health if inhaled. (Supplier No. 20.)

**EPOXY RESINS: ULTRA GLO & ENVIROTEX LITE**—These are similar 2-component, "glass-like" premium high gloss plastic finishes designed for interior use. Both are epoxy-based reactive polymer compounds that will only cure after both their components have been thoroughly mixed together. They may be

poured or brushed, with one coat equal to 50 coats of ordinary varnish or lacquers. The finish will self level after pouring or brushing and requires no polishing to maintain a very high gloss. They can be lightly touched and handled after about 6 hours at 70 degrees and will reach full strength and toughness in about 36 hours when fully cured.

The latest rage for creating crystal clear water scenes, especially "pools," is either one of these two products. Although originally designed for cypress clocks, bar tops, etc., several enterprising artists have incorporated them into habitats. To build one the base must be sealed with fiberglass resin, Ultra-

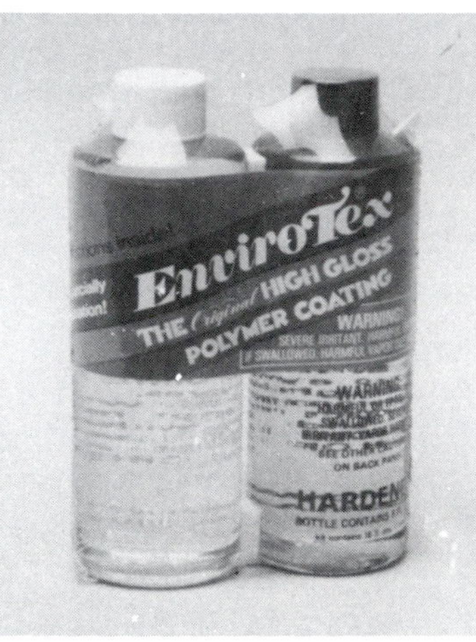


Seal, or similar substance. This will eliminate air bubbles. Once the sealer has dried, mix Ultra-Glo or Envirotex Lite components 1 to 1 as called for in the directions and make the first pour. Don't try to obtain too much depth with one pour. Instead, pour approximately 1/8" to 1/4" and let it cure, and then pour again until a maximum depth of 1/2" is reached. Trying to pour thicker coats than this will result in distortion, etc. Air bubbles may be removed by slightly blowing on them or putting a torch to them as per directions on the container. Additionally, grasses, leaves, duckweed, etc. can be imbedded BETWEEN the layers of resin for very interesting effects. Note: The surface to which you are going to apply these epoxies must be clean and dry. If properly applied, no finer water substitute can be found. The clarity is excellent. Pools of water are as natural as we've seen with the use of these materials. (Supplier No. 60.)

**ULTRA SEAL**—This product has a multitude of uses for the wildlife artist. It looks similar to Elmer's white glue while wet but cures to a transparent, water-clear satin finish. It is safe, non-toxic, and cleans up with water. It will effectively seal wooden bases that are to be coated with epoxies or polyesters.

It is a very good glue that has tremendous adhesive characteristics. Taxidermists use it to back fins for fish mounts, gluing earliners, coating noses for the wet look, and a multitude of other purposes. It is also an excellent decoupage. (Supplier No. 10, Code No. US16.)

**RTV SILICONE RUBBER**—RTV, which stands for *Room Temperature Vulcanizing*, is primarily a silicone based product that is very useful for certain types of molding operations. RTV molds are very rubber-like, and have good tear resitance, which make them good candidates for molding objects with undercut areas. Also, the better grades of RTV are not inhibited by moisture, so that excellent detail can be cpatured even on a damp surface. RTV rubber molds have a relatively long mold life if handled carefully, and can be used repeatedly to make casts of many unusually shaped objects. The unfortunate part of using RTV is the material cost, which at the time of this writing is about $150.00 per gallon. (Available from large chemical companies.)

**MOLD BUILDER**—This is a liquid "brush-on" latex rubber commonly used for making molds. Simply paint several coats on the object to be molded and allow to dry prior to adding another. Once an adequate thickness is reached, simply peel off the mold. This will make a good reuseable mold that can be used for casting wax, plaster, urethanes, and resins for artificial water. (Supplier No. 60.)

**MOLDING PLASTER**—This is a fine white powder that has a creamy smooth consistency when added to water. It will harden after being added to water and is ideal for making molds, making positives from molds, base work, maches, etc. Hardening times will vary according to water temperature and length of time the product was mixed. As a rule, warm temperatures and mixing long periods of time will increase the "set" time. (Supplier No. 10, Code Nos. MP5, MP10, MP25.)

**HIGH FIBER**—This is an ideal synthetic binder (holds ingredients together to make a tough bond) for maches, hide paste, etc. High Fiber, which is totally safe and biodegradable, replaces the extremely dangerous asbestos. It is much more effective than paper pulp or similar products. High Fiber is also excellent for making "beds" for subjects to be molded. Simply add water until the desired consistency is reached. It can be reused many times over if Lysol is added to prevent bacteria growth. (Supplier No. 10, Code Nos. HF5, HF10.)

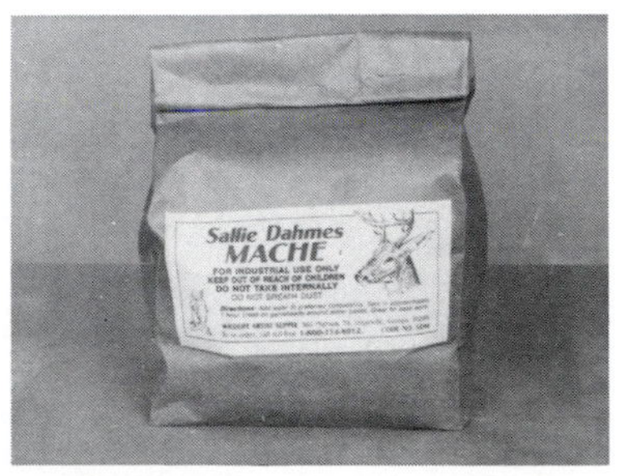


**JIM HALL MACHE AND SALLIE DAHMES MACHE**—These are ideal maches that can be used for a multitude of projects and base work. Sallie Dahmes Mache contains more adhesive and is somewhat stronger than Jim Hall Mache. Jim Hall Mache is slightly lighter in weight and accepts pins more readily. Both can be mixed with Tempera Paints to add color. Hardening times will vary according to water temperature and length of time the product was mixed. As a rule, warm temperatures and mixing long periods of time will decrease the "set" time. (Supplier No. 10, Code No. JHM, SDM.)

**TOM SEXTON FISH FILLER**—This material is similar to mache, however, is much lighter and does not have as much strength. It is used wherever obtaining a lightweight exhibit is a priority. This product will harden approximately 45 minutes to an hour after mixing with water. Hardening times will vary according to water temperature and length of time the product was mixed. As a rule, warm temperatures and mixing long periods of time will decrease the "set" time. (Supplier No. 10, Code No. FF.)

**POTTER'S CLAY**—This economical material can be purchased as a dry powder form which can be mixed with water to make any consistency (viscosity) clay desired or it can be purchased premixed in 5 or 25 lb bags. It is whitish/grey in color and can be used for a variety of basemaking applications. Potter's Clay is easily sculpted and modeled. It dries rock hard and can be carved and/or painted as desired. It can be used as a separator coat in many molding operations that utilize molding plaster. (Supplier No. 10, Code Nos. PC [wet] or DPC [dry].)

**PLASTILENE CLAY**—This is an oil based clay that is used for many moldmaking and sculpture undertakings. Unlike Potter's Clay, Plastilene Clay will not harden and can be used repeatedly. (Supplier No. 60.)

**ALGINATE**—This material is commonly used by dentists to make impressions of teeth. It is an ideal molding material for wildlife artists. Quick molds that hold incredible detail can be made in two minutes time by simply adding water and mixing. Alginate requires no separator and can easily be lifted from the subject that is being molded, as it will not stick to anything.

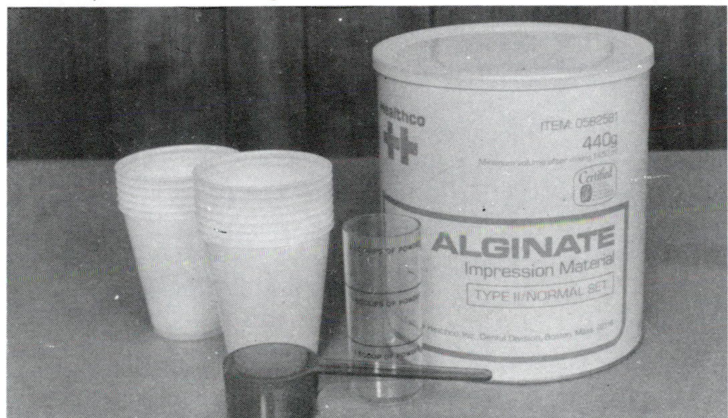


Alginate molds will not hold up to repeated use. The positive part must be poured into the Alginate mold soon after making it as the Alginate will shrink and develop cracks and become brittle after just a few hours. The most common material used to make positives is molding plaster. Moisture sensitive products such as polyesters will not work very well, as the Alginate mold does retain some moisture. (Available from medical or dental supply companies.)

**POLYURETHANE FOAM**—This product contains two components: A (isocyanate) and B (polyol). Shortly after mixing the two components in the recommended proportions the mixture will expand and begin to rise. When fully expanded and cured, it will become rigid. (There are some flexible urethanes but generally rigid foam is used.) The foam is usually poured into some type of mold and when it cures, it will maintain the exact shape of the mold. Care must be taken to determine how much foam to use and adequate ventilation holes must be provided for in the mold to allow excess foam to escape. Compressing too much foam in a mold without "venting" it can destroy a mold.

Urethane foam can be purchased in different densities and weights. The heavier the density, the harder the foam. For example, a 1½ lb. density foam is not nearly as strong and much softer than a 4 or 5 lb. density foam. Compressing the foam into a mold can increase the strength of a low density urethane, however, the artist risks damaging the mold if too much foam is compressed into the mold.

Maintaining a temperature of 70 to 75 degrees is important when mixing the components. Excessive heat will cause the foam to expand too fast and colder temperatures will inhibit the foam from curing properly and slow it down. It is best to mix a small portion, according to the directions, to determine the curing time. Generally urethane foam will begin to rise very rapidly and one must work fast to use it. The foam must be mixed thoroughly and exactly in the proportions called for or it will not cure properly. Always read the instructions carefully as all urethanes are not the same.

Urethane foams are common to the taxidermy industry, as most all mannikins are now being made from it. It is also very useful for base work, artificial rocks, or as a powerful adhesive. (Supplier No. 10, Code No. UF5 or UF20)

**ULTRA LITE BODY FILLER**—There are many brands of auto body filler, each with their own trade name, including Bondo®, Ultra Lite, Dyna Lite, and many more. They are thick consistency types of polyester resin that are mixed with a cream catalyst. They were originally developed to be used in automobile body shops to repair dents, etc. It hardens quite rapidly to a lightweight but strong plastic and once cured may be sanded or ground to a nice smooth finish. It has a myriad of uses in the wildlife artist studio. Here are some of the many uses: making quick mother molds, or positive parts from a mold, adhesive for foam and other materials, strengthen molds, landscape, making mushrooms, and base work. It may be thinned by adding polyester lay-up resin and will cure using the cream hardener (catalyst) from the tube. (MEKP will not be necessary when using Ultra Lite Cream Hardener.) (Your local auto parts or discount store).

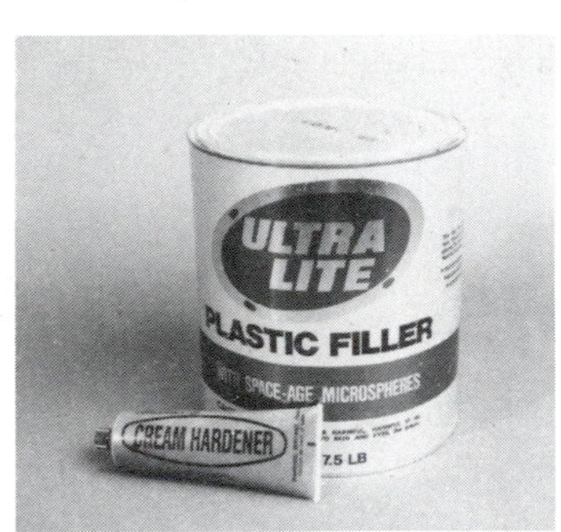


**SCULPTING EPOXIES**—There are several sculpting epoxies (also called epoxy putties) now on the market: Magic-Sculp, Sculpall, All Game, Magic-Smooth, Smoothout, and Ultra Smooth. All are available in two component system with resin and hardener. They all accept paint, dyes, and pigments very well and cure to a very hard finish.

Magic Sculp, Sculpall and All Game are very similar and can be used anywhere a permanent filler is needed. They are not as fine of a finish as Magic-Smooth, Smoothout, or Ultra Smooth which cure to a glossy feather thin edge.

Both can be smoothed with a modeling tool wet with water. To use, take an equal amount of part A and part B and knead together thoroughly. Apply to a clean surface, sculpt, and allow the compound to harden. (Supplier No. 10.)

# Glues and Adhesives

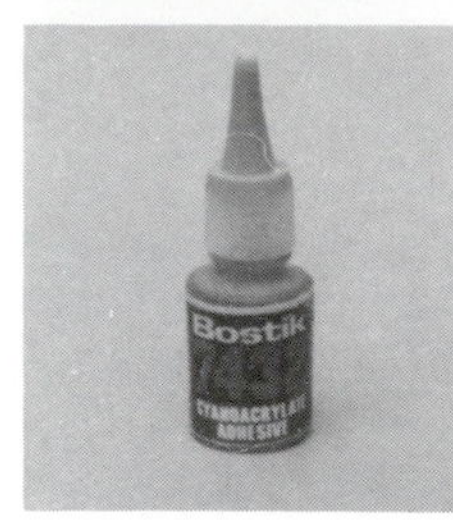


**INSTANT BONDING GLUE**—Also referred to as Super Glue, this is an excellent cyanoacrylate adhesive that bonds in seconds. It is good for gluing plexiglass, leather, bone, rocks, etc., or wherever a strong, fast-drying glue is needed. Just a drop is all that is needed for most operations. (One drop is advertised as holding 2000 lbs.) (Supplier No. 40.)

**SPRAY ADHESIVE**—This glue comes in a 17 oz. aerosol can and can be sprayed with a touch of the button. It is useful for gluing habitat materials to bases and a variety of other purposes. (Supplier No. 10, Code No. HM606.)

**ULTRA SEAL**—This is a good glue for many base making projects. It is a white color in the container but dries completely transparent. It can be applied with a paint brush or thinned with water and sprayed with a hand sprayer or an airbrush. (Supplier No. 10, Code No. US16.)

***POLYTRANSPAR* FIN BACKING CREAM**—This is an outstanding adhesive that has many uses. It is supplied in a caulking tube and has a consistency of Vaseline. It is white in the container but dries transparent. It may be brushed on or thinned with water and sprayed with a hand sprayer or airbrush. It can be used as an adhesive or sealer on plastics, foams, wood, leather, and other materials. (Supplier No. 10, Code No. FC101.)

***POLYTRANSPAR* HIDE PASTE**—This adhesive paste is excellent for base work and gluing open-cell foam. It has a thick consistency and is generally applied with a paint brush. (Supplier No. 10, Code No. FP190.)

***POLYTRANSPAR* EPOXY ADHESIVE**—One of the most powerful adhesives available. Mix equal parts of the A and B components on a paper plate and apply to the surfaces to be joined. The mixture will permanently harden in a few hours (with or without air or moisture). (Supplier No. 10. Code No. EA10.)

**DEXTRINE**—This product is used as a glue in many different formulations. It can be mixed with maches, fillers, other glues etc., to increase the adhesive properties. It can be mixed directly with a binder such as High Fiber, for a stronger consistency to make an excellent glue. Mix dextrine, High Fiber, and Lysol disinfectant (to prevent mold or bacteria growth) with hot water and stir. The more Dextrine added, the stickier the glue. (Supplier No. 10, Code No. DEX5, DEX10.)

**HOT MELT GLUE**—This is one of the most useful adhesives available. Simply insert a glue stick into a glue gun. Once it has heated up, the melted glue can be applied by squeezing the trigger. It dries very rapidly (less than 60 seconds) and will glue just about anything. Unused glue remains in the gun with no waste. It is clean, quick, and economical. (Supplier No. 40.)

# Reinforcement Materials

**BURLAP**—This material is most often used as a reinforcement for plaster molds that require above average strength. It is generally cut into strips that are submerged in plaster. They are then placed on the plaster mold overlapping each layer until a satisfactory reinforcement coat has been built. Available from fabric stores or check with your local feed store.

**HARDWARE CLOTH**—This is a galvanized screen wire mesh. Hardware Cloth with 1/8" square is the normal size used by most artists. It is usually stapled to a base and then bent and shaped to resemble rocks or a landscape of the artist's choice with maches and/or polyesters used to cover it. (Supplier No. 10, Code No. FCS76.)

**METAL LATHE**—This is a screen mesh that has diamond shaped openings rather than squares such as Hardware Cloth. It is used the same way as Hardware Cloth, but this screen meshing is much stronger. It is used by stone masons and is available from most any masonary supply dealer.

# Miscellaneous Supplies

**ARTIFICIAL DUCK WEED**—Small round pieces of paper can be used to represent duck weed. Most any print shop that owns a hole punch machine can provide as much of this material as an artist could want or ever use. Smaller diameter circles can be obtained from anyone that owns a telex or teletype machine. The holes that are punched out are a much smaller diameter than those from a standard hole punch. Using both sizes is probably the best bet. Simply paint the small paper cuttings with *Polytranspar* FP50 or FP60 series airbrush paint using a device as pictured in Chapter 13.

**ARTIFICIAL LILY PAD KIT**—Realistic looking lily pads prepackaged in kit form can be used with or without stems. Perfect for swamp/water scenes. Simply insert stem tube into fastener glued to the lily pad. Wired stem can be inserted into a foam base or the pad can be laid flat in artificial water to resemble floating lily pads. (Supplier No. 30.)

**WASCO ARTIFICIAL SNOW**—This material is offered by Wildlife Artist Supply Company. It is applied by first spraying with Spray Adhesive Glue, misting water-thinned Fin Backing Cream, or Ultra Seal, and then sprinkling the snow over the glue with a can with holes punched in the top. This operation can be repeated several times until the desired thickness is reached. This kit comes complete with a realistic sparkle material that is lightly sprinkled over the completed snow scene. (Supplier No. 10, Code No. HM600.)

**SNOW FLOCKING**—This is applied with a flocking gun or misting water with a handsprayer and sifting or sprinkling the snow over it. It has an adhesive in it that is activated when it is mixed with water. This snow is a finer diameter than WASCO Artificial Snow. (The most accurate habitat scenes contain at least two different textures of snow.) (Supplier No. 10, Code No. ASF.)

**LUSTER PIGMENTS**—These powders are available in a variety of irridescent colors and pearls. They are useful in snow scenes as well as ice and water exhibits. (Supplier No. 10.)

**IRIDESCENT GLITTER**—This material works exactly like the *Polytranspar* iridescent shimmer colors. This product must be applied "sparingly" to artificial snow scenes. If very small amounts are "very lightly" sprinkled over the surface, this glitter will exactly duplicate the refracted light that actual snow produces. (Supplier No. 10, Code No. IG.)

**SCULPEY**—This is a modeling material that bakes in your home oven to permanent hardness in 30 minutes. It is the only sculpting compound of its kind. It works very well for sculpting artificial plants. (Supplier No. 60.)

**GLASS BEADS**—Clear glass beads recreate air bubbles in artificial water scenes. Imbed the beads in the second pouring in order to suspend them. (Supplier No. 60.)

# Protect Your Hands

**LATEX GLOVES**—Due to the hazardous nature of many of the chemicals, paint, (and diseased specimens), etc., that the wildlife artist must encounter on a daily basis, it is always advisable to use latex gloves to prevent unnecessary exposure. Disposable latex gloves are inexpensive and easy to use. They are not only safer but they also keep your hands clean. (Available from your local pharmacy.)

**RUBBER GLOVES**—Chemical resistant rubber gloves are necessary when handling dangerous chemicals and solvents.

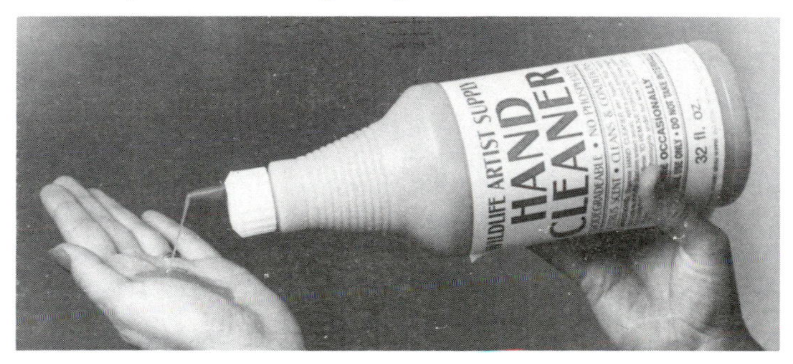


**HAND CLEANER**—Most any shop should have a supply of hand cleaner on hand. The type cleaners that contain grit are generally the best.

# Decoupage

**ULTRA SEAL**—This is an excellent "water" based product that can be used for decoupage work in the same manner as Polytranspodge. (Supplier No. 10, Code No. US16.)

# Paints

***POLYTRANSPAR*™ AIRBRUSH PAINT**—Any habitat scene can be improved by highlighting and coloring different areas with an airbrush. *Polytranspar* has a wide variety of premixed colors from which to choose. It is available in both water acrylic paints and lacquer based paints. (Supplier No. 10.)

***POLYTRANSPAR*™ PLASTICOAT PAINT**—This base coat effectively seals and primes acetate, vinyls and plastics. It can be tied in a knot without cracking and once primed with Plasit-Coat most any paint can be used on top of it. Works great with *Polytranspar* or Oil Paints. (Supplier No. 10.)

**ACRYLIC PAINTS**—Tube acrylic paints can be thinned with water and applied with a brush, handsprayer, or airbrush. They dry quickly and clean up with water. (Supplier No. 60.)

**LATEX PAINTS**—Latex pains can also be thinned with water and applied with a brush, handsprayer, or airbrush. Latex paints will brush easily from hair. (Supplier No. 60.)

**WINSOR & NEWTON OILS**—These are premium oil coulors. They are used in combination with artists' brushes. They are slow to dry, therefore, they afford excellent opportunity to blend colors exactly as desired. (Supplier No. 60.)

**TUFFILM**—This crystal clear matt finish is applied from a 12¾ oz. aerosol can. It leaves a gentle, satin finish that is ideal for final spraying and sealing habitat scenes. It can also be used as a fixative for drawings and charcoal sketches, effective as a sealant against dust and water. (Supplier No. 60.)

**TEMPERA PAINT**—These pigments are perfect for many habitat scenes. They can be mixed directly to color epoxies such as Sculpall, All Game, plaster, maches, etc., to give them a pleasing-colored base coat. Water can be mixed with the pigments to make paint suitable for a variety of base work. They are often applied and textured using a natural sponge, sprayed from a handaprayer, or spattered on by flinging with a brush. (Supplier No. 60.)

# Chemicals

**NAPTHALENE**—This is the same material that is in moth balls. These are used as extra protection from insects for exhibits. It works extremely well for subjects enclosed in glass cases. Available from most large discount stores.

**BORAX**—An excellent desiccant, borax is great for removing moisture from natural stems or limbs. Simply cover the material to be dried with borax powder for a few days to remove the moisture for preserving. (Supplier No. 10.)

**GLYCERINE**—This product is often used to preserve various plants, pine needles, etc., by soaking them in a solution of 50% glycerine and 50% isopropyl alcohol. The solution replaces the moisture in the plant and effectively preserves many species. (Supplier No. 10, No. GL1.)

**ACETONE**—This is a very fast evaporating solvent that is used to clean up equipment that has been used with polyester resins. It is also one of several thinners contained in lacquer thinner. Acetone is also used in "hand molding" acetate operations. This solvent is extremely flammable and the fumes are hazardous. Use only where adequate ventilation is provided, and where the threat of excessive heat, sparks, or open flame is non-existent. "Never" smoke when using this solvent, and always wear a charcoal filtered respirator. Available at auto supply stores.

# Tools, Equipment & Supplies

Most studios (or homes) probably already have on hand the majority of the tools necessary to create a wide variety of quality habitat exhibits. The following list details a variety of tools that are commonly used to create them. Keep in mind that it is not necessary to own all of these tools to create a good habitat; however, using the proper tool for the job certainly eliminates a tremendous amount of lost motion and valuable time. ("Time" translates to money!)

**DeVILBISS TOUCH-UP GUN**—This is a moderately sized paint sprayer, commonly used by taxidermists for sealing and glossing fish. It is good for evenly covering large areas with color or finishes. (Supplier No. 60.)

**PAASCHE AIRBRUSH**—Airbrushes are great for achieving detailed blending of colors by controlling the flow of a fine paint spray. There are numerous brands on the market; however, we have had good luck with the Paasche VL series such as the Paasche VL3.

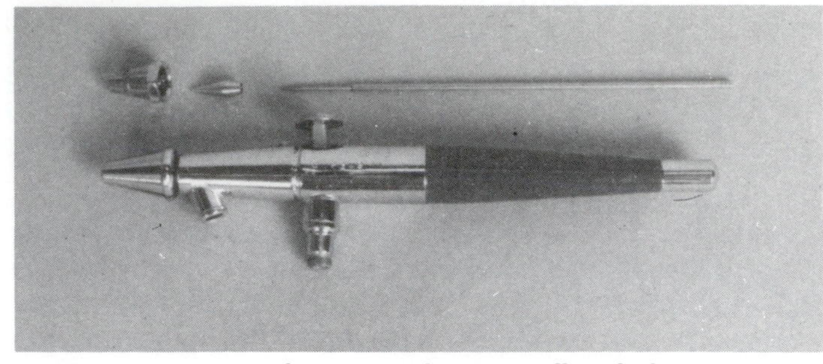

Basic accessories for a Paasche VL3 will include conversion kits for changing a VL3 to a VL1 (very fine detail) or VL5 (wide spray), extra airbrush bottles, caps, extra needles and tips, siphon assemblies, 6' or 12' –1/8" airbrush hose. Accessories for a DeVilbiss could include extra cups and a 12' hoses. The secret to having success with an airbrush is to keep it extra clean and don't drop it on the floor. (If you do drop it more than likely you will need to replace the needle and tip.) (Supplier No. 60.)

**AIR COMPRESSOR**—An air compressor that is capable of delivering at least 80 lbs. PSI will be needed. Tank types are best. The continuous running diaphragm compressors are not only noisy, they don't deliver enough air pressure to operate a DeVilbiss airbrush and cannot produce enough air pressure to drive air tools. The air compressor should be able to operate an air stapler, air grinders, sanders, etc. They are also handy for blowing loose debris from exhibits.

An excellent time saving system for interchanging different air tools or airbrush equipment is quick-disconnect couplers and plugs. These require a simple push on the coupler to remove an air tool. The coupler locks onto the plug to reattach another tool. These save a tremendous amount of time and effort unscrewing hoses and reattaching.

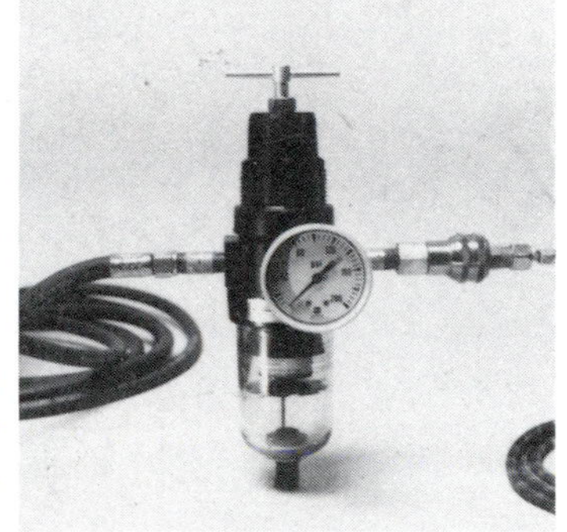

Moisture is a problem with compressors as condensation will develop and water will build in the tank which, in turn, will be pushed out the line when the compressor runs. The tank should be drained daily. Be aware that moisture will adversely affect painting operations and will ruin air tools. A moisture trap/air regulator (see photo) is a necessity and no studio should be without one. (Supplier No. 40 and/or 60.)

**ARTISTS' BRUSHES**—Much of habitat and exhibit building relies on painting artificial rocks, plants, trees, etc., to recreate nature. Additionally, background paintings are often desirable in this field. Basically you get what you pay for with artists' brushes, and as a rule, the more expensive brushes last longer and perform better than their inexpensive counterparts. It is not advisable to buy a $75.00 Kolinsky sable brush to paint a large mache artificial landscape; however, intricate background paintings or subtly highlighting small details on an artificial mushroom or background painting would be well suited for this type of a brush.

To obtain good results with artists' brushes, it would be wise to invest in two items: the Silicoil brush cleaning/thinning system and the Winsor & Newton Brush Jar. These items protect and prolong the life of your valuable artists brushes.

The Silicoil brush cleaning/thinning system (see photo above) is a "necessity" as it effectively cleans and conditions brushes, keeping them soft and ready for use. As a pleasant surprise, it also helps to blend colors together and actually makes it much easier to obtain desired results. The inexpensive jar has a metal coil and the jar contains a special cleaner/conditioner. Simply brush the coil after dipping it in the fluid and it will both clean and condition the brush.

The Winsor & Newton brush jar organizes your artist's brushes and provides a convenient place to keep them. No longer will you have to search throughout the studio looking for artists' brushes. As a by-product the artist has a very professional looking container to store them in. (Supplier No. 60.)

**TONGUE DEPRESSORS AND STAINLESS STEEL SPATULA**—Thoroughly stirring resins, epoxies, polyurethanes, airbrush paints, etc., is absolutely essential. Tongue depressors make good disposable stirrers. Stainless steel spatulas can easily be cleaned and used time and time again. (Suppliers No. 10 and 40.)

**DISPOSABLE PAINT BRUSHES**—Every artist that uses epoxies, resins, and the like should have on hand a good supply of inexpensive disposable brushes. (These are sometimes called acid brushes.) It's just not pracical to take the time to clean brushes when using these type products, especially since these disposable brushes are available at such a low price, (19¢ each at the time of this writing). (Supplier No. 40.)

**PARRAFIN WAX**—Many artists use parrafin wax for mold making. Often a wax impression is used when clay or other materials could cause an adverse reaction to sensitive RTV silicone molds. Parrafin wax is easily heated to a liquid state, then poured over the suject to be molded where it solidifies rapidly. Once it is removed from the subject, it forms a good mold. Wax can also be heated and easily removed once other materials have been cast from its impression. This is commonly referred to as the "lost wax method." (Supplier No. 40 or available from your local grocery store.)

# Cutting Tools

There are a variety of tools commonly used for cutting the many different materials that will be encountered on a day to day basis. The wildlife artist will find that having the right tool for the right operation will save both time and money. In many instances, a particular tool will more than pay for itself in saved labor in a very short period of time. most of these are available at local hardware stores.

**BAND SAW**—This handy tool is perhaps "the" most useful tool in the wildlife artist's studio. It will easily cut contours of all shapes and sizes. Carvers and taxidermists rely heavily on it to cut wood bases, foam blocks, and other materials. When obtaining a band saw try to purchase one with as big of a clearance between the base and the head of the saw as affordable. It is often desirable to saw large blocks of foam for habitat work and the extra clearance will be appreciated. (Supplier No. 40.)

**JIG SAW**—An alternative to a band saw is a jig saw. In fact, a jig saw will work in many situations where a band saw cannot be used, such as large bases that cannot be picked up and moved to the band saw. (Supplier No. 40.)

**KEY HOLE SAW**—This hand saw will come in handy for sawing holes and squares out of bases. (Supplier No. 40.)

**COPING SAW**—This fine bladed saw is used for trimming miter cuts on molding. (Supplier No. 40.)

**HAND SAW**—Self explanatory.

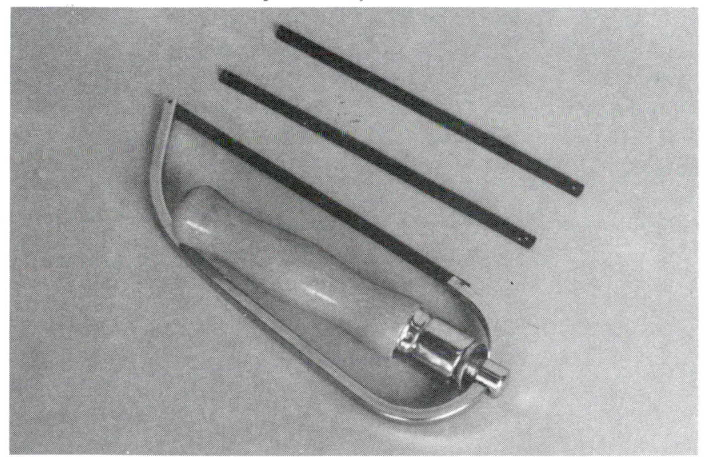

**HACK SAW**—Used for cutting metal (the Olson pocket hacksaw is pictured above). (Supplier No. 40.)

**CHAIN SAW**—A chain saw is necessary for collecting and trimming driftwood, stumps, etc. (Supplier No. 40.)

**BONE SAW**—Bone saws are excellent for many purposes besides the obvious "sawing bones." They can be used to cut wood, masonite, and other materials as well. (Supplier No. 10, Code No. KL17.)

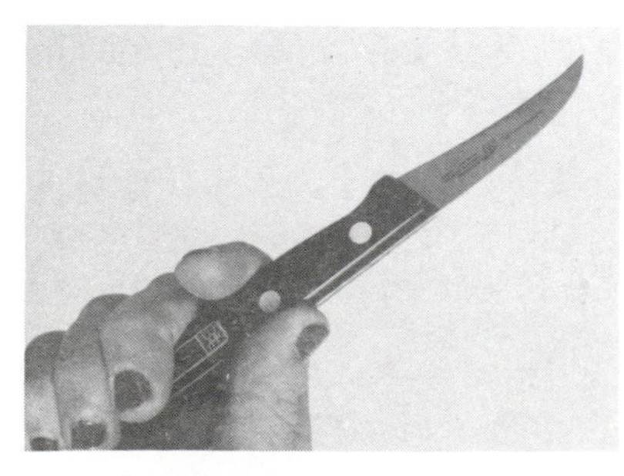

**KNIVES**—Every wildlife artist will have a good selection of knives on hand. A fisherman's fillet knife is a necessity when carving foam, and the "Perfect knife" (photo on left) is perfect for just about anything. (Supplier No. 10, Code No. MR120.)

**KNIFE SHARPENERS**—There are several excellent sharpeners available; the Paper Wheel Sharpening System is one of the best. Ceramic sharpeners, crock sticks and sharpening steels work great for touching up a razor edge. (Suppliers No. 10, 40.)

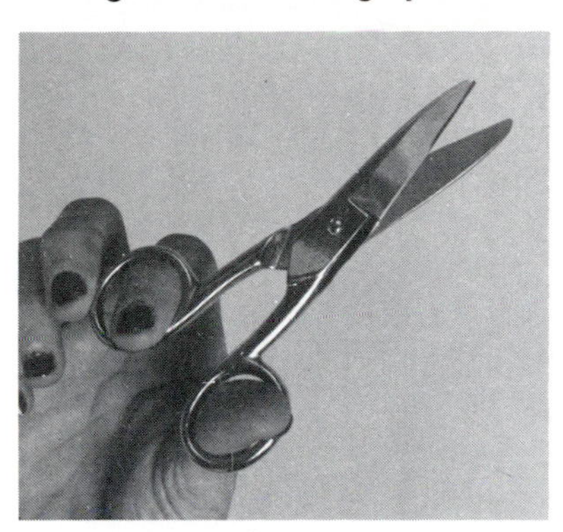

**SCISSORS**—The Ultimate Scissors (pictured at the left) are the absolute best that we can find. They cut virtually anything and retain a good lasting edge. (Supplier No. 10, Code No. MR110.)

**SHEARS**—Used for trimming sheet metal, metal lathe, and hardware cloth. (Supplier No. 40.)

**WIRE CUTTERS**—Good for cutting metal lathe and hardware cloth. Also handy for cutting wires for mounts that are installed into habitat scenes. (Supplier No. 40.)

**CHISELS**—Wood chisels come in handy for shaping driftwood, roughing out wood shapes, etc. (Supplier No. 40.)

# Measuring Devices

**PLASTIC BEAKERS**—These handy, solvent-resistant beakers are graduated in both fluid ounces and ml (ccs), which makes them ideal for making volume measurements. (Supplier No. 40.)

**VOLUME CUPS**—These are disposable, heavy paper cups that come graduated in fluid ounces for easy mixing. (Suppliers No. 40, 60.)

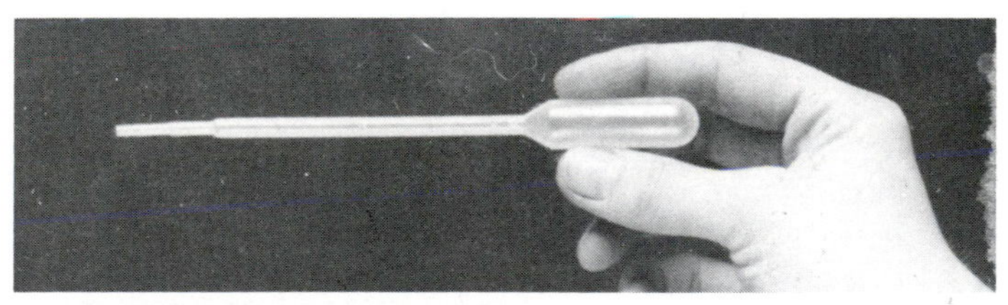

**PIPETTES**—These are good for transferring small amounts of liquid. They are generally graduated in ml (ccs). (Available from medical supply companies or pharmacies.)

**SYRINGES**—Glass syringes are ideal for transferring many liquids. These are also graduated in ccs for easy, but accurate, liquid measurements. (Avaible from your local pharmacy.)

**BEAM SCALES**—For measurements by weight, a beam scale is ideal. Measurements are in grams. (Supplier No. 50.)

**TAPE MEASURE**—The artist will need both cloth and metal tape measures. (Supplier No. 40.)

**MEASURITE**—This handy tool measures inside and outside diameters. Gives accurate readings in both inches and milimeters. (Supplier No. 10, Code No. MR100.)

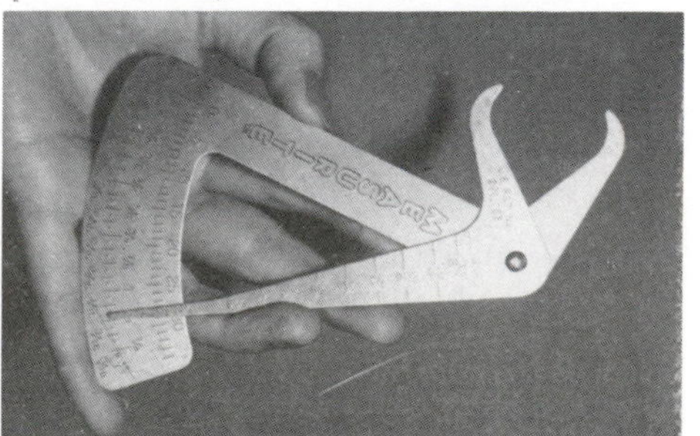

**CALIPERS**—Thse are ideal for measuring inside and outside diameters. They come in varied sizes. (Supplier No. 60.)

## Fastening Devices

**SCREW GUN**—This item is a necessity. Power screw guns will save a tremendous amount of time. They are geared for driving screws quickly and effortlessly. Although attachments can be had for phillips or flat head; phillips head is the preferred screw. (Supplier No. 40.)

**STAPLER**—An industrial grade air or electric stapler is ideal for studio work. The inexpensive hand staplers or small electrics are not nearly as sufficient for driving staples deep enough. This item will pay for itself. (Supplier No. 40.)

**HOT MELT GLUE GUN**—This handy tool heats up glue sticks to a melting stage. Squeezing the trigger pushes the melted glue out of the gun onto the surface that is being glued. It cools and hardens in 60 seconds, bonding just about anything. (Supplier No. 40.)

**HAMMERS**—Self explanatory. (Supplier No. 40.)

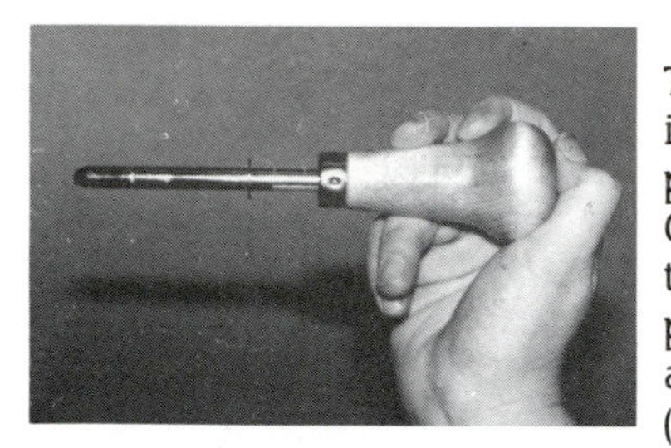

**BRAD & PIN SETTER**—This is a good tool. Simply insert a galvanized nail or pin into the magnetic tip. Once loaded simply push the sliding handle in and it pushes and locks the brad all the way to surface level. (Supplier No. 40.)

## Material Removal & Grinding Equipment

**DRILLS**—An assortment of drill bits with a reversable power drill is necessary for almost any project. Variable speed drills are more versatile. (Supplier No. 40.)

**FILES/RASPS**—Heavy duty industrial grade file/rasps are great for quick removal of material. Works on bone, foam, wood and a variety of other materials. (Supplier No. 40.)

**FLAPWHEELS**—These come in assorted sizes and are excellent for lightening-fast, rough sanding of surface areas. Also good for removing the flashing from mannikins. (Supplier No. 40.)

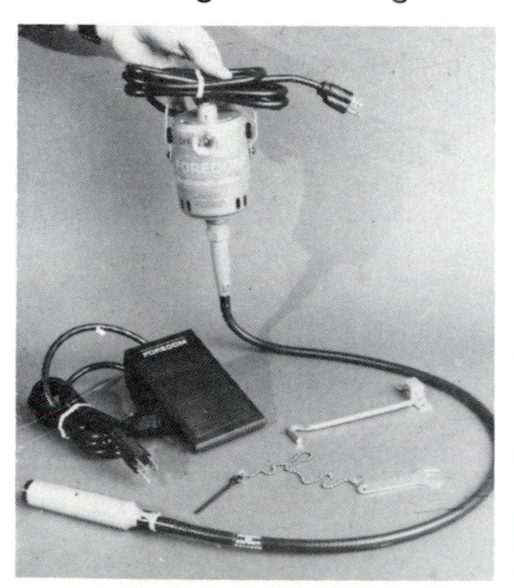

**FOREDOM TOOL**—This flexible shaft tool is very useful. It can be used for grinding, carving, sanding, deburring, drilling, shaping, polishing, routing, engraving, and more. Dremel tools, which are handheld motors, are also very useful. (Supplier No. 40.)

**BENCH GRINDERS**—These are excellent for heavy duty grinding. Always use safety goggles with this tool. (Supplier No. 40.)

**SANDING SCREEN**—Useful for sanding foam and other similiar materials. It is reuseable, self-cleaning, and doesn't wear out easily. (Supplier No. 40.)

**SANDPAPER**—Keep a variety of sandpaper on hand to shape foam, sand wood, and smooth rough edges. (Supplier No. 40.)

## Safety Equipment

**FIRE EXTINGUISHER**—No studio should be without at least one fire extinguisher. Locate them in convenient, easy-to-get-to places and know how to use one. (Check your Yellow Pages for a local supplier.)

**EYE WASH STATION**—Many chemicals commonly used by wildlife artist could cause blindness if accidentally splashed into their eyes. Locate an eye wash station near your work area and know how to use it. (Yellow pages under safety equipment.)

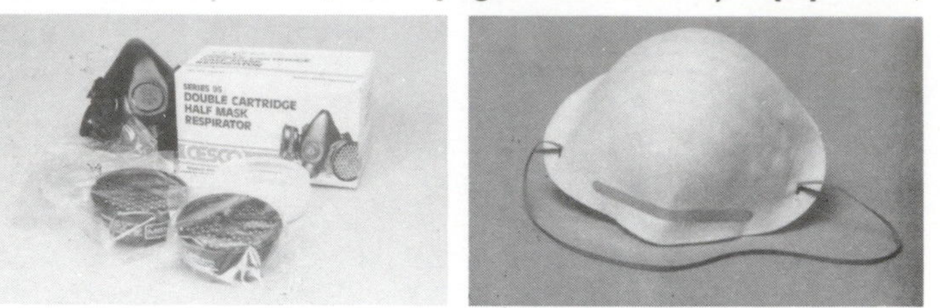


**RESPIRATOR**—There are different types of respirators that are designed for different hazards. Active charcoal filter masks (above left) are used for many types of dangerous fumes such as lacquer, polyester, etc. Dust masks (above right) should be worn when using pigments, powders, etc. (Supplier No. 40.)

**PROTECTIVE EYEWEAR**—Protect your eyesight with protective eyewear. Use chemical resistant goggles for chemicals that could splash or for protection against flying objects. Don't become a blind artist. (Supplier No. 40.)

**RUBBER GLOVES**—Many chemicals can be abosrbed directly through the skin. Others can cause dermatitis and rash. Avoid these hazards with chemical resistant gloves. (Supplier No. 40.)

**CHEMICAL RESISTANT APRON**—Some chemicals can burn, cause skin rash, etc. Protect yourself with a chemical resistant apron. (Supplier No. 40.)

**SPRAY BOOTH**—A booth with a powerful exhaust fan to vent harmful fumes. A necessary item for airbrush painting lacquers, using polyesters, etc. Good ventilation is a must! Don't breathe anything but fresh air. (Supplier No. 40.)

## Texture Tools

**SPONGE**—Natural sponges make great texturing tools for mache bases. They are also handy for cleaning up.

**GROOMING BRUSH**—Brass, bristled grooming brushes are good for texturing habitat scenes as well as grooming mammals. (Supplier No. 40.)

**WOOD BURNER**—This unique tool allows the artist to burn detail rather than draw or paint it. The tip can be shaped for very fine texture work. (Supplier No. 60.)

**DRYWALL TEXTURING BRUSHES**—These tools are available from most any drywall supply company. They are good for texturing mache habitats. (Supplier No. 40.)

**FLOCKING GUN**—This gun is a good applicator for flock. They are available from auto supply shops in a hand sprayer as well as auto sprayers.

## Miscellaneous

**CAULKING GUN**—Necessary for using Fin Backing Cream. Eliminates waste. (Supplier No. 40.)

## More Information

The *Breakthrough* How-To Library's "Commercial Fish Reproduction in Fiberglass" is an excellent booklet with a step-by-step guides to producing fiberglass fish. It contains a wealth of information concerning making molds. Much of habitat and exhibit building is molding, therefore, this book is invaluable. (Supplier No. 10, Code No. BP2003.)

# CHAPTER 3

*Every accomplished artist in every medium understands the value of good reference. Habitat construction is no different. The closer one comes to recreating an actual habitat, the more natural the scene will be depicted, and the more pleasing it will ultimately appear.*

# The Importance of Good Reference Material

Creating realistic habitat exhibits is no accident. It requires careful study, planning, and a thorough knowledge of the artist's subject.

One of the most helpful tools in a wildlife artist's studio will be a reference library. Any successful artist will have good reference materials within an arm's length while working and will refer to them often. Reference materials are not restricted to wildlife or fish—they are just as important for recreating the subject's environment accurately. For the best results, study the subject and study its environment. Don't for example try to recreate ripples on water moving out in front of a swimming duck without a first hand knowledge of exactly how they should look. Should they be rings, or waves, or "exactly" how should they look? Often the total effect (and all your hard efforts and materials) is ruined by a lack of knowledge of the subject and or environment.

Before actual construction begins, always obtain good reference materials. Photos are an excellent source of reference. Collect photographs from outdoor and nature-oriented magazines, or use a camera and take your own. There is absolutely no excuse for not being able to obtain good reference photos of natural habitat scenes. Duplicate your reference as closely as possible. If you don't have reference material for a particular scene obtain it prior to creating it.

Perhaps nothing is more important than obtaining and "using" reference materials for creating natural and realistic scenes. Don't overlook this important subject!

## Building a Reference Library

Building a reference library of environments is far easier than building one for wildlife and fish. Taking photographs of environment components requires very little effort and is time well spent. Take a systematic approach when building a reference library. One way to do this is first organize the reference library according to classification, i.e., **Rocks**, **Vegetation, Water (liquid), Water (frozen),** and **Groundwork.** Then, subdivide each classification to scenes that will likely be recreated in your studio.

**ORGANIZING A REFERENCE LIBRARY**

| ROCKS | VEGETATION | WATER (Liquid) | WATER (Frozen) | GROUNDWORK |
|---|---|---|---|---|
| Granite | Forest | Splash | Dry Snow | Sandy Bank |
| Sandstone | Swamp | Stream | Wet Snow | Dirt |
| Shale | Desert | Pond | Ice | Mud |
| Lava (volcanic) | Oak Leaves | Ripples | Icicles | Desert |
| River Rocks | Grass | Droplets | Frost | Forest Floor |
| Pebbles | Pine Needles | | | Creek Bed |

If a classification changes with the seasons, (summer, fall, winter, spring), be sure and have reference for each one. Almost any artist would know better than to portray a summer plumage bird in a winter scene. The same holds true with the environment. The artist wouldn't want to use winter foliage in a spring environment nor would they portray blooming spring flowers in a winter scene.

Many elements of the environment such as rocks, plants, grass, etc., can actually be brought right into the studio and the artist can use the subject itself as reference.

VCR recorders with stop action capability are useful for studying moving objects such as the splash created from a leaping fish or the ripples made from a swimming duck.

Photographs are the most widely used reference tool. They can be easily filed and stored and take up very little room. They are readily available and can be obtained with relative ease. The serious wildlife artist should own a quality 35mm camera and know how to use it. A few hours spent taking pictures in the forest will yield a wealth of reference that can be used for years to come.

## Tips for Photographers

- By holding an empty 10 gallon aquarium in a stream, one can shoot pictures of stream bottoms (or purchase an underwater camera).
- Color print film works great for color reference shots. It's inexpensive, and can be developed overnight.
- When shooting splash shots, use a high ASA/ISO rated film (400 minimum, 1000 works better although the photo will be somewhat grainy) and a fast shutter speed (1/500 or more) to freeze the action.
- Placing a yardstick or ruler beside the subject being photographed helps to establish the size of the subject.
- A Macro lens for a 35mm camera is a valuable asset for taking close-up pictures. (Close-up photographs will be more beneficial as a rule than general scenery shots for habitat recreation work.
- When studying photographs of reference subjects, study "fine" details. Notice areas that are shadowed versus highlighted. Notice color patterns, shapes, size, etc. Reproducing intricate detail into your rendering can make the difference between an average exhibit and a superb rendering.
- When shooting a plant, flower, or leaf, place a piece of paper behind the subject to eliminate background clutter.

# Miscellaneous Water Reference Photos

## Moving Water

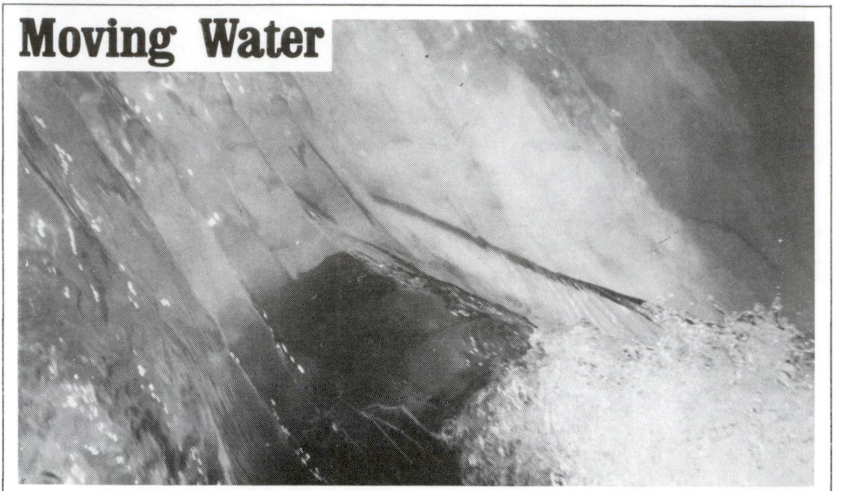

*Waterfall—falling and churning water*

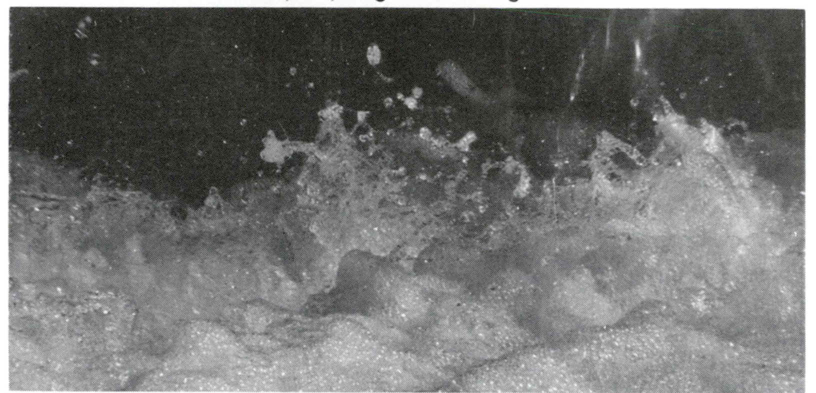

*Severe turbulence—frothing water*

*Waterfall—turbulence*

*Ripples from a swimming duck.*

## Splash

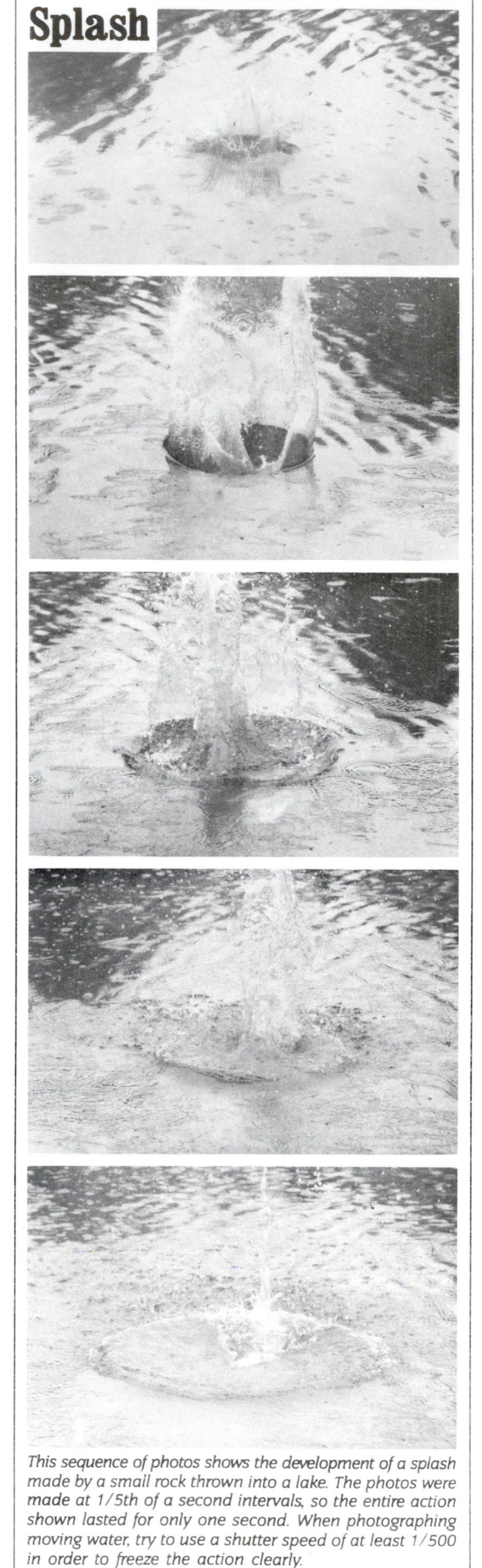

*This sequence of photos shows the development of a splash made by a small rock thrown into a lake. The photos were made at 1/5th of a second intervals, so the entire action shown lasted for only one second. When photographing moving water, try to use a shutter speed of at least 1/500 in order to freeze the action clearly.*

Snow
Stream Banks

# Desert Habitat Reference Photos

It occurred to us as we were photographing material for this section, that tne word "desert" was perhaps misapplied. True desert is really devoid of habitat, and about all we could show would be various formation of sand, garnished perhaps with a cap belonging to a long departed French Legionaire!

The "desert" we are most familiar with is really "semi-arid" desert, which contains all sorts of plant and animal life, and is the type found in the western part of the United States.

One question that comes to mind when discussing any type of habitat is whether to use the natural materials found in nature or to reproduce these materials synthetically. Habitat materials found in the east, for example, are usually very rich in moisture and very green and healthy appearing, which presents the problem of proper preservation. Semi-arid habitat is different, mainly because the moisture level is so low, and everything has a more or less dried-out look. In the spring, the semi-arid desert also blooms green, but stays that way only a short time. Because of these conditions, western habitat can be used basically "as-is," with no deteriorating effect whatsoever. Various western cactus plants, sagebrush, grasses, weeds, and flowers fall into this category. The following series of photographs should stimulate many ideas in creating a semi-arid habitat.

## Rocks

*These photographs display typical rock scenes found in semi-arid country. Notice how the rocks are almost always accompanied by lichen and grass. The top two photos show lava rock formations. One of the most difficult problems in creating any type of habitat is to show what we wish to show in as small a space as possible. To increase the visual impact of a habitat base, try to incorporate vertical height in a manner such as can be seen in the above examples.*

## Earth

*The basic ground effect of semi-arid country will almost always have a dried appearance, sometimes even parched looking (as shown at upper left). Ant hills (above center) are very common in western habitat, and would be quite easy to construct. Gopher holes (above right) can also be incorporated in semi-arid scenes. Badger diggings (at left) can be used to increase visual interest in a habitat setting.*

# Vegetation

**CACTUS**
*Notice how the cactus shown in photos above is always accompanied by other types of natural habitat, such as grass and rocks.*

**SAGEBRUSH**
*The most common plant-life in semi-arid country is sagebrush, and can be used either alive (top photo), or in the dried, weatherbeaten condition.*

**WEEDS, WILD FLOWERS, AND ROCK LICHEN**
*Virtually any type of weed or wild flower can be used to add interest to your scenery (top). The botom photo shows typical moss or lichen found in the western part of the country.*

# Habitat Artifacts

*While photographing material in the field, we came across many items that we had to classify simply as habitat artifacts. These items encompassed many things, such as the rusty tin can in (top left), or the remains of a steer (center left). Almost any item that offers a hint of man's influence on his environment is a habitat artifact. The log and rail fences at the lower left should suggest many ideas for your habitat arrangements. Sometimes just the end of a log or rail is all you need to add more impact to your scene. Barbed wire is also very useful in many western scenes. Something as simple as an old piece of pipe (upper right) can be used in some habitat bases. Many "habitat artifacts" can be located by searching around old buildings. The number of ideas to utilize artifacts is endless, and may even stretch to include a feather on a cow pie (center right). Our favorite artifact photo for our whole trip was the broken wagon wheel shown at the lower right. This photo really set our minds to wandering, and it almost tells a story out loud. Incorporating this scene with a group of quail or small animal would be extremely dramatic and appealing also. If a large mammal were used in conjunction with the broken wheel, for example a black bear with one foot resting on the hub of the wheel, a very emotional visual appeal would be created, and the bear and the wheel would become the center of interest. This type of habitat is what helps win Best of Show awards. One enterprising taxidermist, Vic Heincher, incorporated this concept into a mountain lion scene and walked away with a gold medallion in the 1986 Master of Masters World Taxidermy Championships, again indicating how well received such a concept is.*

# CHAPTER 4

# Design & Composition

***by Bob Williamson***

Successful wildlife artists realize that the outcome of each rendering is heavily influenced by the initial effort put into the design and "composition" of the piece. Webster's dictionary defines *composition* as "the uniform arrangement of artistic parts." Systematic forethought and careful planning are prerequisites of building an artistic exhibit that is worthy of putting on display. Keep in mind that fine art doesn't just "happen." It is carefully planned and will require tedious research and study. Size, shape, and color coordination (blending and/or contrast) must all be taken into accord. Creativity and artistic expression are vital to the success of this initial planning stage. There are two major types of exhibits or bases that are generally incorporated into renderings: *natural* and *abstract*.

Obviously, an artist could incorporate *both* types into a scene, but most prefer one or the other. Natural scenes require a "thorough" knowledge of one's subject and careful research of the natural habitat that is being recreated. Knowing the habits, characteristics, and environment of the subject especially during the time of year that is being depicted is elementary for successful habitat scenes. Abstract settings are not hindered by strict scientific recreations and interpretations, therefore, they require less attention to exact and natural detail and rely more heavily on artistic expression, design, composition, and *impact*. "Dyed-in-the-wool" conservative artists often turn up their noses at abstract scenes. However, many artists, especially some of the younger, more "daring" types, have created some abstract renderings that are undeniably masterpieces.

*These are excellent examples of both "natural" and "abstract" bases by the same artist. Both renderings by Larry Barth won Best in World at the Ward Foundation World Carving Championships Lifesize Decorative Division. The snowy owl and Bonaparte's gull on the left (1985's winner) exhibited a very natural scene while the pair of terns on the left (1986's winner) were displayed in more of an abstract setting.*

*Here is another fine example of both types of artistic expression by the same artist (Bob Berry). The porpoises (above) are an example of an abstract base and the butterflyfish (at left) is shown in a realistic natural scene. The porpoises won a blue ribbon at the Taxidermy Review competion and the butterflyfish won Best in World Fish carving at the 1985 World Taxidermy Championships.*

The addition of some type of base or background to a finished piece greatly adds to the visual appeal and potential monetary value of the overall rendering. A base may be as simple as a small piece of oval panel or driftwood to "frame" the subject, or it may be an extensive background that completes the story of the specimen. An example of this would be a snow scene for an otter or an Alaskan ptarmigan complete with background painting. Abstract bases on the other hand, can only be limited by the artists' imagination and can be constructed of virtually anything.

Some subjects, such as very large fish, could probably get by without "any" additional base or scene as their size alone creates sufficient viewer interest; however, it is the opinion of this writer that some sort of base, driftwood, etc., will always add to the impact of virtually any piece. A pedestal or two-sided fish will, of course, have to rest on something, therefore, a base or scene of some type will be required.

Lifesize large animals also require some type of base to physically support them and a natural base can be very easily included. Small subjects may be displayed with or without a base or habitat of some kind, but will almost always have a better viewer appeal if some type of base is added. For example, if you wished to pose a raccoon or squirrel lying down, they would need to be positioned on top of "something," and a small imitation rock or stump would add immensely to the visual appeal of the overall piece.

As artists, it is part of our job to explain to a customer the advantages or possible disadvantages of various types of bases or habitat additions to a rendering. Try to have several examples of habitat types for them to look at, and let them decide which would add the most to their particular piece.

Pricing also becomes important when discussing various bases or habitat additions. If one intends to mount or carve a squirrel or bird for a customer, the cost of a small corkwood or driftwood base is usually included in the overall price. On the other hand, if the customer likes a snow or forest "floor" scene, the price must be adjusted in your favor to include the extra work. In many instances, the additional cost of a nice snow scene or a small creek bank overhang will be as much or greater than the basic cost of the mount or carving. An artist should realize that by displaying and promoting some of the nicer types of habitat bases, that they can add hundreds or even thousands of dollars to their annual income.

The biggest bonus to promoting habitats perhaps is the fact that they are fun to build, and always break up the monotony of routine studio work. Habitat scenes are not (at least at the time of this writing) commonplace; therefore, they are good advertising tools for your business—especially to customers who have never seen mounts or carvings displayed in such a manner. Many artists have established sideline businesses with local gift shops, etc., with quail or pheasants in a natural setting complete with glass dome or plexiglass case, where the selling feature is primarily the habitat, not the bird!

# The Design/Composition Process

*by Jim Hall*

## To Begin the Project

To begin a project and properly design it, preliminary sketches should be made long before the actual work can start. These sketches will answer many questions and help to resolve problems that will inevitably arise when the actual work begins. It must be emphasized that one doesn't have to be an accomplished pen and ink artist to make these sketches. Even crude sketches will suffice. Primarily the purpose is to establish a good design and formulate a plan (just as a builder needs a plan). Often unforeseen problems can be solved ahead of time on paper rather than costly time-consuming experimentation, wasting material on a poor design that has not been completely thought through.

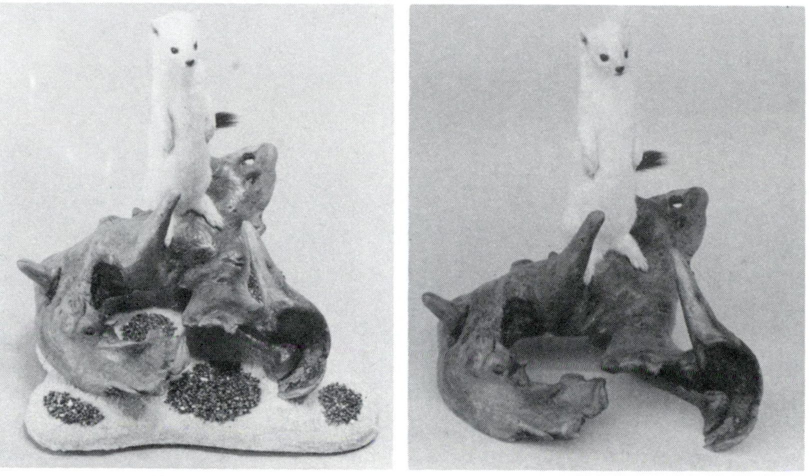

*Figure 1* *Figure 2*

Don't be reluctant to try more than one approach. Make several sketches and drawings of varied poses, positions, etc., until completely satisfied with the overall look. It is often advantageous to get other opinions as to which design is most liked.

Once the final design composition and pose has been selected, carefully analyze the piece to determine the best and easiest way to construct it. Many, many hours can be expended in designing and building a habitat base, particularly the first time, as many things must be considered. Some of these are:

1. The overall size of the base in relation to the actual mount and the physical size of the overall piece.
2. The color of the base, again in relation to the color of the subject.
3. Exactly how complicated or simple the design.
4. Deciding the type or style of the piece.
5. Determining the pose or attitude of the subject.
6. Projecting and actually costing of the project.
7. Arrangement.

Other considerations the artist feels are important can be made and then, once the design of the habitat base is well thought out, it can be produced. If so desired, it can be duplicated over and over quite easily; but it is vitally necessary to first plan the original concept thoroughly and completely.

## 1. The Size of the Base

The physical size of the base in relationship to the size of the mount is an important factor to consider. Bear in mind that the center of interest is the mount or carving, and that the base should only add to the visual appeal. For example, creating a habitat base four feet long to display a small muskrat or mink won't work well because the viewer's eye will be detracted *AWAY* from the center of interest because of the large size of the base. This is easy to illustrate. Notice the ermine mount on the combination driftwood/rock base in figure 1. In this situation, the small ermine is really overwhelmed by the base. Notice how difficult it is to keep your eyes on the ermine when viewing the overall piece. A smaller or less complicated habitat, such as the simple piece of driftwood as seen in figure 2, might be better for the ermine. Note how the two pine marten mounted on a combination driftwood/dirt base (figure 3) display a nicer overall balance between the size of the specimens and the base.

*Figure 3*

*Figure 4*

A reverse situation can be seen concerning the lifesize black bear in figure 4. In this case, the viewers attention is focused almost entirely on the bear, mainly because the base is rather small and plain. A larger, more detailed base may have been better for this mount, but this type was utilized to direct attention to the bear and minimize showroom space requirements.

Another consideration is the overall size (and weight) of the completed piece. Considerations such as the average size of door openings, transportation problems, display area size, and so forth need to be pre-planned. Large dioramas, such as many museums require, must be built in sections and later the sections assembled on site. Weight is always something to consider and whenever possible, "lighter" is always better. (Take it from someone that has hauled many a "heavy" exhibit.)

The customer will generally decide whether they want a rendering to be a table or wall display, and the artist should be prepared to discuss the merits of each basic type. Table displays always require the most work as they can be viewed from any direction; also, more perfect specimens are generally required. Hence, one should charge accordingly for these more complicated models. Another thing to advise the customer of, is that table displays require more room space and are more easily damaged, especially if there are small children around. Wall displays on the other hand are, more-or-less, out of the way and can adequately display the subject very well. Generally wall displays at least partially conceal the back side of the renderings making them easier to produce.

The physical size of the overall piece will be governed by the size and the type of specimen and the habitat desired. These are a few of the things that a customer should consider when purchasing your sevices. As you are discussing these various problems with the customer, you are actually starting the "mental composition" of the mount.

## 2. Color

The color of the base is also important. As a "general" rule, place a dark colored specimen (mink, crow, fisher, etc.) on a light colored base; and a light colored specimen on a rather dark colored base. As can be seen in figure 5, the dark colored fisher is much better displayed in a snow scene. Mink and otter also show exceptionally well in snow or snow-covered forest floor habitat.

*Figure 5.*

The light colored ranch mink mount combined with the darker base background in figure 6 illustrates a nicely balanced contrast of color. Refer to the photograph of the single pine marten mounted on the forest floor base below (figure 7). Notice how the visual image of the rust colored marten tends to blend with the background, defeating much of the contrast in color. The snow patch helps some, but not quite enough.

*Figure 6*

*Figure 7*

*Figure 8*

The white ermine mounted on the forest floor base (figure 8) again reverses this color contrast technique. Unfortunately, the taxidermist created one error in this mount in that the ermine is only white in the winter months and looks somewhat out of place when no snow is present. Always remember to place the subject in a setting that would be natural to it; i.e., do not place a summer plumage bird in a winter scene or a winter coat ermine in a summer scene such as this one was. A light dusting of artificial snow over the fall scene would have made this mount more convincing.

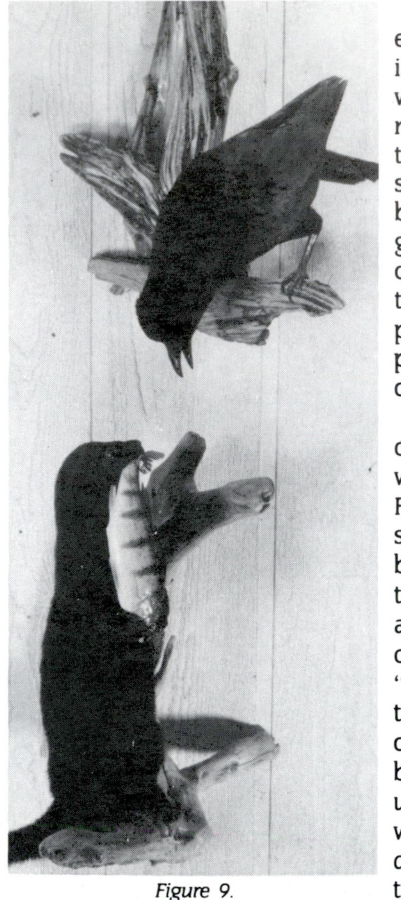

Figure 9.

Color considerations can even be carried as far as to include the color or type of wall upon which the finished mount will be displayed if this fact is known. Figure 9 shows a black mink and a black crow displayed on a grey colored wall. The color contrast is obvious. If these two pieces were to be displayed on a dark colored paneled wall, a great deal of color contrast would be lost.

A good knowledge of colors can help tremendously when composing a piece. For example, cool colors such as certain shades of blue can often lend authenticity to a cool scene such as a snow scene that would otherwise not capture the "feeling" that the artist is trying to express. Warmer colors such as reds, yellows, browns, etc., may be more useful in a summer scene or wherever a "warm" effect is desired. (Most viewers prefer the warmer colors.)

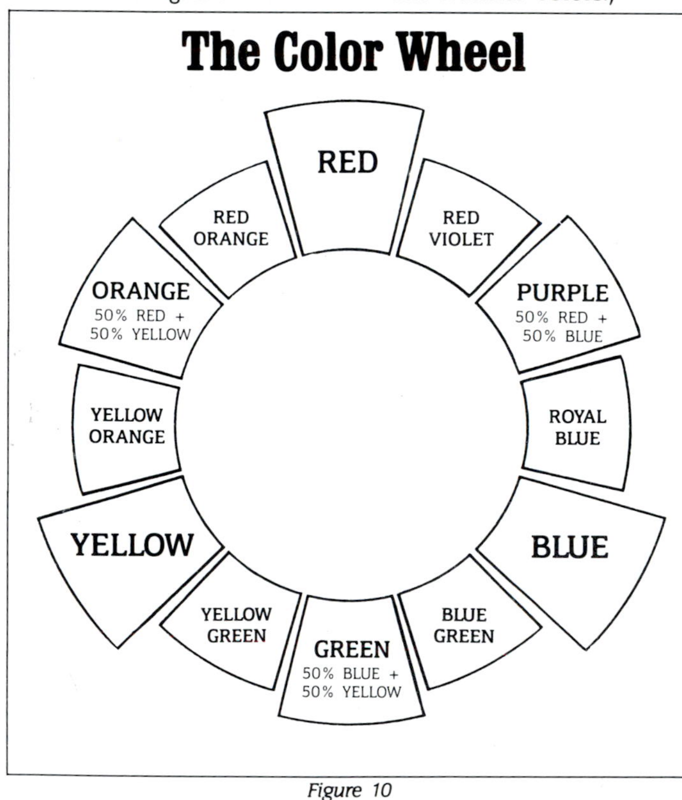


Figure 10

Knowing complementary colors and opposites will also be of considerable help when composing scenes. A color wheel is illustrated above. Complementary colors are colors that are positioned next to each other on the wheel. Opposite colors are situated directly across from each other. A good book and/or an art class that covers this valuable subject is definitely recommended to aide the artist in color-coordinating renderings.

# 3. How Complicated or Simple?

Should the habitat base be complicated or simple in design? This is a question only the artist can answer but perhaps we can offer some guidelines. At a major competition years ago, there was a largemouth bass display that hung on a wall that contained so much plastic plant material that it was difficult to even "see" the bass much less to judge it fairly. This piece was so cluttered with plant life that it could have won a botany competition possibly, but not taxidermy competition! The taxidermist simply got carried away with his habitat ideas. On the other hand, we have seen entries in competitions that had no habitat or base work associated with them at all which is just as bad an alternative. Obviously, a well-balanced specimen/habitat relationship falls somewhere in between and would be a better choice. This well-balanced relationship is illustrated in the carvings below.

*Figure 11. Tan Brunet's 1981 Best in World Decorative Pairs—Canvasbacks*

*Figure 12. John Scheeler—Mourning doves.*

If we were to establish rules for just how much habitat to add to a base, Rule Number One would have to be to use the "KISS" technique, described by Onno van Veen years ago as meaning, "Keep It Simple Stupid." The word "Stupid" really doesn't belong in the rule because designing a base or habitat to reflect simplicity, taste, and enhancement of the specimen requires much thought and experimentation, but the basic idea is sound. Look at how some former World Taxidermy Competition winners have utilized the simplicity technique.

*Figure 13. Arctic char, Best of Show by Jeff Compton, WTC 1983.*

*Figure 14. Largemouth bass, Best in World by Tom Sexton, WTC 1985.*

*Figure 15. Walleye, Best of Show by Alan Gaston, WTC 1986.*

Notice in all three of these fish mounts the simplicity of the habitat base, and how nicely the base compliments the mount. Similarly, carvings that reflect a simple yet elegant base design offer the finest appearance. Note how the base complements rather than detracts from the focal point on these fine renderings.

Rule Number Two might read: Don't attempt to create a habitat base environment for a customer or competition that cannot be done CONVINCINGLY! Nothing detracts from a nicely mounted specimen and habitat base more than a good idea that was carried out poorly. If your artificial water *REALLY* looks like a warped punch bowl, then practice with resins until it truly looks like *WATER*. If your icicles *REALLY* look like "sick" candles hung upside down, then practice with resins and colors until they truly look like icicles. After you have developed the "techniques" required to create these effects CONVINCINGLY, then utilize them in conjunction with your mounts.

Figure 16 illustrates a very convincing "splash" created from artificial water by award winner Rick Lauriente of Denver, Colorado. The actual construction methods and techniques for creating various scenes are discussed elsewhere in this book. Here it would be wise to mention that regardless of proficiency with technique, obtaining and using reference is by far the most important aspect of creating a convincing habitat. One must know "exactly" how something should look in order to recreate it convincingly.

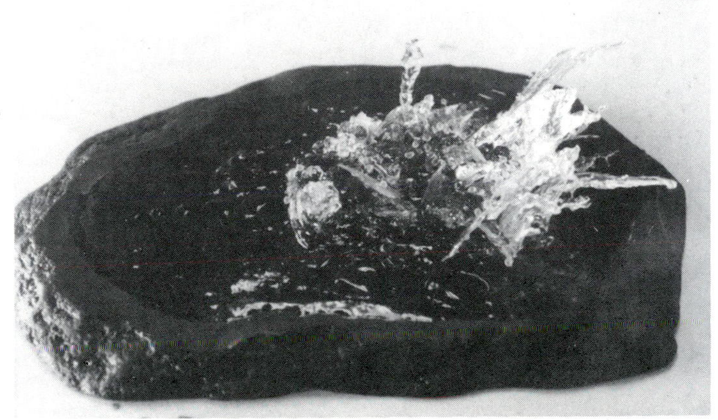

*Figure 16. Artificial splash by Rick Lauriente.*

There are hundreds of examples of good design and composition that have been used in various competitions over the years, and the techniques used by these winners will be as good tomorrow as they were yesterday. For example, if one wishes to display a lifesize black bear feeding in a garbage dump, then by all means add recreations of the garbage and junk that would appear in a dump to the habitat base! If you wish to recreate nature as God made it, perhaps you should use more conservative ideas that incorporate some of the techniques used by many of these previous winners.

*Figure 17. Frank Newmyer's 1983 Carl E. Akeley Winner*

*Figure 18. Vic Heincher's 1986 WTC Master of Masters—Mountain Lion*

*Figure 19. Wendy Christensen-Senk's 1984 WTC Master of Masters—Wolf*

# 4. Determining the Pose & Attitude of the Subject

This important aspect should "dominate" the entire piece. The "feeling," sight, smell, and emotion derived from the primary focal point (the pose and attitude of the subject) is what carries the piece to eventual glory or lackluster.

I have witnessed several competitions where an excellent idea was ruined by lack of attention to posing the subject correctly. I have seen beautiful habitats and bases with "ho-hum" subjects that all but destroyed the effort in the other areas. "Remember"—don't try to make the base or habitat "carry" the subject. The subject should *dominate* and capture the viewer's attention.

*Figure 20. Roger Smith's 1986 WTC 2nd place Carl E. Akeley—Red Fox.*

One of the best examples of creative posing and effectively transmitting an attitude is this red fox rendering by Roger Smith of Leslie, Michigan. Roger effectively and brilliantly posed the fox on a simple but elegant brass and hardwood base and captured his subject perfectly. The viewer is tempted to yawn with it but certainly not from boredom.

Put some thought into the pose and attitude of your subject. It goes without saying (but I'll say it anyway) that one must be "thoroughly" familiar with one's subject in order to effectively pose it. Study, research, and study more.

# 5. Projecting and "Actually" Costing the Project

It is essential to "cost" each project. Often enthusiasm and the fact that it's fun and interesting to create various bases and exhibits dulls the business sense of the artist. Creativity and artistic expression are great but many artists don't want to spend their lives as a "starving artist."

Artists need to make a profit also, and one essential ingredient in making a profit is possessing a firm knowledge of costs. Primarily (without getting into a business manual), all that is required is a knowledge of direct and indirect costs and profits. See Chapter 14: *Financial Considerations* for more information on this subject.

One must project the anticipated costs of the project after the concept has been developed. Hours needed to complete the project and required materials can only be estimated. If this is a commissioned job for a specific customer rather than a speculative job intended to be sold later, estimate "high." Speculative jobs are less crucial as the selling price can be increased if cost overruns are incurred.

When the actual job begins, very carefully keep up with the amount of time, materials, etc., spent on the project. Write it out upon completion and compare it with the projected figure. After just a few times with this exercise, your estimates will become very accurate. Once costs are under control profit is almost assured.

# 6. Arrangement

Once one has decided upon the basic plan, some thought must be given to "arranging" the various items within the setting. As an example, figure 21 shows three imitation stones

*Figure 21. Three separate stones.*

*Figure 22. Three stones in a "group."*

placed in a row, and figure 22 shows one stone placed on top of the other two. Both photos contain the same three stones, but the geometric placement of one stone upon two others is much more appealing to the human eye. Much of this type of arranging is simply a matter of trial and error, and by shifting various items, one arrangement will appear better than the others. As a rule, if a particular arrangement appears better than any other to you, it will also appear better to others. It is just this type of trial and error arranging that helps transform a craft item to elevate it to a piece of "fine art."

The arrangement of the three rocks in figures 21 and 22 also follows the basic Rule of Three. Five stones could be used for the center of interest, however, care should be used to ensure that the artist doesn't make the scene look too "busy." Some artists attempt to add too much to a scene, something like a Walt Disney menagerie where all sorts of animals live in a tiny area. This type of approach is usually unconvincing to the viewer, and should be avoided. We all know that "beauty is in the eye of the beholder" and this axiom is true of the artist as well as the customer or judge.

When constructing a particular habitat, always bear in mind the over-all theme of the piece. For example, if one wishes to create a habitat scene that is to be an excerpt from nature, as it were, one might consider a mother duck and several small ducklings. If the theme or title of the piece was going to be "Getting Your Ducks In A Row," then by all means the ducklings should be strung out behind the mother in single-file fashion. If the theme or title was going to be "Over Worked" or "Under Foot," then the ducklings should be piled over, under, and milling around the mother duck.

Your library of reference material will generally contain many habitat ideas. Pictures from magazines such as *National Wildlife*, *National Geographic*, or any of the outdoor sporting periodicals will aid you immensely in developing habitat ideas. It will be surprising how many "new" ideas will appear if you keep an open mind and just look for them. In many instances, the specimen or specimens will be shown in an ideal habitat, and the time required in designing a piece will be eliminated. All the artist will have to do is recreate what they see.

# The Rule of Three

Most of the work an artist produces will deal with only one specimen, such as one bird or one small mammal, etc. On occasion, however, a customer will desire that more than one specimen be included in his rendering, and very interesting ideas for "group" subjects will be presented. Virtually all state, regional, national, and world taxidermy and carving competitions have group categories, and they are highly popular since they offer the artist greater opportunity for expression. These are specific instances where the *Rule of Three* comes into consideration.

What is the Rule of Three? It is my opinion that there is a phenomenon that occurs within the mind of the viewer when observing various numbers of objects, regardless of what the objects are. For some reason, the human brain prefers to view "off" numbers of items. For example, figures 23, 24, and 25 show two, three and four stones, respectively. As you study these pictures, note how much more pleasing to the eye the photo of three stones appears as compared to two or four. This same reasoning will apply to wildlife art as well. A mother mallard hen and two chicks will be more visually appealing than if three chicks are used in a display. Three fish mounted in a group setting instead of two or four will also be more appealing. The Rule of Three may be a rather debatable point but is something to consider when designing that next group setting. As an added bonus, this phenomenon may enable the artist to convince the customer to include "three" specimens in a group

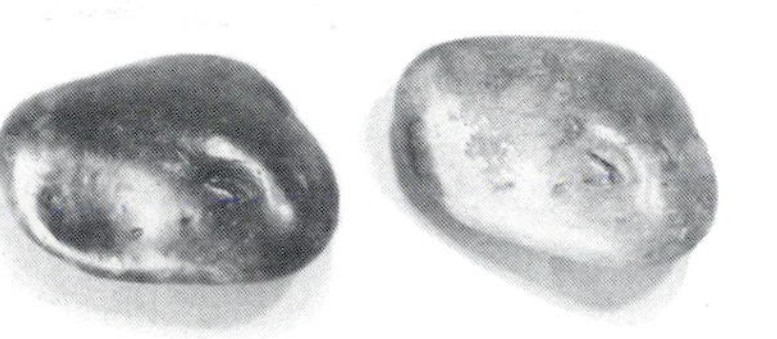

*Figure 23*

*Figure 24*

*Figure 25*

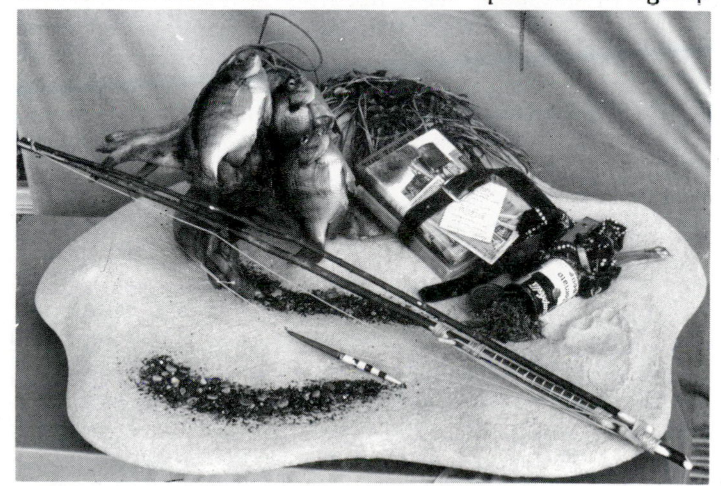

*Figure 26. "School Days" by Jim Hall.*

setting rather than two, and the value of the additional income provided is certainly *NOT* debatable.

The Rule of Three can, of course, be used with just one specimen and designing the shapes and components of the piece to reflect the same idea. Simply use three basic objects as focal points in the scene.

# Composition and Visual Balance

***by Ken Edwards***

Do most viewers even *notice* when a rendering has a good composition? Is composition really as important as its cracked up to be? Does composition actually play an extremely big role in the impact of the final result (as big as, say, technical ability or anatomical accuracy)?

The answers to these questions are yes, yes, and yes! Composition is every bit as important as any other phase of wildlife art, and a good (or a great) composition makes an enormous impact in the final piece. Even a totally non-artistic viewer will be able to see the difference between a well composed rendering and a poorly composed piece. They may not be able to put their finger on *exactly* what the difference is, but if pressed for an answer, they would probably say that the good composition looked more *professional*.

What *is* a good composition? The answer to that question could fill a dozen books. Basically a composition is the positioning of visual elements within a field. In a drawing, painting, or a photograph, the composition consists of all of the elements and their relationships to each other, as well as their relationships to the borders (edges) of the field. In photography, composition would include *framing*, which is basically where the subjects are placed within the picture frame. The examples below show the same subject framed in different ways.

The example on the left puts the center of interest right in the center of the picture frame (this is similar to the way most amatuer "snapshots" are composed). The result is a poor composition that would be unlikely to get the viewer involved with the subject. The example on the right shows the same subject framed differently (better) with composition in mind. In this example, the entire area of the frame is integral to the subject. The entire result is more interesting, more pleasing to the eye, and more professional.

When working in three dimensions, the relationships between the elements of a composition become even stronger and more complex. Not only does the artist have to be concerned with all the lines, shapes, colors and textures, but also with mass, depth, and a rendering which produces a pleasing composition when viewed from *more than one angle*. A good example of an outstanding 360° composition would be Larry Barth's 1986 Best in World Decorative Lifesize Wildfowl Carving Championships rendering of two terns flying low over the ocean. While photographing Larry's sculpture for an article in *Breakthrough* magazine, the photographer (who always tries to find the "best side" from which to photograph any three-dimensional artwork) quickly discovered that the sculpture was beautifully composed from every viewing angle and that there wasn't a "good side" or a "bad side."

*Larry Barth's Best in World terns sculpture didn't have a "bad side."*

Larry Barth has previously stated that when he is working on the planning stage of a carving, he tries to forget about the representational aspect of his subject. While composing the piece, he is no longer working with *birds*, but merely working with *shapes* in space, as an artist creating an abstract sculpture would. Being in this frame of mind (in which composition is of supreme importance) is the key to creating an artistic masterpiece, versus creating merely a technically correct rendering. As to the importance of the design and composition process, Larry Barth has this to say:

"All of the work that comes *after* this point of composing and designing the piece is important, but superficial. It has been said that bird carving involves a thorough knowledge of birds; the ability to draw, carve, and paint them; and *most importantly* the ability to bring these elements together three-dimensionally into an artistic composition. There are many people who can draw birds, paint birds, carve birds, and know about birds, but these abilities alone won't necessarily make any carving come "alive" artistically.

"Bird carving is on the rise. We've come to a point where the technical skills are fairly evenly spread out. The only thing that ultimately separates the artists from the craftspeople is their sense of *design* and *composition*."

That statement, made in early 1985 (before Larry Barth had won back-to-back Best in World titles), sums up the philosophy which gave him the edge to twice win one of the toughest artistic competitions in the world. He knows that there are plenty of artists who can realistically paint a feather or carve a bill—so he puts extra emphasis on the "art" of composition. And he continues to win.

Is there a formula for creating a good composition? In general, this can't be done. Composition is a highly *intuitive* process, and although it can be learned to some degree, it is a talent some people seem to be lucky (or blessed) enough to be born with, while others may never be able to develop an artistic "eye." It is easy enough to point out good compositions, but difficult to find the words to explain *why* they are good.

The best test for a good composition is to look at it with a critical eye and try to imagine *what would have made it better.* In a good composition, everything works. The entire piece is a complete artistic statement, totally self-contained. Every individual part of the rendering should add to the total effect of the whole. Look at the composition and ask these questions:

**Is there too much going on?** Is the composition cluttered with unnecessary details which detract (rather than enhance) the subject matter? This is probably the most common problem with habitats in general—the artist gets caught up in the fun of creating and keeps adding "one more pine cone" until the rendering appears cluttered with debris. Know when to quit—and keep it simple.

**Does the composition reinforce the concept?** The composition should *enhance* the concept of the rendering, not work against it! If the concept of the piece is "peace and solitude," don't have an overly dynamic habitat with dramatic angles and sharp edges. Choose a serene, simple composition with predominantly horizontal lines and large areas of uncomplicated space. Conversely, if the concept is "a fight to the death," by all means use a dynamic composition with strong diagonal lines, jagged rocks, bent trees, or anything else to support the concept. In the case of Larry Barth's terns, the concept might have been "graceful flight," which is perfectly defined by the strongly vertical composition with its sweeping overlapping curves.

**Is every part of the composition in the BEST possible position?** This is referring to the basic arrangement of the elements within the composition. Look at every individuall element and ask is this in the "right" place? Should this rock be a little to the left or right? Should it be a little higher or lower? Is it too big or too small? If all the elements are in the best possible position, the composition is a successful one.

**Do the colors work together?** This is not a question of scientific accuracy (not yet, anyway, because we are dealing with composition only at this point). Are the colors pleasing? Do they clash? If the colors are ugly (even if they are accurate) the piece will be ugly, so plan color schemes like an artist and analyze them with an artist's eye.

**Do the values enhance the composition?** The *values* are how light and dark a particular color are. If a black bear is sleeping on a pile of coal, the dark values will obviously be working against the composition. Imagine a black and white photograph of the piece to determine the way values appear. A good way to see the values of a composition is to look at it with squinted eyes.

**Is it truly a three-dimensional composition?** Does it look good from more than one side? With the exception of dioramas or other fixed exhibits with only one viewing point, a good 3-D composition should continue to "work" from many different angles (not only from the front).

**Is it balanced?** When we say a composition is balanced, we mean that it is at a state of equilibrium—that it looks "just right" and that nothing more could be done to improve it. The balance itself can take many specific forms; the balance of light and dark areas (*value*), the balance of smooth and rough surfaces (*texture*), the balance of coordinated hues (*color*), the balance of big things and small things (*mass*), the balance of active and inactive zones (*dynamics*), and the balance of important things to unimportant things (*interest*). All of these balancing acts are related to each other within the whole. In a successful composition, they all balance with each other as well.

In order to explain visual balance, here are several examples of works of art to illustrate different aspects of visual balance. For clarity, the fundamentals of these compositions have been greatly simplified—please be aware that in all the following examples, many more relationships are working in the compositions than merely the ones illustrated.

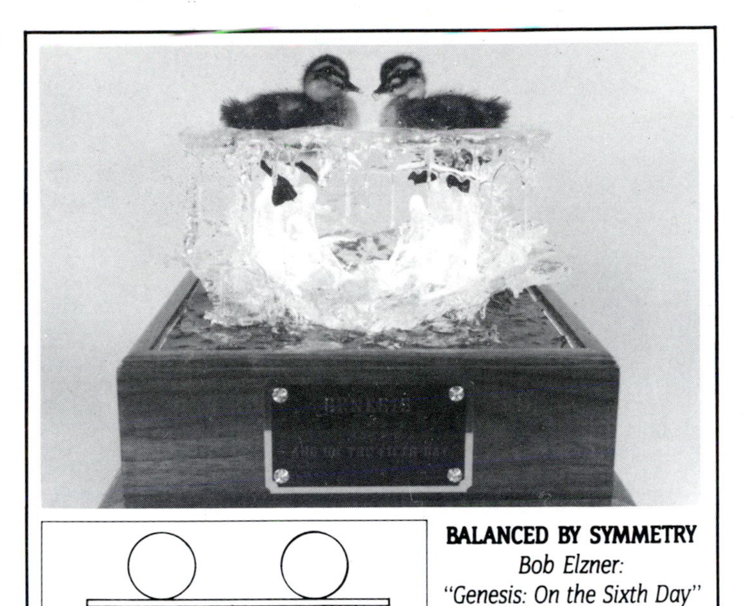

**BALANCED BY SYMMETRY**
*Bob Elzner:*
*"Genesis: On the Sixth Day"*
*1986 Carl E. Akeley Entry*

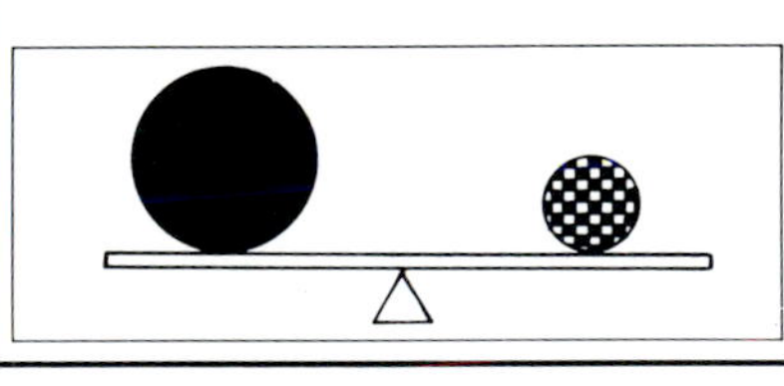

**BALANCED BY COLOR (VALUE) AND TEXTURE**
*Garry Senk:*
*Cappercallie Grouse*
*1985 Carl E. Akeley Entry*

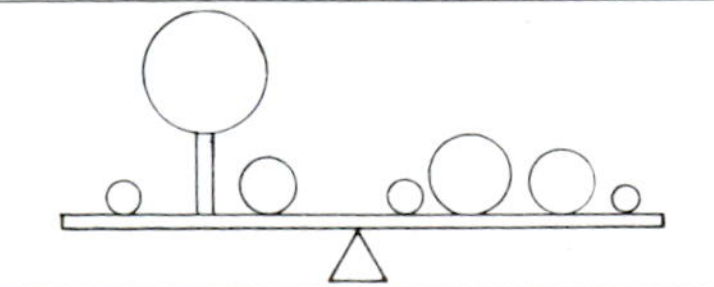

**BALANCED BY POINTS OF INTEREST (IMPORTANCE)**
*Sallie Dahmes:*
*Green-Winged Teal & Decoy*
*1985 Carl E. Akeley Winner*

**BALANCED BY DYNAMIC IMPACT**
*Tom Sexton: Largemouth Bass 1985 WTC Best of Show*

Using Tom Sexton's 1985 WTC Best in World/Best of Show Largemouth Bass mount as an example, let's examine the visual balance of the piece more carefully.

Tom's original composition was exquisitely balanced and composed for maximum impact. Although the design of the rendering indeed "worked" from all angles, for illustration purposes, we will use this frontal view. Tom's concept was of a leaping bass that had just broken the line as it leaped from the water. Instead of creating a large "exit" splash, Tom chose to depict the fish as it was re-entering the water. A small splash under the lower lip shows the begining of an "entry" splash, and concentric ripples at the left of the base point to the spot where the bass leaped out of the water. The bass has a nice curve to its body (not apparent in this view) and it is falling into a stand of wiregrass growing through the surface of the water. The wiregrass is leaning from the impact of the fish. When viewing this mount in person, the dynamic energy is overwhelming. It is as if an exciting, active, and dynamic moment in time was frozen for all to see. In short, the composition "works" to reinforce the concept.

A balanced composition doesn't necessarily mean equal weight on both sides. This design is balanced by the blank space (a dynamic zone) through which the fish has traveled immediately before striking the water surface.

Let's examine what would have happened if Tom had made minute changes in the positioning of the subject.

If the bass had been moved any further into the corner of the base, the composition would have become "cramped" and unbalanced.

If the fish had been moved 2 inches closer to the center, the composition would have lost its dynamic impact. There is no feeling as to "where the fish has been" in this composition and the overall effect is much more peaceful than exciting.

A less self-confident artist might have even centered the fish within the base. While the result is certainly not a "bad" composition, it doesn't ring with authority like the original mount.

The danger in analyzing a particular composition like this is that the analysis is taken as an axiomatic truth. This analysis, like this composition, is only for *this particular mount*. Yet it is almost certain that, after this is published, the very next taxidermy competition will have several mounts with the subject similarly positioned on the right side of the base.

# Thoughts from One of the Best

*One of the finest wildlife artists to come along in many years is Bob Berry of El Cajon, California. Bob tries to capture what he calls the "spirit of nature" in his art. Here is how he does it:*

***by Bob Berry***

To evoke the spirit of nature—that is the essence of wildlife art! It is up to the artist to relate his knowledge of nature in a clear statement to the viewer. Appreciating the beauty of the natural world through art takes one to that special moment or spot through imagination.

Is it art? Is it nature? No, it's a blend of man-made shapes and forms that suggest nature. It is impressionistic, rather than an exact duplicate of nature as photography might be.

I strive to make my carvings a brief contact with nature's great calm. A carving composition should suggest an entire scene to the viewer—with sounds, smells and the feel of the environment.

How do you achieve all the elements that make a successful piece? The desire and inspiration is usually there—but does the end product lack a little something? Be assured, good compositions don't just happen! They are totally planned and executed. All factors of design and composition are considered, from the mental concept or idea to the final paint stroke. This planning facilitates execution and controls or minimizes guessing errors.

Composition is essential to good art. Composition is the positioning of all parts in a space. It is the rhythm or repetition of parts. It is the arrangement of different shapes to unify everything around the center of interest and focal point of the work. The center of interest is one object, shape or point of interest that is more important than the other parts.

Plato, the Greek philosopher explained composition as *diversity with unity* or *unity with diversity*. Diversity, as we use the term, can be defined as variety. This includes variety in line, color, shape, size, and position of all the elements in the given space. Be aware, diversity can be overdone. It should not detract from and destroy the work's appeal. The *whole* must be more important than the parts, yet the parts must be consistent in degree of detail. Unity can be described as order. We enhance order by combining all elements of a work—line, size, shape, color, and position—into a harmonious whole.

Good composition is achieved by balancing some masses against others. Balance comes from a combination of factors—size of masses, distance between masses, value of masses, all in relation to one another.

Some basic rules should be your guidelines. These rules are time-honored art principles, but not necessarily etched in stone. Remember, art is not an exact science. The rules are only a guide. Your expression and originality can modify the rules.

- Be original, sometimes the hardest part of a composition.
- Plan a balanced composition. The key words here are *plan* and *composition*.
- Develop repetition of colors, shapes, textures, etc.
- Develop contrast of colors, shapes, textures, etc.
- Position shapes and space relations to each other. Shapes should show balance of masses in symmetry and asymmetry.
- Develop a rhythm or flow. Add that element of understanding that tells the story.
- Heighten the theme through materials and techniques.
- Add and develop drama through emphasis of a focal point.

That's still not the whole answer. As wildlife artists, we must learn about nature by getting out and studying nature, not by copying photographs or other works. You can learn from others by copying but only as a beginner imitating technique. To be original is to be unique, to start fresh, to create something without copying anyone or anything. Remember, copies are never as good as the original.

The process of understanding nature is endless. You must set aside an amount of time for field studies and R&D (research and development). I personally allow 20 percent of my time for R&D. That's one day out of a five day work week. Be it photography, sketching, observing, collecting, reading or compiling data, this time is by far the most important 20 percent that goes into my artwork. The success of the other 80 percent depends upon it.

For my decorative bird and fish carvings, I've also incorporated some time-honored rules from the Japanese art of Bonsai. It's very much like wildlife art or taxidermy, manipulating nature as an art form. There are a variety of basic Bonsai styles, cascade, formal upright, slanting, windswept, grove, clump, etc. I find inspiration for design, shape and form by organizing my carvings like a Bonsai. Using Bonsai rules for my carvings has resulted in more successful renderings. Each Bonsai is painstakingly manipulated and cared for over many years. The tree itself is the center of interest with the trunk the focal point. The branches, foliage, exposed roots, rocks, mosses, soil and container are all carefully considered for a total composition. The size and shape of the container is as important as any element of the tree itself. It is a harmonious part of the whole planting. In a sense, it disappears—that is, it's not especially noticed when you are viewing the tree. I use this rule for my bases for each carving. I carefully select hardwood for color and grain and then individually turn them on a lathe myself. This gives me total control of the entire piece.

All parts must work together to achieve a total unified effect. I've departed from trying to capture nature's "absolute" reality. It's far too time consuming. Instead, I try to represent my subjects by working on design and composition, not minute nonessential detail.

Don't let failure or "success" hold you back, always renew your efforts. Once a piece is finished, I'm never totally pleased with it, but what's done is done! Then, on to the next original that is sure to be my best yet!

# CHAPTER 5

*Armed with a basic knowledge of mold-making techniques, the wildlife artist will be able to reproduce almost anything they wish. This chapter describes some of the most common materials and procedures used by wildlife artists for creating molds from original specimens.*

# Mold Making: Techniques and Applications

***by Jim Hall***

As the wildlife artist progresses within his or her field, sooner or later it may become desirable to duplicate or reproduce an idea or form that they encounter. Duplication of various habitat materials may prove profitable for a variety of reasons, i.e.; to reduce the weight of an exhibit, to be able to make and have available several objects that are identical in order to be able to recreate the same scene more than once, and also to save valuable time in searching for that "ideal" rock, grass, driftwood, or other habitat material.

In many instances, reproduced items may be preferable because: they are more durable; they are insect proof; they will last longer; or they will project a better appearance than the original. Often, using natural materials will not work simply because they will dry and shrink. Others materials may create a variety of problems. Cattails, as used in various habitat settings, are a good example. The natural cattail has the nasty habit of blossoming and breaking open long after it has been installed in a nice background, creating a terrible mess, and the need to replace it. If an exhibit needs an item such as a cattail, then creating a reproduction is the only way to go (see Chapter 9).

Moldmakers base their work on one principle; namely, that virtually *anything* can be molded and reproduced. This chapter will deal with some of the more common molding techniques that are easy to utilize by the wildlife artist, and are easy to learn. Many of these mold-making techniques are described in detail in other sections of this book.

## Types of Molds

The best type of molding technique to use for an item is determined by the type of material that will be used for the final "positive;" the complexity of the shape to be molded; and the materials used for the actual mold making operation.

The simplest of all molds is the single piece mold. This is the type commonly used to cast very simple, one-sided objects such as wall plaque figurines, etc. They can be made from just about any casting material, but number one molding plaster is usually selected because it is inexpensive and easy to use. The half-cast method of mounting fish, where the shape of the fish is cast in plaster, is a typical example. A one-piece mold is the easiest to make, and is recommended for beginning mold makers.

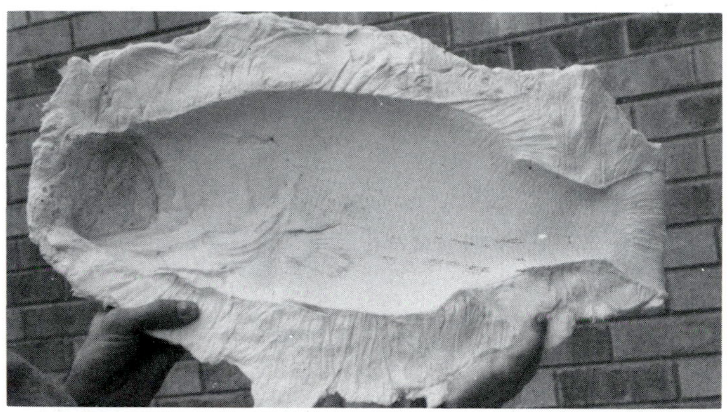

*A one-piece mold.*

*A two-piece mold.*

Two-piece molds are used to cast and reproduce objects where *both* sides of the model will be displayed. They are constructed just like a single piece mold except that two halves are made separately and then combined to cast the final part or positive. Again, molding plaster will prove satisfactory for this type of molding, especially when only a few positives need be made. After the first or front half has been cast, a separator is brushed or sprayed along the plaster flange to keep the cast of the second or back side from sticking to the first. In the case of plaster, liquid soap, Vaseline, or a thin clay solution will work very well. Other specialized separators are required when working with urethanes, fiberglass resins, RTV compounds, and the like. Always follow the manufacturers recommendation in the use of these special materials.

Multiple piece molds are just a continuation of the single and two-piece type, and are only used for molding very irregular-shaped objects that would present difficult or impossible demolding problems. They are the most complicated of all molds to produce, often requiring as many as fifteen separate pieces.

A glove mold.

A glove mold is a type of very thin and flexible mold that is usually used to reproduce some irregularly shaped objects, such as a fish tongue, or horn, or antler. It is made by applying several coats of latex mold building compound or RTV silicone rubber to the surface and permitting it to dry and cure. It is then peeled off (like a glove—hence the name) and can be used "as is" for some operations, or it can be supported in a container of loose sand or something similar to offer additional support. After the positive (or cast part) has been poured into the mold and allowed to dry, the latex or flexible mold is simply peeled from the reproduced part, like a glove.

For more detailed molding, such as applications in which duplicating the exact shape is required, a mother mold can be fabricated. This is a secondary type of supporting mold that is used to hold the flexible latex mold in its correct shape, and is usually made of plaster.

# Release Agents

A release agent is a product used to keep any part from sticking to another. There are dozens of agents available to use, and a selection is made based on the mold materials that will be used. A very good release agent to prevent plaster from sticking to plaster is a thin slurry (cream-like consistency) of dry potter's clay and water. Johnson's Paste Wax also works well when working with plaster.

Fiberglass molding operations require more sophisticated release agents. The best agents are those containing very high concentrations of carnauba wax. Some waxes, such as Ceara Mold Release Wax, are 100% carnauba.

As further assurance that a mold will "release," sometimes two parting systems will be used. A wax layer is first applied and buffed lightly with a soft cloth, and then another release agent is applied. The most common of these secondary release agents is PVA, which stands for poly-vinyl alcohol, and is applied with a brush or spray gun. In theory, the release actually takes place between the PVA and the wax. This is an almost foolproof system, and is usually used in fiberglass molding operations. This technique is described in detail in the *Breakthrough Publications'* Wildlife Art How-To Library Booklet *Commercial Fish Reproduction in Fiberglass* by Jim Hall.

One specialized type of release agent is called barrier coat, and is used in conjunction with RTV rubber molds. A light coating of barrier coat is sprayed into the rubber mold and permitted to dry. Two-part urethane foam can then be mixed and poured into the mold. The barrier coat will not stick to the rubber, but will bond chemically to the urethane, providing a near-perfect system. Manufacturers handling RTV rubber compounds are familiar with and usually stock barrier coat.

Common release agents.

# Making Positives From Molds

A positive, or the finished part that has been reproduced, can be made from any material suitable for the final product. If your final product was to be a candle, for example, then obviously it should be made out of parrafin wax, or a combination of parrafin and beeswax. Other small items can be reproduced in wax as well.

Some latex materials, such as Chicago Latex, No. 613, can be poured directly into dry plaster molds to reproduce parts. The plaster must not be sealed as it draws the moisture from the latex. When cured, these parts can be painted by various means and are quite attractive. Manufacturers selling these types of products are all very helpful in teaching individuals how to use their products, and usually will provide instructional sheets or brochures.

# Molding in Plaster

Perhaps the simplest and most economical material commonly used in mold-making is No. 1 Molding Plaster. This material is sold in small quantities at many drug stores as Plaster of Paris, and is available at some supply companies (such as Wildlife Artist Supply Company) in 10, 25, 50, or 100 pound bags. Since molding plaster is relatively inexpensive, it lends itself well to mold-making experiments, and to many of the more basic types of molds. It is commonly used for casting fish in the half-cast mounting technique; for making body casts of various animal or bird carcasses; for the casting of very large fish; for creating "back-up" or "mother" molds that are used to reinforce flexible mold-making materials such as latex, silicone rubbers, and others for casting parts from molds.

# Molding in Wax or Clay

Molding with the use of wax or clay is another very inexpensive technique often used to reproduce small items such as mushrooms, plants, rocks, and many one-shot operations.

Wax has the advantage that it can be warmed and softened, so that it can conform to small details easily. It can then be chilled to form a more or less rigid material. Plaster or some of the two-part rigid urethanes can then be poured into the wax impression to create the image of the desired object. Many waxes will be found to be satisfactory for the simple types of molds, such as parrafin, beeswax, or the brown sculpturing wax used by many miniature sculptors (available from Wildlife Artist Supply Company). One distinct advantage to using wax in molding is that it has its own "built-in" release agent, which makes demolding much easier.

Moldeling clay can be used in much the same manner. We recently desired to replace the open-mouth nose of a whitetail mannikin with a closed mouth. Rather than go to the bother of altering the lower jaw of the deer, we simply pressed oil clay upon the nose and lower jaw of a closed-mouth mannikin, building it up to about a one inch thick layer of clay. We then gently twisted and pulled the clay mold from the head, applied one coat of PVA (Poly-Vinyl Alcohol) release agent to the mold, and poured 3 lb. density urethane foam into the clay mold. When the foam had cured, we stripped off the clay. It was then a simple matter to cut off the unwanted open mouth and nose from the mannikin and glue on the new section. The total time for this simple molding and casting operation was about 25 minutes, which was significantly less than the time that would have been required to alter the open-mouth form. This same technique can be used on a variety of habitat materials including rocks, fence posts, etc.

## Alginate Mold

Alginate is a very unique molding material that is actually derived from some types of plant life, particularly seaweed. It is the soft, spongy type of molding material that is used by dentists and dental laboratories to make mouth impressions in dental work. It is used in conjunction with molding plaster to produce the final product. Its main advantage as far as wildlife art is concerned is that various molds can very quickly be made of objects and then immediately cast with plaster. Alginate molds are useful because they produce excellent detail for reference study casts in a minimum amount of time. It is a moderately expensive material to use. One negative aspect of Alginate is that it is easily destroyed by only 4 or 5 casts from the mold (and is sometimes destroyed by the first cast). Since Alginate cures without any heat build-up and will not stick to hair, flesh, trees (bark), or anything, it has a definite place in every wildlife artist's studio.

## Polyester Resin (Lay-Up Resin)

The heading of polyester resin covers many products, some of which include lay-up and scenery resin, gelcoat, and most of the auto-body filler resins such as Bondo or Ultra Lite. All contain the same basic polyester resin and all require the same basic catalyst (methyl ethel ketone peroxide, or MEKP) in either liquid or paste form. Various fillers and chemicals are added to the basic resin to obtain certain properties such as high flow rates, low flow rates, slow or fast cure cycles, high or low viscosity, or different mechanical properties.

When an object is said to be made out of "fiberglass," the basic materials are usually polyester resin with fiberglass added for tensile strength. When used in conjunction with fiberglass, cloth, mat, or chop is added to the polyester resin to give it great strength, and is widely used for both small and very large molds. Because of the strength of the mold, mold life is very long, often capable of producing literally thousands of parts if the mold is well constructed.

Moldmaking with polyester resins is not easy, and requires considerable training and experience before consistent good results can be obtained. Also, specialized equipment and materials are necessary, and it is absolutely essential to use very good ventilation and respiratory equipment.

Polyester resin and fiberglass probably has the greatest range of usefulness of any common mold making material. It can be used to make molds of just about any item, and with the use of the proper release agent, one can use it to cast the positive or final product. The *Breakthrough* Publications booklet entitled *Commercial Fish Reproduction in Fiberglass* (BP2003) details this process in depth. In the form of paste, such as Ultra Lite filler, it can be used as a glue for form alteration, quick repairs that may be needed on urethane forms, as a glue used for the bonded ear liner method of treating ears on big game, and dozens of other uses around the shop. Fiberglass, or polyester resin, is also one of the most economical, fast-setting materials available today.

## Epoxy Resins

Epoxy resins are similar in appearance to some polyester resin compounds, but are different in many ways. They include such products as Sculpall,™ Smooth-Out,™ REN 1253, 5 minute epoxy, epoxy paint, and many others. Epoxies must be catalyzed to cure and usually consist of one unit of resin and one unit of hardener. When mixed properly, epoxies exhibit great strength. Some of the epoxy puttys, such as Sculpall, are very useful to the wildlife artist. They can be used to recreate the thin barbell feelers on catfish; to set eyes in mannikins, for wood filler repairs, basemaking, and many others. When fully cured, epoxies can be filed, sanded, drilled, and painted. The only drawback to the use of epoxies is their rather high cost.

## RTV Silicone Rubber

RTV, which stands for *Room Temperature Vulcanizing*, is primarily a silicone based product that is very useful for certain types of molding operations. RTV molds are very rubber-like, and have good tear resistance, which make them good candidates for molding objects with undercut areas. Also, the better grades of RTV are not inhibited by moisture, so that excellent detail can be captured even on a damp surface. RTV rubber molds have a relatively long mold life if handled carefully, and can be used repeatedly to make casts of many unusually shaped objects. The unfortunate part of using RTV is the material cost, which at the time of this writing is about $150.00 per gallon.

Moldmaking tips, specifications, and ideas are available from the various manufacturers of the RTV compounds at no cost and it is a good idea to request such information when placing an order.

## Latex Rubber

Latex rubber, or Mold Builder,™ is a common material useful in reproducing small-to medium-sized objects. Small objects, such as stones, can easily be molded in latex and then cast with fiberglass (this procedure is described in detail in Chapter 8). Larger molds will have to be supported with reinforcing or mother molds, which often can be made of plaster. This material is moderate in cost and is very easy to use—just follow the manufacturer's instructions.

## Vacuum Molding

Vacuum molding is an interesting process and is useful for reproducing items such as leaves and grass from a thin vinyl sheet. The molds (usually made from high density plaster) are reproduced with a vinyl sheet that is placed in a heating box until limp. It is then draped over the mold or die containing the shape desired, and a vacuum is used to pull the limp plastic sheet tightly over the mold, where it becomes rigid in the exact shape of the mold. When cool, the plastic sheet is removed from the machine, trimmed to size, and painted with a special primer. Due to the initial cost of the machine (about $1,000.00), this method is generally not very practical for a small shop owner.

# Mold Making Materials and Considerations

| MATERIAL | ADVANTAGES | DISADVANTAGES | SAFETY CONSIDERATIONS |
|---|---|---|---|
| | | | *NOTE: These are offered as general guidelines only and should not be taken in any way as complete instructions. Always read and observe all manufacturer's warnings.* |
| **No. 1 Molding Plaster** | •Inexpensive<br>•Easy to mix and use<br>•Can be used for large molds<br>•Easily reinforced<br>•Provides excellent "back-up" or "mother" molds | •Brittle molds<br>•Short mold life<br>•Sometimes can only be used one time<br>•Requires careful storage in a dry environment<br>•Are very easily damaged<br>•Curing can be inhibited by moisture on mold | •Very safe and non-toxic when in wet form<br>•Avoid breathing any plaster dust when mixing |
| **High Density Molding Plaster** (Hydrocal, Hydrostone, Densite K-34, etc.) | •Same as Molding plaster but more expensive<br>•Molds will last for years | •Although stronger than standard plaster molds, care in handling is still required | •Same as above |
| **Molding Wax or Clay** | •Inexpensive<br>•Easy to use<br>•Ideal for small impressions<br>•"Captures" detail fairly well<br>•Releases from part easily with no release agent<br>•Material is reuseable | •Low in strength<br>•Usually good for only one cast | •Very safe to use non-toxic<br>•Avoid breathing dry clay dust<br>•If hot wax is used, be very careful as severe burns can result |
| **Alginate** (dental impression material) | •Easy to mix<br>•Will not stick to hair or skin<br>•"Captures" detail very well<br>•Requires no release agent | •Moderately expensive<br>•Limited to 4 or 5 casts<br>•Easily damaged<br>•Mold will shrink, crack, and be unusable after 24 hours | •Very safe to use and non-toxic |
| **Polyester Resins** | •Relatively low cost<br>•Useful for any size mold<br>•Very long mold life<br>•Produces very strong molds | •Requires considerable training to use effectively<br>•Requires the use of specialized equipment<br>•Reduced shelf life | •Requires adequate ventilation and respiratory equipment<br>•Polyester resins present storage problems concerned with temperature and age<br>•Flammable |
| **Silicone Rubber (RTV)** | •Extremely fine detail<br>•Relatively long mold life<br>•Very flexible molds possible<br>•Useful for undercut pieces<br>•Will cure over damp materials | •Very expensive material<br>•Often requires special mold design<br>•Materials must be weighed very accurately<br>•Takes a moderate amount of time to fully cure | •Avoid excessive contact with skin.<br>•Clean up with lacquer thinner and paper towels<br>•Safety glasses are recommended when handling RTV compounds |
| **Latex Rubber** (Mold Builder) | •Usually requires no release agents<br>•Moderate cost<br>•Material "goes a long way" | •Making molds is slow<br>•Often 6 to 10 coats are applied and each coat must be allowed to dry prior to adding another | •Very safe to use<br>•Non-toxic<br>•Wash hands after use |
| **Vacuum Molding** | •Rapid reproduction of thin objects such as leaves and grass | •Costly initial set-up<br>•Not practical for the small shop owner | •Non-toxic<br>•Very safe to use<br>•Heating elements are very hot |
| **Epoxy Resins** (Sculpall, Smooth-Out, All Game, REN 1253) | •Very strong<br>•Good detail<br>•"Sculptable" when wet<br>•"Carvable" when dry | •Relatively expensive | •Wash hands after use |
| **Urethane Foam** | •Relatively inexpensive<br>•Lightweight yet strong | •Limited shelf life<br>•Quality of foam is affected by the ambient temperature and humidity | •Components are flammable<br>•Generates high heat<br>•Produces toxic gas—proper ventillation is a MUST |

# Making a Plaster Mold of Leaves

Mozelle Funderburk, of Stone Mountain, Georgia, is a well-known museum habitat artist. This section shows her procedure for molding leaves in plaster as well of several methods of reproducing leaves from the mold.

Before making any mold of plants and leaves, Mozelle makes a special point of selecting *only* the finest shaped plants or leaves she can find. Always be sure that the plants are naturally indigenous species found within the environment that is being duplicated. For the sake of accuracy, be sure the plant is of the wild variety rather than a domesticated "exotic." Also keep in mind that some plants prefer shade while other require direct sunlight and won't survive long in opposite conditions. Mozelle always makes certain that her habitat and exhibit constructions follow these botanical laws, just as serious wildlife artists follow anatomical laws.

An important point to remember regarding the collecting of plants is that there are "rare and endangered" species in the plant world as well as in the animal world and care should be taken not to violate regulations governing them. If you *must* duplicate a protected species, do so with plants purchased from nurseries or arboretums. Or as an alternative, use artificial materials and rely on good references to help you construct the plants by hand.

To begin molding leaves in plaster, a "bed" or molding surface must first be prepared. Styrofoam trays, the kind meats are sold in at grocery stores, make excellent, reuseable surfaces on which to lay beds of "High Fiber" (the asbestos substitute from Wild Life Artist Supply) or mache on which to arrange leaves for molding. Mix the High Fiber with water until a stiff consistency is reached. The tray is covered with this mixture and smoothed out evenly.

Arrange the leaves, one at a time, in the bedding material. Press them into the High Fiber in such a manner to hold them securely but not so much as to cause the High Fiber to fold over the upper edges of the leaves. It is important to consider which side of the leaf should be the "show side." Even though the bottom surface of most leaves exhibit more detail than the top surface, Mozelle generally makes molds of the tops of the leaves. Since most artificial leaves in a habitat are viewed from the top, this will give the most realistic impression for a one-piece mold. Eventually, even the best mold will begin to wear out, so it is important to start with as much detail as you can possibly achieve.

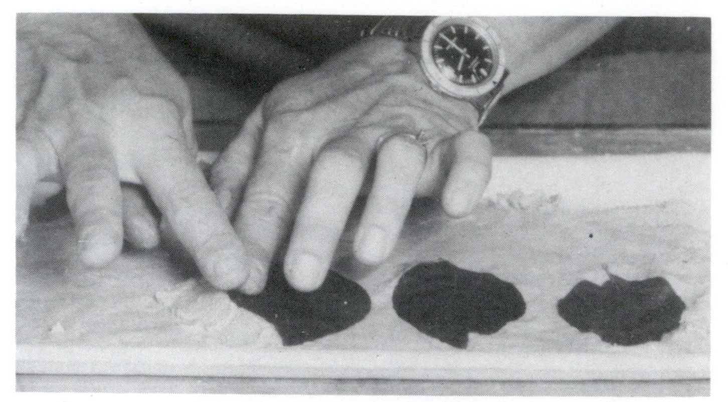

When positioning the leaves in the bedding, be sure the edges keep a nice undulating rather than flat surface. If necessary, place additional High Fiber under the edges of the leaf at different points and press the edge firmly into it to remove air pockets into which the plaster may run. This will give the leaves a more natural rather than "pressed" appearance.

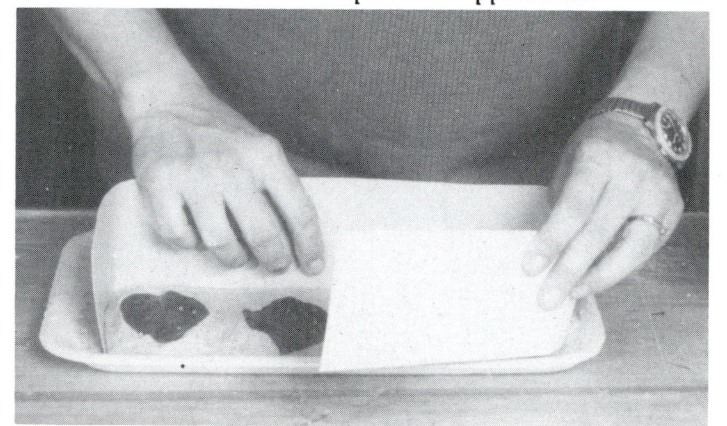

Once the arrangement of leaves on the High Fiber bed has been completed, a cardboard dam is placed around them by pushing the bottom edge of the cardboard firmly into the High Fiber. Staple the upper edges of the two ends together to prevent the plaster from running out.

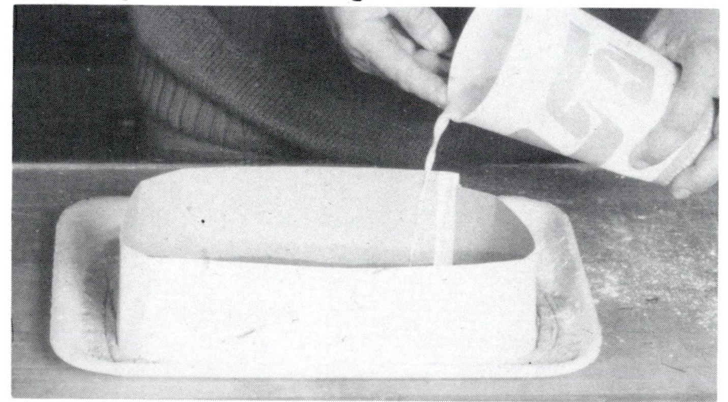

Because the life expectancy of a "plaster" mold depends on the materials being poured or cast in it, it is important to use a plaster with a very high density (or P.S.I.). Number one molding plaster will make an adequate mold for most purposes, however, if an extremely durable mold is required, a high density plaster would be recommended. Mozelle prefers to use Hydrocal to make her molds as it has a very high density giving it a longer life, and won't rehydrate like plaster. (Hydrostone and Densite K-34 are two more plaster-like products recommended for making tough molds).

If the molds are to be used only for making latex, acetate, or vinyl leaves, a one sided mold is sufficient. But if wax is going to be used to cast the plants, it will be necessary to make a second side to the mold. Do this by leaving the leaf in the mold after the plaster has hardened and turn the mold over (plaster side down). Clean away all of the bedding material from the leaves and apply a coating of Vaseline over the entire exposed plaster surface to act as a seperator. Repair or replace the cardboard dam if necessary. Then make a second pour of plaster over the leaves to create a two piece mold. (With most leaves, the detail of one side is sufficient to make satifactory duplications in latex, vinyl, or acetate.)

## Vacuum Molding Vinyl Leaves

If you're fortunate enough to own or have access to a small vacuum-molding machine, you can produce a limitless supply of accurately reproduced leaves from a roll of clear vinyl stock. Vinyl comes in thicknesses from 4 to 20 mil (1 mil = .001 or 1/1000 of an inch) and for most leaves and plants, 4 to 10 mil works fine.

The mold should be a one-piece or one-sided plaster mold, or use one side of a two-piece mold. The mold may have impressions of several leaves; size being limited only to the capacity of the vacuum-molding machine's ability.

Begin by preheating the vacuum's heating element for about five minutes. Before installing the vinyl sheet to its frame assembly, Mozelle heats the mold for a few minutes also. This helps to give better definition and sharper contrast to the pull than it would if done on a cold mold.

When the mold is warm, the clear vinyl sheet is secured in the frame and locked securely in place.

Swing the heating element over the vinyl sheet. Within a few seconds, all of the wrinkles in the vinyl will disappear as the sheet begins to heat up. When the sheet begins to soften and sag, it is ready to make the vacuum pull. Lower it onto the mold and activate the vacuum.

Once the pull has been completed, the heating element is moved away from the vinyl and the sheet is allowed to cool. The frame is released and the vinyl sheet is simple lifted off the plaster mold.

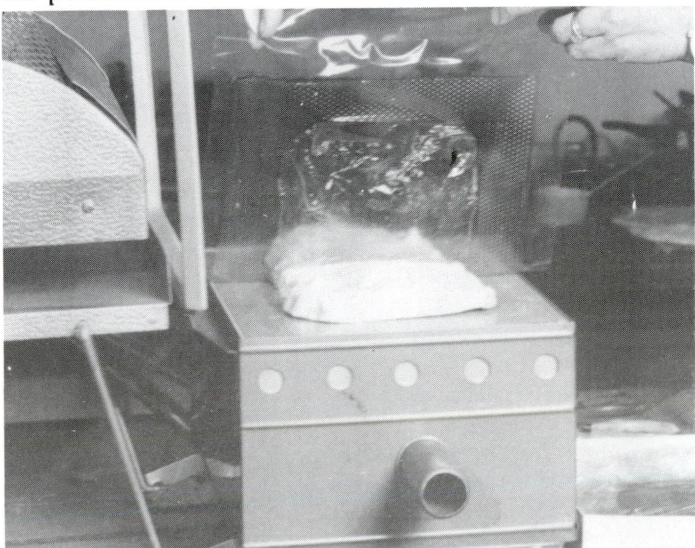

The vinyl sheet has now picked up all of the detail of the original mold.

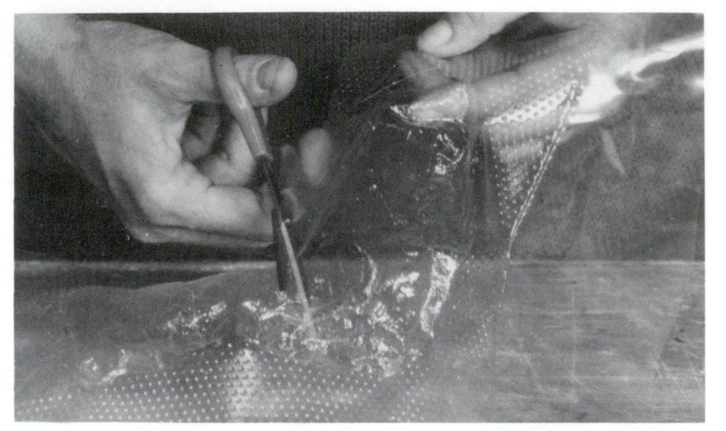

Using a sharp pair of scissors, cut out along the immediate edge of the leaf. The actual stem of the vinyl leaf may be cut away as it will be replaced in the next step. While in the process of cutting out the vinyl leaves, Mozelle turns on her hot glue gun so it will be ready for attaching the stems.

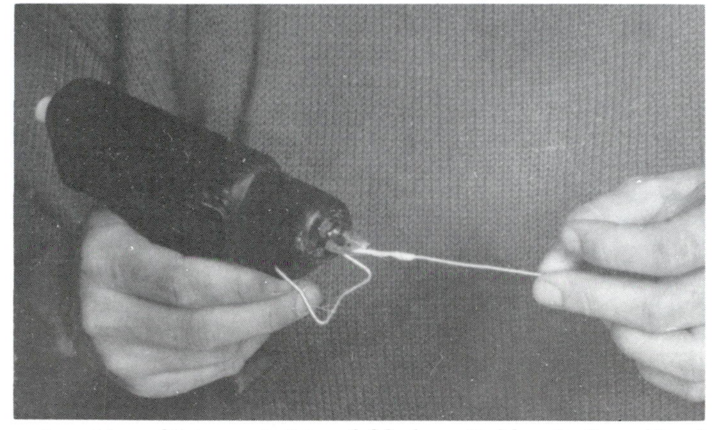

A section of "stem wire" (available from Hobby, Craft, or Floral Supply shops) is attached to the channel of the stem on the bottom side of the leaf using hot melt glue. It may be easier to actually insert the wrapped wire *into* the tip of the glue gun about an inch or so (completely coating the wire), than to try to eject hot glue out of the gun onto the wire. It's not a good idea to apply the hot glue directly to the vinyl as the heat may cause distortion of the leaf.

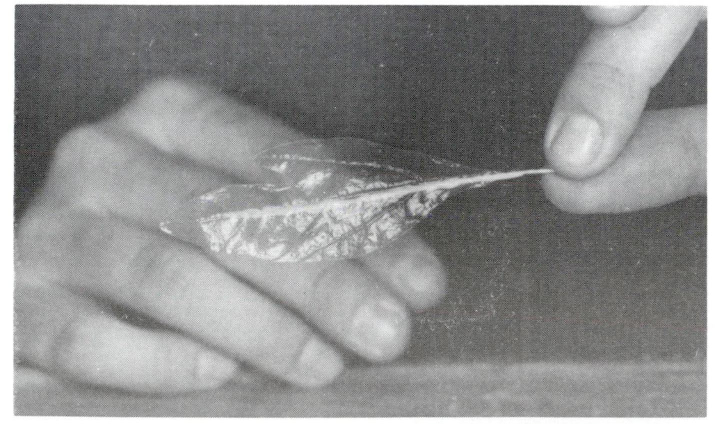

Once an assortment of leaves have been assembled, it is most efficient to paint them all at one time. Arrange the wire stems of the leaves in styrofoam or cardboard blocks and spray in appropriate colors with an airbrush. *Polytranspar*™ PlastiCoat airbrush paint (available from Wildlife Artist Supply Company) will prime and paint plastic leaves in one coat. The paint adheres to the surface of vinyls, plastics and acetates and once applied, will not chip, crack or peel, even if the plastic is tied in a knot. *Polytranspar* PlastiCoat is pre-colored to match most habitat and plant requirements in a realistic assortment of four greens and one brown. If other colors are desired, *Polytranspar* Water/Acrylic or Polymer/Lacquer paints may be applied over the PlastiCoat paint as it creates a "primed" surface on which to work. (Tube oil and acrylic colors may also be applied over a PlastiCoat base coat.)

## Acetone/Acetate Pressed Leaves

In the absence of a vacuum pulling machine, a slower but similar reproduction can be achieved by using thin acetate sheets cut slightly larger than the leaves being duplicated. These sheets can either be "pressed" *into* a one piece plaster mold of a leaf, or *onto* an actual leaf itself; both methods produce comparable reproductions.

Caution: Acetone is an extremely fast solvent. It is highly flammable and explosive. It is probably the most flammable solvent that the average wildlife artist will ever use. Handle acetone with *extreme caution!* The tiniest spark in the vicinity of acetone fumes could cause an explosion and fire. Use only with adequate ventillation. Keep away from all sparks, heat sources, or open flame. Make sure everyone in the vicinity knows when acetone is being handled so they won't attempt to light a cigarette, plug in an appliance, or create any other hazard.

In a metal pie plate or cake pan, mix 75% acetone and 25% water. Cut a piece of acetate slightly larger than the area to be molded. Holding both ends of the acetate sheet, dip it into the solution to completely cover it.

When the acetate begins to soften and shrink (about 15 seconds), quickly lay it over the mold and press it down firmly into place using a block of soft modeling clay large enough to cover the entire leaf.

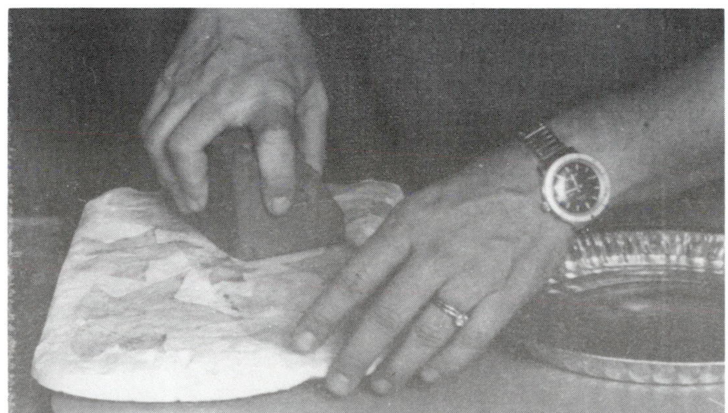

Hold the clay in place momentarily until the reaction of the acetone stops. Then remove the sheet from the mold and examine the quality. If it hasn't reproduced the detail adequately, the acetate may need to remain in the acetone/water bath a bit longer. If a number of leaves are being produced with this method, additional acetone must be added to the mix periodically as it evaporates away and becomes less effective.

Another method which is extremely quick and easy is to use the acetate/acetone procedure to make *direct casts* from natural leaves. No plaster mold is required for this method, making it ideal for one-time projects.

The same mixture of 75% acetone and 25% water is used. A natural leaf is placed on a flat surface (top of leaf is up). An acetate sheet slightly larger than the leaf is cut. Holding the sheet by the edges, it is immersed in the acetone solution for about 15 seconds (or until it becomes soft and limp).

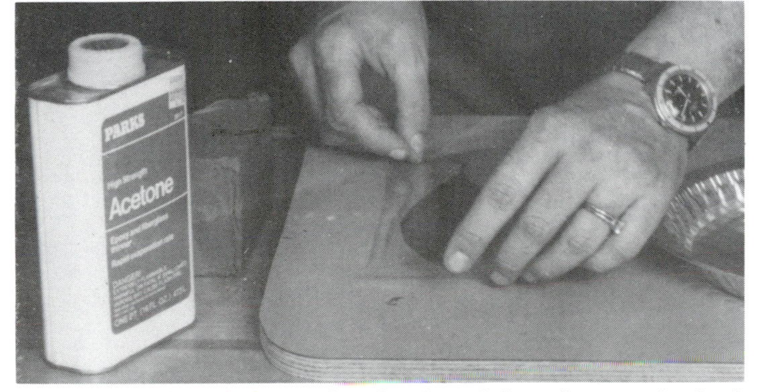


The limp sheet is immediately placed directly over the leaf.

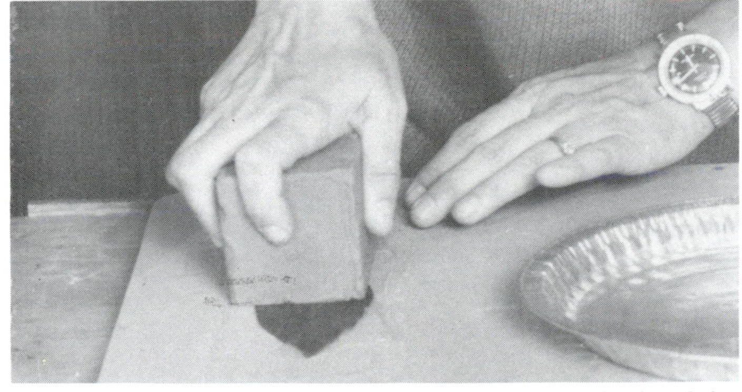

A block of soft modeling clay is placed onto the leaf and the acetate sheet. Steady downward pressure is applied, forcing the limp acetate to conform to the contours and detail of the leaf.

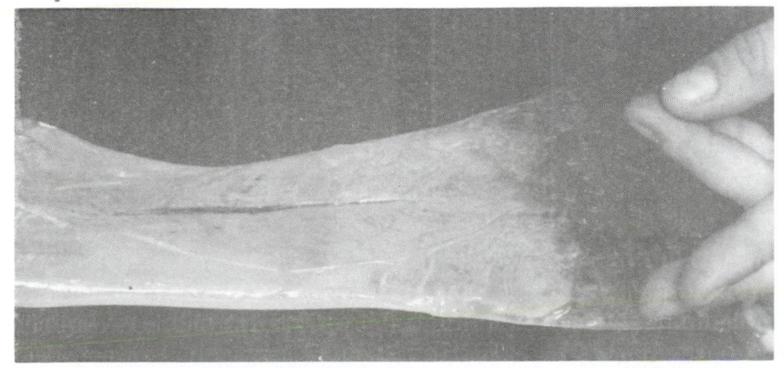

As the acetone evaporates, the acetate sheet will regain its rigidity while keeping the detail of the leaf. The leaf is now cut out with scissors. A stem is added with hot glue (in the same manner as leaves made on the vacuum press using vinyl). The leaves are then airbrush painted with a coat of *Polytranspar*™ PlastiCoat paint and once dry are ready to be used in a habitat or exhibit.

# Latex Leaves from Plaster Molds

The same plaster leaf molds used for the vacuum-formed and the acetate/acetone methods may also be used in making leaves from latex rubber. Mozelle recommends Chicago Latex No. 613 as a good all-purpose latex for the wildlife artist.

Using a brush, the liquid latex is applied to the plaster mold. When using latex in a plaster mold, the plaster surface must not be sealed and no separator should be used. The plaster actually draws moisture from the liquid latex, allowing it to cure and harden.

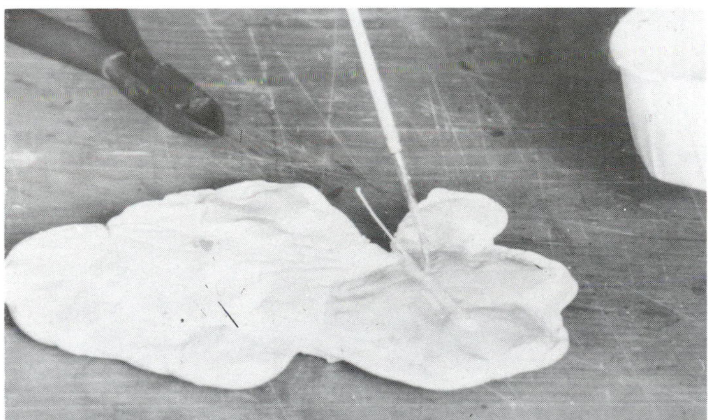

After a sufficient thickness of latex has been built up, a wrapped wire stem is placed in the central groove of the leaf and additional latex is applied over it. If a two-piece mold was to be used, the second mold piece would be assembled at this point and set aside to cure.

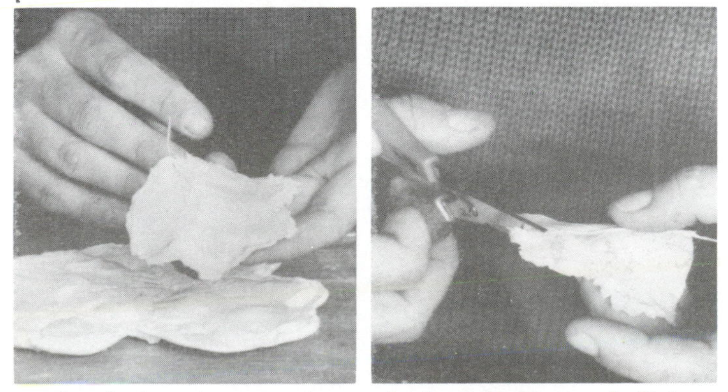

After drying (about one hour), the latex leaf is "peeled" from the mold, trimmed with scissors, and finished with *Polytranspar* PlastiCoat airbrush paint.

# Reproducing Driftwood In Urethane with an RTV Mold

*by Jim Hall*

Driftwood, either in its natural or reproduced form, provides the wildlife artist with a very attractive base or background for many of his or her products. It can be very effectively used both for wall mounted items or for three dimensional table pieces. In many areas of the country driftwood is in good local supply and one or two days of gathering will provide the artist with an ample supply. Many areas, however, do not have a local supply and driftwood must either be purchased and shipped, or it must be reproduced.

Reproducing driftwood, which may appear to be bothersome at a first glance, does offer many advantages. Once a mold has been made, your supply of "driftwood" is assured, and the cost of the foam material is very low per piece. Also, you never need to worry about any tiny critters eventually working their way out of one of your backgrounds. Once you have selected a choice piece of natural wood to reproduce, you will always be able to use that same size and shape, and you will never run out of pieces at an inopportune time. You will also be able to color the new "driftwood" in any manner you or your customer may choose.

## Making the Mold

Because of the very irregular size and shape that will be encountered in a piece of driftwood, an RTV (Room Temperature Vulcanizing) rubber mold is required. From the various types of RTV compounds available, we have found the following to be satisafactory for this purpose.

Silastic HS (High Strength)
Dow Corning, Midland, Michigan 48460

Mold-It II RTV Compound
Wildlife Artist Supply Company, Loganville, Georgia 30249

To begin, we selected a nice, clean piece of driftwood that displayed good character and was of a size and shape to suit our needs. Be sure to select a piece that is firm with no loose pieces that may come off during the molding process.

We then sprayed the driftwood with a very liberal coating of *Polytranspar*™ Sanding Sealer (WS-100) and set it aside to dry. This sealer provided a nicer surface to cast and also sealed in any loose particles of wood and dust. As a further guarantee that our mold would come apart easily, we applied a coat of mold release wax with the use of a sponge and a small brush (below).

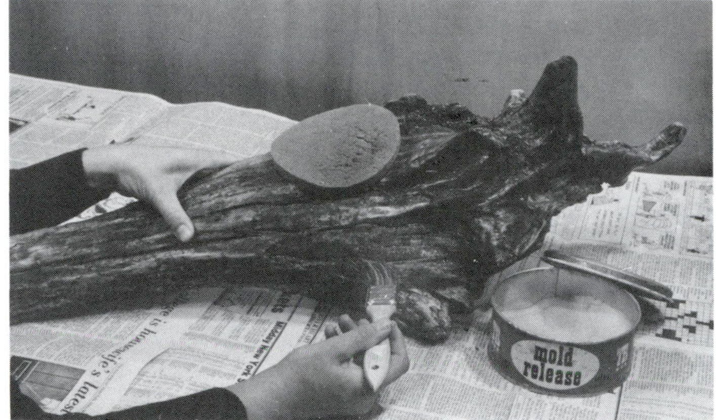


A piece of formica-covered plywood was then cut out to provide the base to hold the driftwood during the molding process.

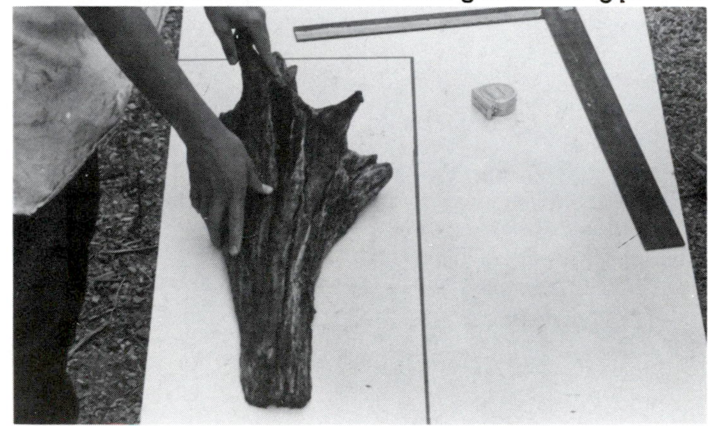

A screw hole was drilled through this base and the driftwood was attached with the use of a single Grabber screw.

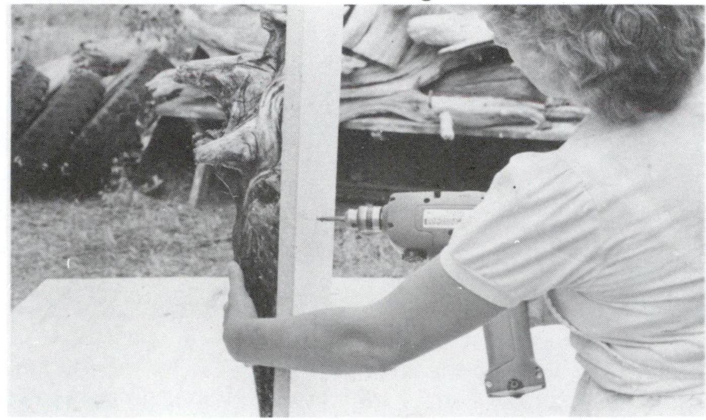

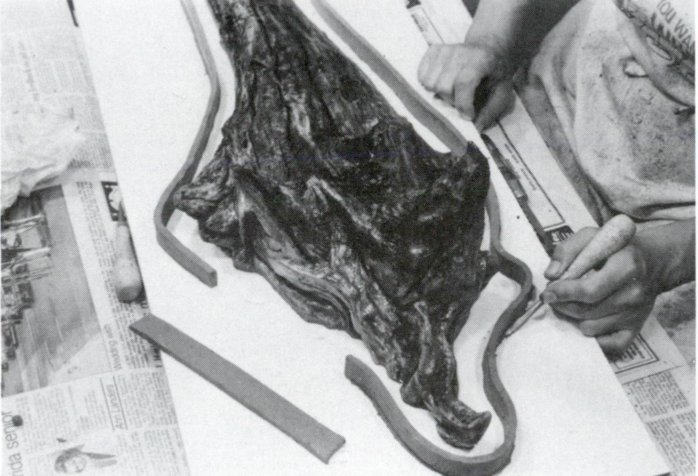

In places where the back of the piece of driftwood did not touch the back board, clay was added. When this was complete, a clay dam about 1/2" high was erected around the piece of driftwood to help contain the molding compound.

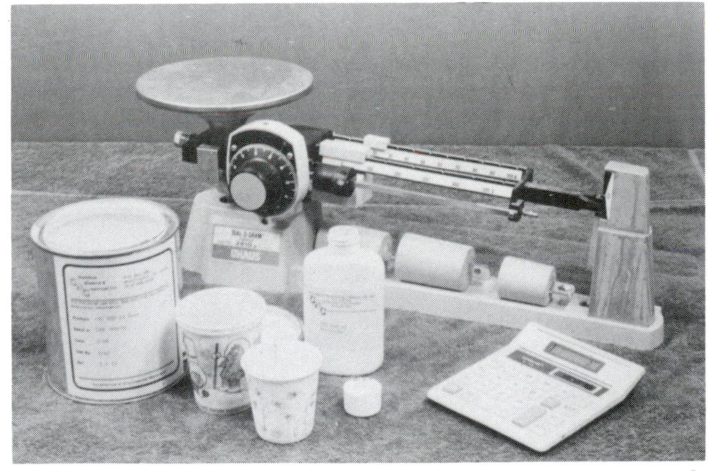

We selected Mold-It RTV Compound for this series. As with any of the specialized (and expensive) rubber compounds, we recommend weighing each of the components accurately and exactly as the manufacturer specifies. A three-beam balance is particularly suited for this purpose.

After the Mold-It II Compound and catalyst were thoroughly weighed and mixed, a layer was brushed over the driftwood model. The first layer of rubber is the most important and it must be worked into the cracks and crevices carefully to avoid trapping air bubbles, etc.

This process produces a relatively thin (or glove) mold, and the rubber compound is applied in several layers. It is very important to watch the compound as it is curing to determine just when the next layer should be applied. Reapplying additional compound should be done just after the previous layer has begun to cure and has ceased to run or sag. Do not wait until the previous layer has completely cured because you will achieve a very poor bond between the layers.

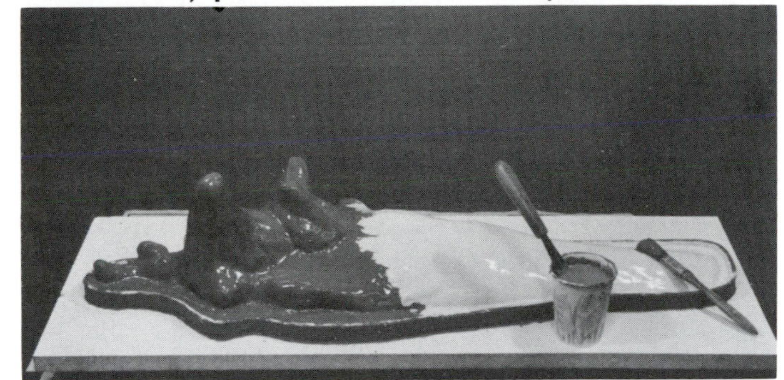

In the above photo, a small quantity of black tempera powder has been added to the rubber compound used for the second layer. By changing color of the compound, it will be easier to determine how heavy a coat is being applied.

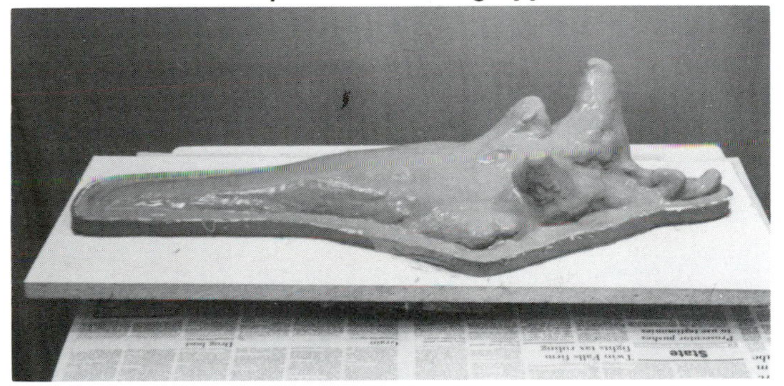

The construction of the mold should continue, layer by layer, until a thickness of 1/4" is obtained "at the thinnest spot." The actual number of layers will be determined by just how heavy each coat has been applied. This mold required six layers to obtain the desired thickness.

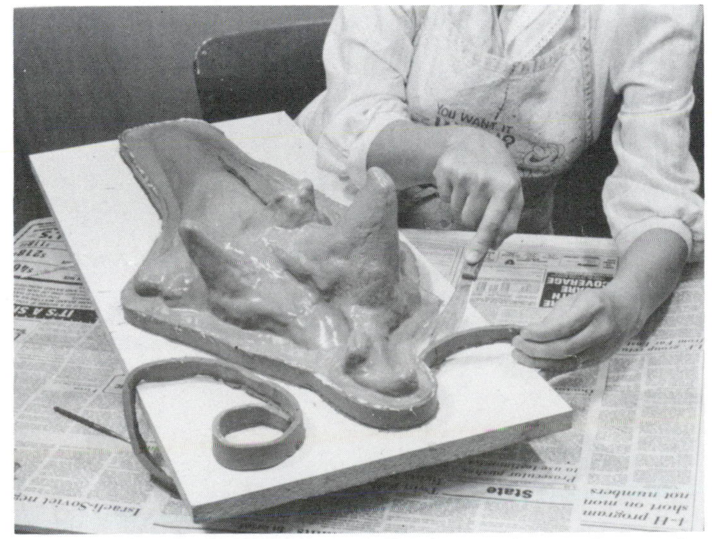

After the rubber mold has completely cured according to the manufacturers instructions, the clay dam material can be removed.

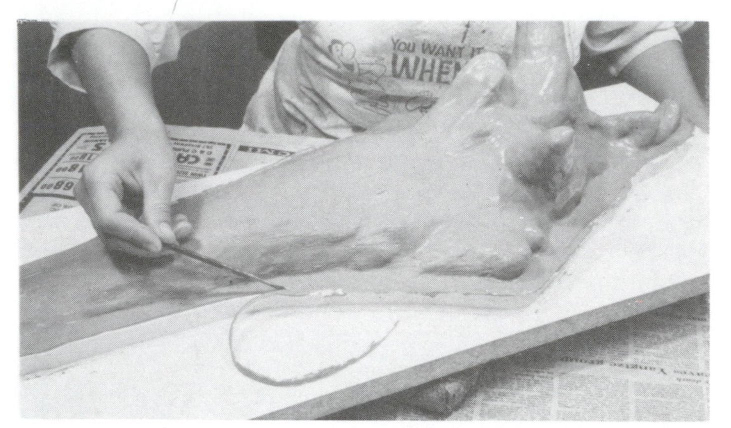

Excess rubber along the mold edges can easily be trimmed.

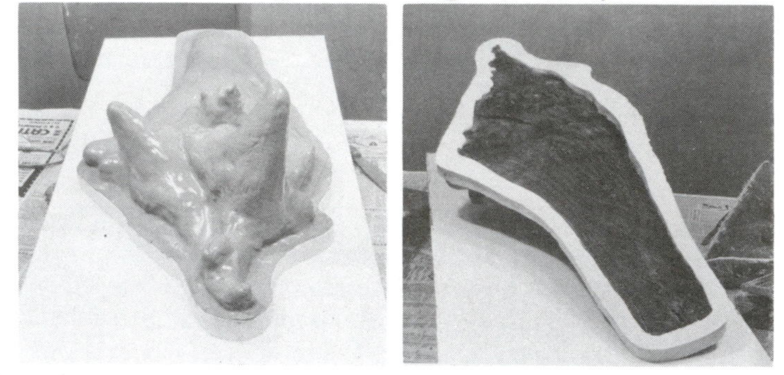

The two photos above show the front and back sides of the finished rubber mold.

Once the rubber mold has been completed, one has several options to continue. The simplest option would be to position the mold open side up in a bed of dry sand, which will offer an improvised mother mold. In this case, two-part urethane foam would be poured into the mold and allowed to free-rise to fill the cavity. When cured, the excess foam would be trimmed flush with the straight side of the mold.

Another option to using the rubber mold would be to construct a plaster mother or supporting mold. Since the rubber mold has such an irregular shape, a multiple piece mother mold would be required, but could be constructed quite easily.

For this article, we constructed a mother mold out of fiberglass, as can be seen in the following photographs.

# Casting The Driftwood Mold

The first step in actually producing the driftwood reproduction is to assemble the various parts of the mold. If a rigid type of mother mold is used, either plaster or fiberglass, it is clamped together with "C" clamps or vise-grip pliers. Then the rubber mold is placed into position within the rigid mold.

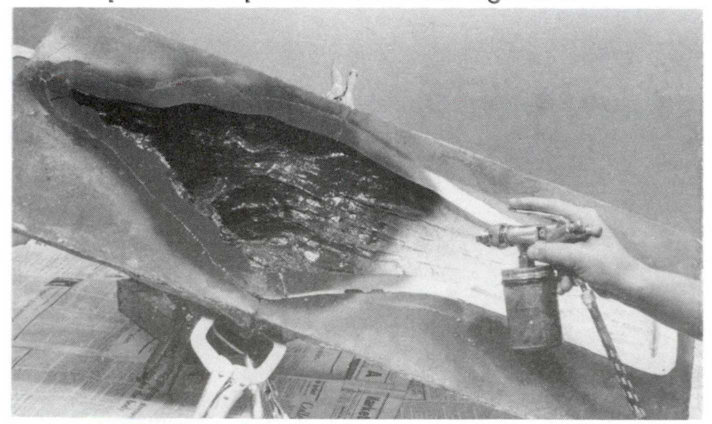

The surface of a rubber mold of this type is best prepared by spraying in a layer of a product called barrier coat. Barrier coat acts as a release agent for RTV rubber molds, and can be colored to provide a base color of the driftwood. We tinted our barrier coat to a charcoal color with black tempera dry pigment powder.

Our next step was to gather the A and B urethane foam components and mixing equipment. Our urethane foam was measured in equal quantities, poured into a convenient mixing

container, and thoroughly mixed with a beater attached to a 1/4" electric drill motor. When the foam was mixed, it was

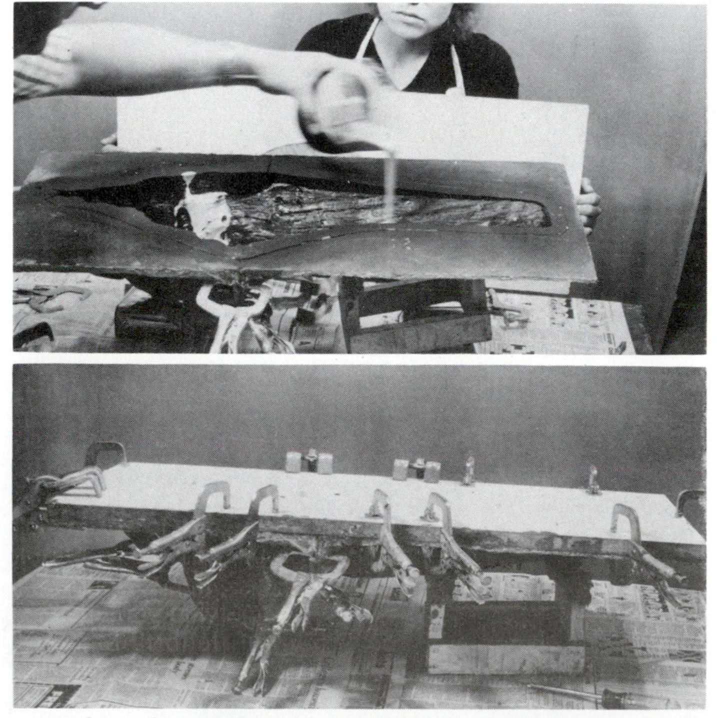

poured evenly into the rubber mold, and a backing board with a plywood mounting board screwed to it (see next photo on following page) was secured to the fiberglass mother mold.

The foam was allowed to expand and cure for about 30 minutes, after which time all of the clamps were removed and the mold sections separated.

When all of the supporting mold pieces had been removed, the rubber mold was gently "peeled" from the new piece of "driftwood."

After the remaining flashing had been removed, and the edges lightly sanded, the foam driftwood was ready to paint.

## Painting the Foam Driftwood

Since we had colored our barrier coat material a charcoal color, much of the coloration of the driftwood was complete, requiring only some additional highlighting.

We only used four colors to tint the driftwood, namely, *Polytranspar*™ White WA-10, Black WA-30, Sienna WA-200, and Chocolate Brown WA-70. We used the Water-Acrylic paint system because it extended the drying time and permitted any amount of blending of the colors we desired.

The four colors were mixed with small quantities of Ultra-Seal (US 16) to give them a little more "body" and then brushed and stippled on the surface of the driftwood. We had our best luck when we applied dark colors to the cracks and crevices, and light colors to the high spots. If a mistake was made, it was very easily changed by shading in another color. After about 15 minutes, we were satisfied with the results. No other finish was required.

Above displays three pieces of driftwood. Which is the original? If you can't tell, we should be able to call the project a success.

Above, a reproduced piece of driftwood is used as a background for a fish mount.

# CHAPTER 6

# Base Building

***by Bob Williamson***

Essential to any habitat scene is a base. There are several different styles of bases that can be built with relative ease or specialized bases can be purchased.

In the above photo an excellent quality hardwood base is shown that can be used for a variety of scenes. This particular base is available from Wildlife Artist Supply Company and is ideal for fish, bird, or small mammal renderings. It is recessed on ther sides to accommodate artificial water pours without using molding to conceal the shrinkage from the edges.

All that is required when using a base such as this is to seal it prior to making a water pour. This will eliminate leakage and/or air bubbles. Ultra Seal is a good product for this step. It is best to seal the base with at *least* two heavy coats allowing one to dry prior to adding another.

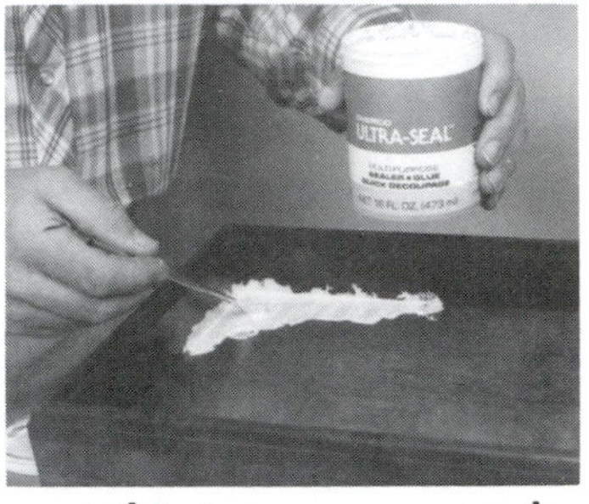


Once it has been sealed the habitat can be built, adding rocks, vegetation, mud, and artificial water for a marsh scene. Forest scenes can be built with leaves, dirt, peat moss, etc., and it is but a simple matter to convert this to a snow scene by simply adding artificial snow, ice, and icicles.

A less expensive route for average commercial work that does not require a fancy hardwood base is to construct one. There are four types of bases that are commonly used. **(1)** Hardware cloth/metal lathe; **(2)** Urethane poured; **(3)** Styrofoam carved; and **(4)** Polyester (fiberglass) Lay-up Resin (described in Chapter 8). Often the artist will decide to use a combination of methods according to different needs.

## Hardware Cloth/Metal Lathe

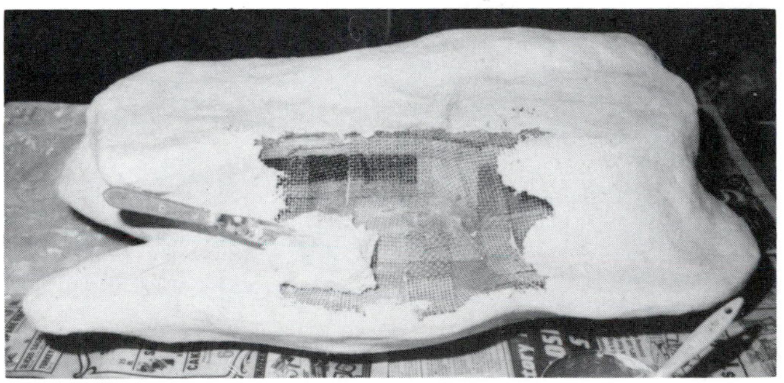

To construct a scene using this method, the artist will need to assemble the following supplies and tools:

1. Plywood base of the desired size
2. Assorted blocks of wood or foam
3. 1/8" hardware cloth or metal lathe
4. Jim Hall or Sallie Dahmes Mache
5. Tempera colors, *Polytranspar*™ Water/Acrylic paints
6. Assorted leaves, peat moss, and habitat materials
7. Ultra Seal
8. Staple gun
9. Sponge
10. Hot melt glue gun

### Step 1

Hot melt glue blocks of wood or use Ultra Seal to glue blocks of foam to the plywood base.

### Step 2

Staple hardware cloth or metal lathe (used by stone masons) to the plywood base. Shape the scene to the desired landscape.

### Step 3

Apply either Jim Hall or Sallie Dahmes Mache (that has been tinted to the proper base color with tempera pigments) to the hardware cloth and smooth it out. NOTE: Polyester resin and fiberglass mat or chop can be substituted for mache in this step. Dirt, sand, rocks, etc., can be added while the polyester resin is still sticky. When it cures, simply blow the excess off with an air gun.

### Step 4

When the mache has dried use Ultra Seal and/or hot melt glue to glue dried peat moss, dirt, leaves, etc., to the base.

### Step 5

Use Polytranspar™ Water Acrylic paints and a sponge to paint, highlight, and texture any rocks in the scene.

# Urethane Poured Method

Materials needed:

1. Plywood base
2. Urethane foam parts A & B
3. Tempera pigments
4. Assorted leaves, peat moss, and habitat materials
5. *Polytranspar*™ Water/Acrylic airbrush paints
6. Hot melt glue gun

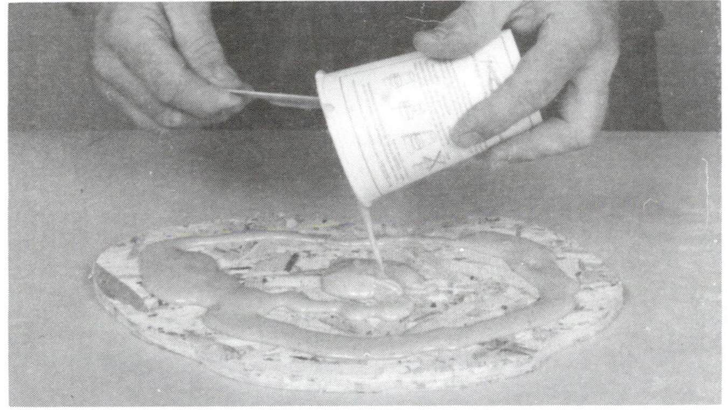

### Step 1

First, add tempera pigment to the 'B' component urethane. Then, thoroughly mix together small quantities of A & B component urethane foam according to the directions on the container. (Be sure and use adequate ventilation and use the recommended safety procedures.) Pour a quantity on the plywood, spatulate or brush it all over the plywood, and let it free rise (foam will expand).

### Step 2

Once it has begun to foam, peat moss, dirt, etc., can be sprinkled over it while it remains sticky. Once it has completely cured and the foam has hardened, it may be carved or sanded to shape, painted, and then finished with rocks, vegetation, etc. Note: If a piece of driftwood, rock, etc., is to be added, place it in the scene immediately after applying the foam and let the foam surround it and stick to it. This will secure it solidly to the base. Areas that are to be a higher elevation on the exhibit need additional pours of foam until built to the proper height.

### Step 3

Now, apply any finishing touches with Polytranspar™ paint as needed.

# Foam Carved Method

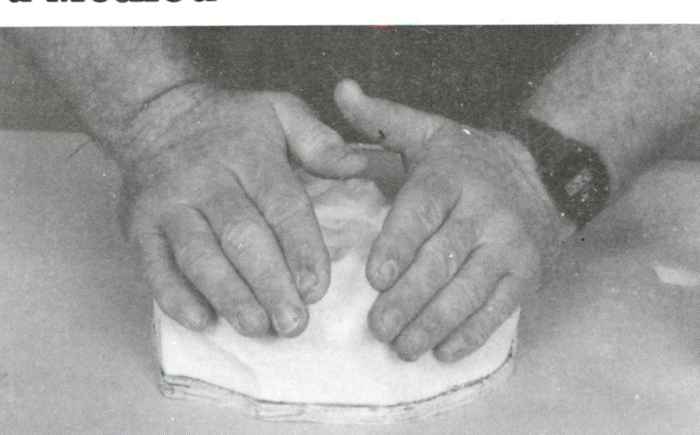

Materials needed:

1. 3/4" plywood base
2. Carved foam
3. *Polytranspar* Hide Paste or 2 component urethane foam
4. Dirt, peat moss, and habitat materials
5. Hot melt glue, Ultra Seal, or Elmer's Glue

### Step 1

Carve a piece of foam to the desired shape. Trace the foam image onto a piece of plywood and bandsaw it to the desired shape.

### Step 2

Attach the carved foam base to the plywood with urethane foam. Simply mix the urethane foam per the directions and paint a thin coating onto the plywood base. Now place the carved foam on it. As it "sets," the urethane will permanently glue the carved foam piece to the plywood. *Polytranspar* Hide Paste, Ultra Seal, Hot Melt Glue, Elmer's Glue or similar products can also be used for this step.

### Step 3

Once the plywood is attached, apply Ultra Seal or similar product to the foam, then lightly sprinkle sand, treated dirt, etc., onto the surface. Once dry use an airhose to blow the excess off, and touch up with *Polytranspar* airbrush paint.

Once a habitat base has been built, it is a good idea to enclose it in a glass or plexiglass case to protect it from dust, children, and other threats. (Snow scenes and water scenes will be ruined by dust if not covered.)

# 1. The Artistic Base: Simple and Effective

*by Jim Hall*

The abstract design of this base completely breaks away from the more traditional natural scenes such as snow, dirt, forest floor, etc. In fact, this base doesn't really resemble anything, and yet it is extremely pleasing to the eye. This base is the brainchild of Master Fish Taxidermist Rick Laurienti of Denver, Colorado. Rick has been very successful with it, in fact, it has enabled him to win several impressive awards. The list includes:

1st Place Single Fish and Best of Show—1985, Colorado Taxidermist Association Annual Convention

1st Place Single Fish and Best of Show—1985 Taxidermists International, Canada 1st Place (2) Fish Category

Best of Category—1986 NTA Convention

1st Place Fish Category and Best of Category—1986 Colorado Taxidermists Association Annual Convention

As can be seen in the photo above, the water in this scene doesn't really look like water, and the rocks don't really look like rocks, however the overall effect is most pleasing. This rendering of a steelhead trout won Rick a Second Place ribbon in the Best of World Fish Category at the 1986 World Taxidermy Championships in Lawrence, Kansas. This base can be made from mostly scrap materials very easily, and does not require a great deal of time to create. Rick begins construction of his artistic base in this manner:

An 18" piece of particle board was selected to make the base. The raw material for this base is extra thick particle board, and we found the easiest way to obtain it was by buying a six foot length of 1" or 1¼" stair tread, which is available at most any lumber yard.

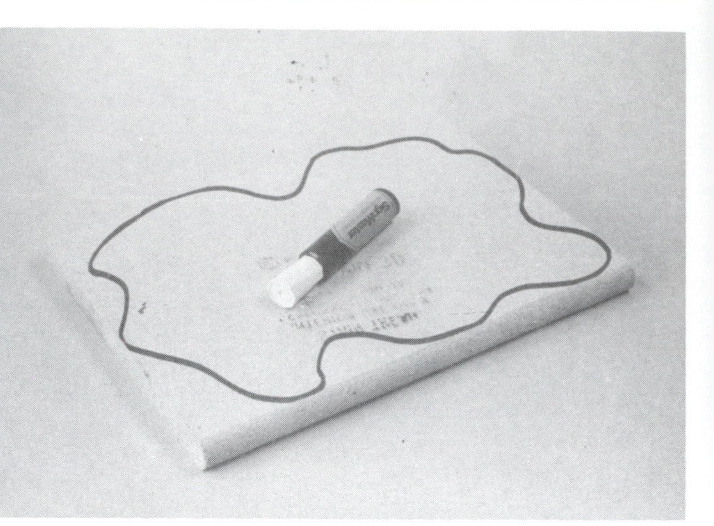

First, an irregular outline was drawn on the board with a felt marker and then it was bandsawed to shape.

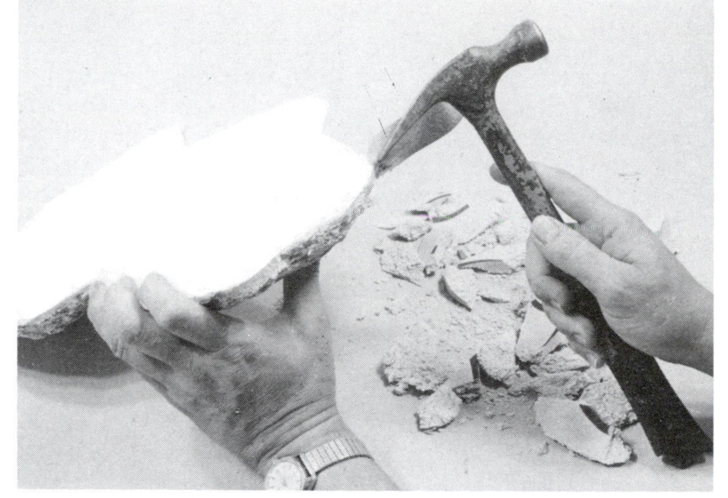

Next, the edges of the particle board were chipped and broken away as shown in the photo above. Strike the board with the claw side of a hammer at about a 45° angle, which will result in a very ragged edge that tapers in on the top or "show" side.

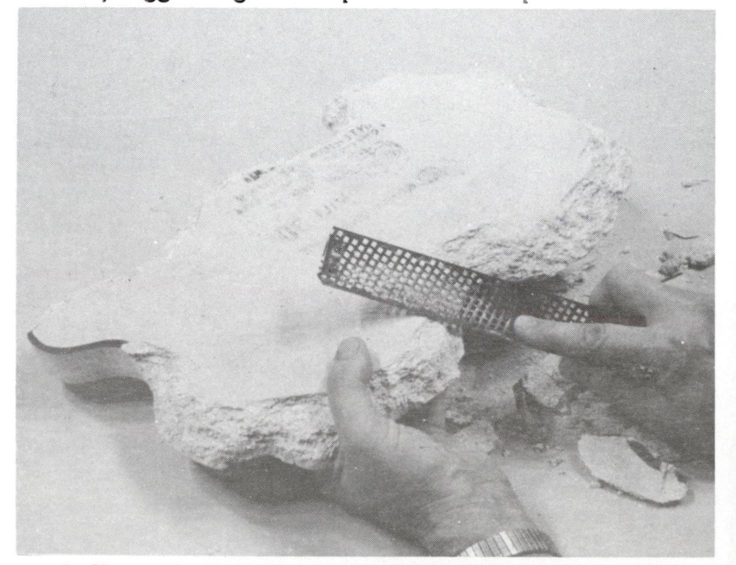

A half-round, Sur-form rasp is then used to smooth some of the rougher edges, and to help "blend" everything together.

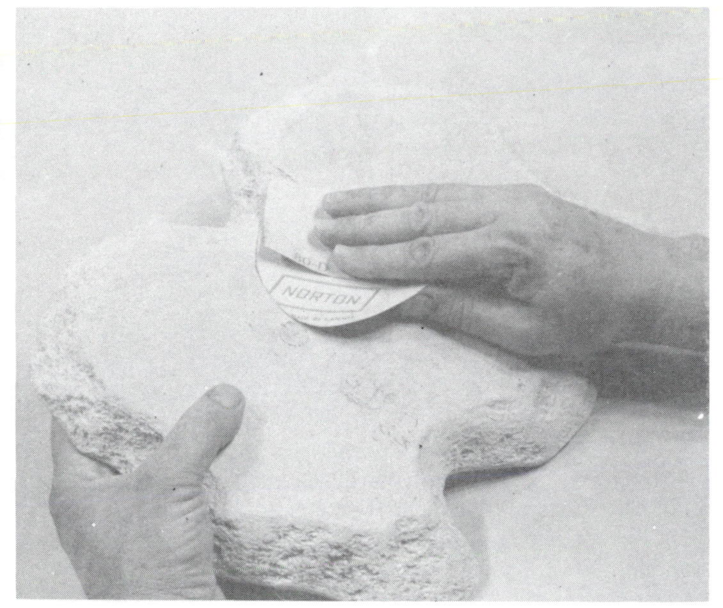


The edges of the board and the top were then lightly sanded with a medium grit sandpaper, such as number 100.

We then applied a liberal coat of *Polytranspar*™ Wood Sealer WS-100 to the base, and permitted it to dry thoroughly.

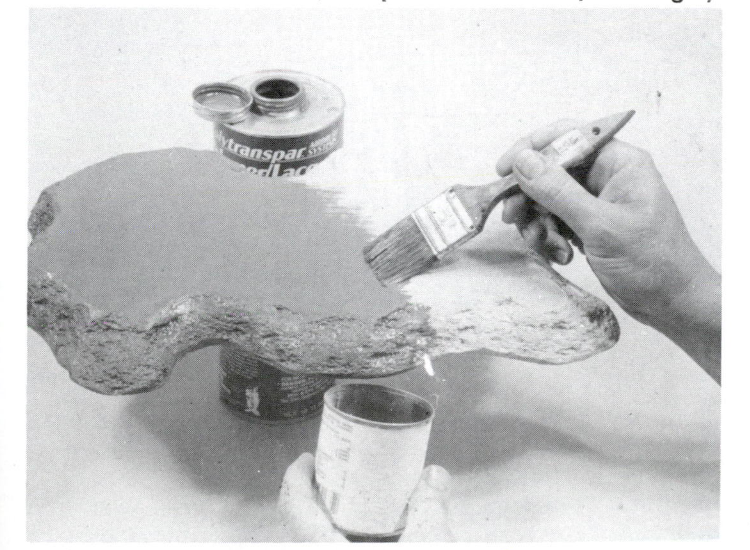

The next step was to apply two coats of *Polytranspar* Gray Fiberglass Primer FP-195G to the entire surface with a 1½" bristle brush.

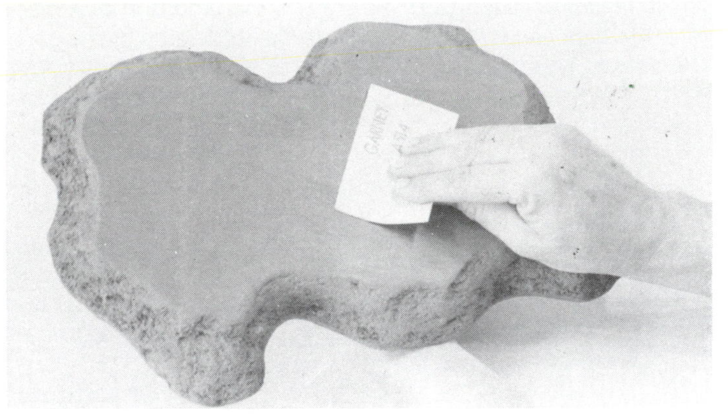

The primer was lightly sanded with fine garnet paper after each coat had thoroughly dried.

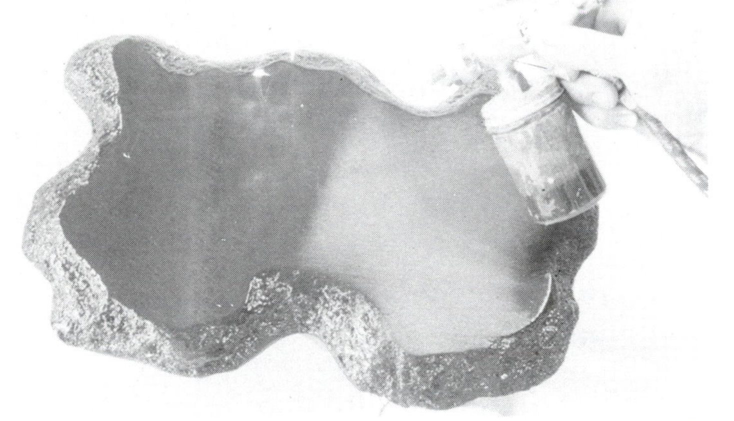

When the sanding was complete, we sprayed the particle board with a heavy coat of *Polytranspar* Black (FP30).

At this point, just about any color could be used to spray the base because, as we said, the base doesn't have to resemble "anything" in particular, and some of the "water" colors such as green, green/blue, or coral would yield entirely different effects. Black is the color preferred by Rick (and was used here because it helped us with the photography). We sprayed two heavy coats of Wet Look Gloss to achieve a very high gloss.

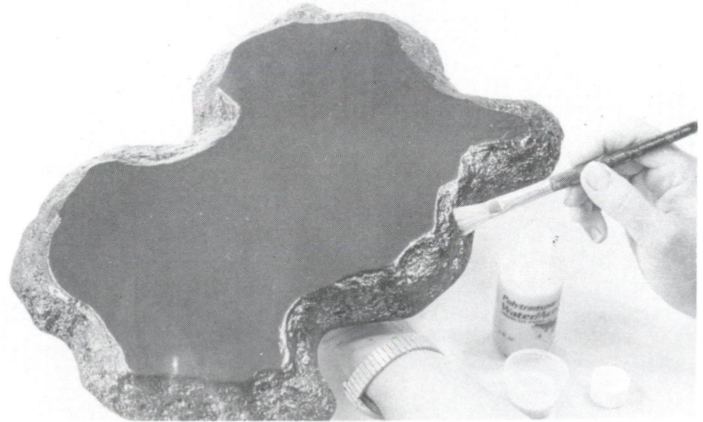

To add additional detail and interest to the base, we stippled Water Acrylic Sparkling Gold WA-423 into the rough edge and wiped off the excess with a paper towel.

Sparkling Gold has a relatively large particle size, and after being applied in this manner, it left a "rich" gold color in all of the deep impressions along the edge. Water Acrylic was used because it permitted us to apply the paint and remove the excess several times until we achieved a nice "goldy" effect. Again remember that this base isn't supposed to look like anything, so any contrasting color could be stippled to the base.

The finishing touch to this base was the addition of a piece of 1/4" thick black acrylic plastic to the underside of the particle board. This plastic is available in many colors through your local window glass supplier, and can be purchased in 4' × 8' sheets, or cut to size pieces. We used a brand called Acrylite, which had a protective paper that was fastened to each side.

All we had to do was lay our finished particle board base upon the Acrylite and trace a rough general outline.

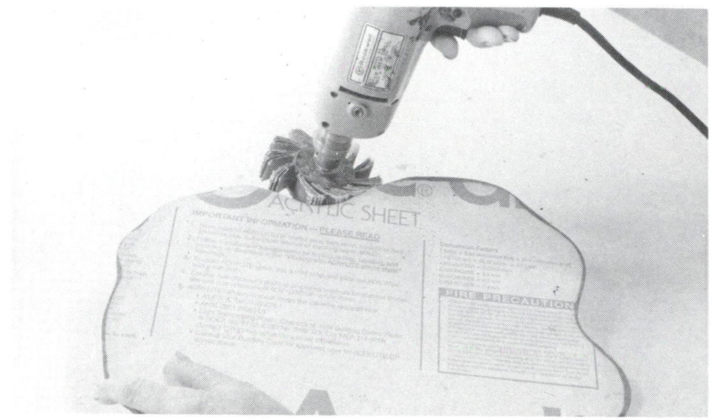

This outline was then cut to shape on a bandsaw. The rough edges of the plastic were then sanded smooth using a "flap-wheel" containing coarse, medium, and fine sandpaper.

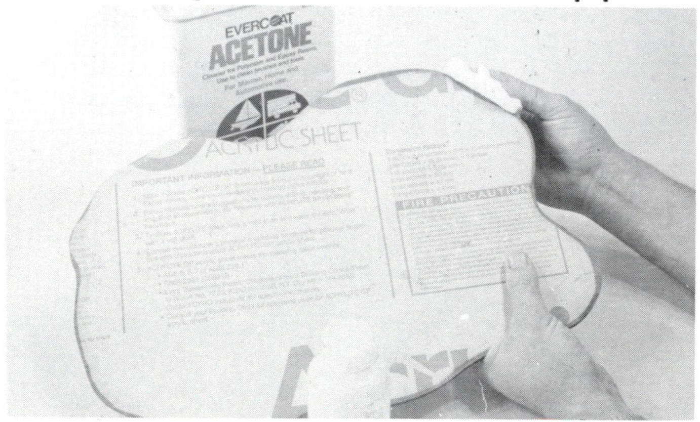


The edges of the plastic were then "polished" by wiping them with a paper towel dampened with acetone.

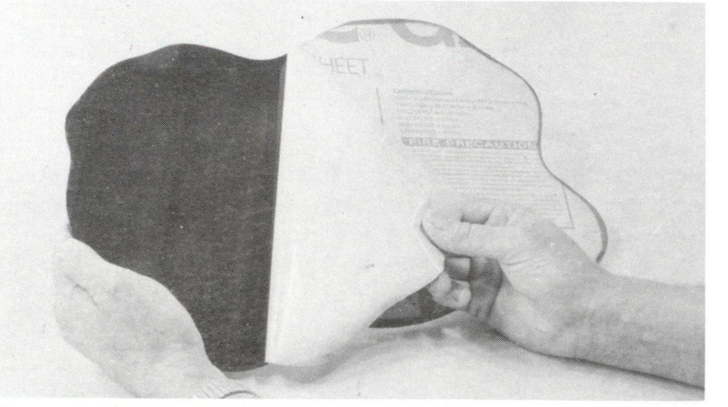

The protective paper was then peeled from the plastic sheet of Acrylite.

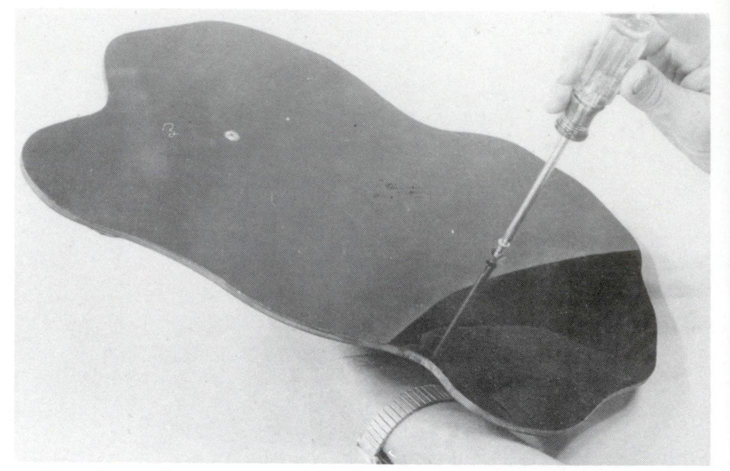

The plastic base was secured to the particle board base with two wood screws. Remember to always drill clearance holes in acrylic plastic sheets to eliminate any chance of the plastic cracking when the screws are installed.

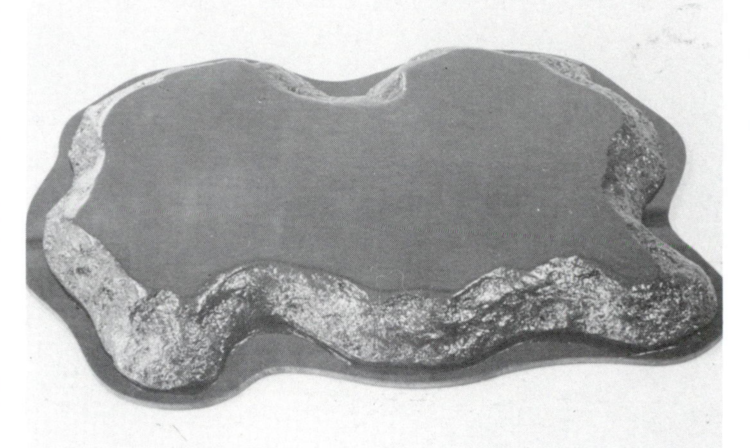

The above photo shows the finished base. At this point, the base could be used for the backing of a wall mounted fish.

With additional scraps of particle board, a table base can be easily constucted for a pedestal mounted fish as shown in the close-up photo of the base above.

Rick has offered us all a very unique way to break away from traditional ideas concerning habitats and bases by showing us this somewhat abstract approach. His technique has served him very well, both for his commercial customers and the critical eyes of competition judges.

# 2. The Realistic Rock Base: A Winning Style

*by Alan Gaston*

My idea was to do a scene of a pedestal-mount walleye feeding on a leech being pulled along the rocks on a Lindy Rig. I felt that this would be a combination that anyone who had been walleye fishing could relate to. Two areas which I wanted to create realism in, were the leech and the rocks. I have never seen a freeze-dried leech that I thought looked realistic, and I was not satisfied with the rocks that I had been making up to this point. I did not want to enter another competition (which was only the fourth time I had ever entered a competition) until I came up with a better method of making realistic looking rocks.

I decided to try casting carefully selected, authentic rocks with a latex mold. In doing this I used a latex molding material called Rub-R Mold. The rocks were collected from Minnesota's most famous walleye lake, Mille Lacs Lake. The process I used went as follows: The rocks I chose were mostly round with a flat side that would make casting easier.

After washing and drying them I built a shelf around each with plastiline oil based clay (above photo on left) and painted about 10 coats of Rub-R Mold on the rocks (above photo on right).

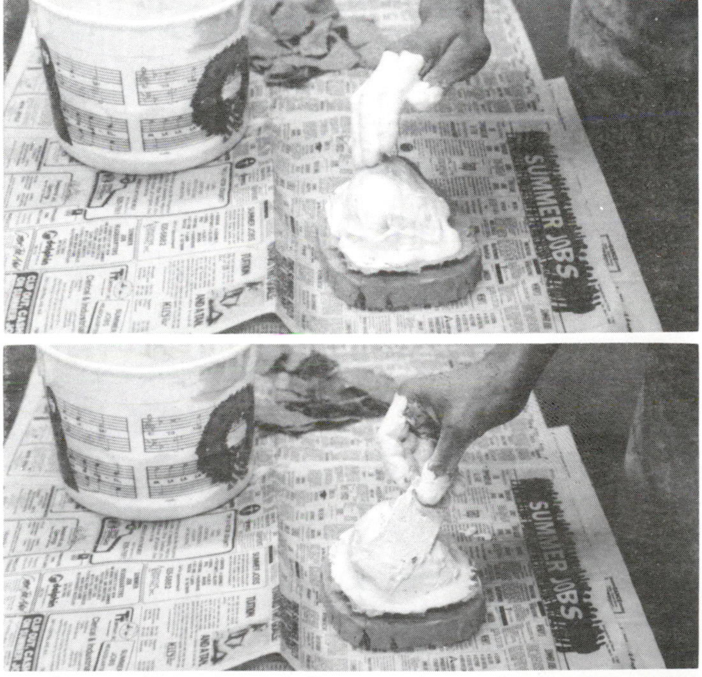


I then made back-up shells or mother molds for the rubber molds with plaster and burlap laminations. When the plaster cooled (which took about one hour), I removed the shell and glued it to a base so that it would stand level for pouring.

Next, I peeled off the latex and ended up with a completed mold.

The weight and color of the rocks were both important considerations. To reduce the weight, I carved a core for each rock out of scrap foam about 1/2" smaller than the mold, and cut it flush with the top of the shell. The original rocks had a subtle sparkling effect. To create this, I simply dropped a few snow sparkles into the bottom of the chosen molds. The plaster was dyed prior to pouring while in its powder form with a small amount of black powdered dye. Now, I was ready to pour the plaster. I filled each mold up about half way and pressed the core into the dyed plaster. Since the foam wants to float, weight

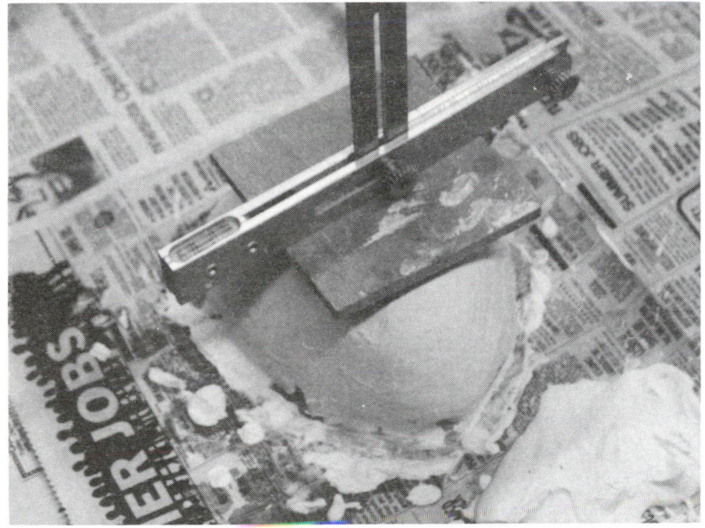

must be added over the foam to hold it down. As soon as the plaster cooled (about one hour), I popped them out.

At this point the detail was near "perfect." I allowed a 24 hour drying period to pass before continuing to work on the rocks.

Before a rock is finished it will need to be sealed and detailed for a realistic appearance. I seal my rocks with polyurethane gloss. The plaster drinks up a lot of gloss so three or four coats are necessary. The final step is to darken the pits. Water colors work the best for this. Spray the entire rock lightly with black or dark brown and quickly wipe the rock down with a clean rag.

Before I had finished, my dad, Master Taxidermist Marv Gaston, came upstairs and asked if the rock that I was wiping was real. His arm rose quickly as he picked up my cast rock. At this point, I knew that they were good enough.

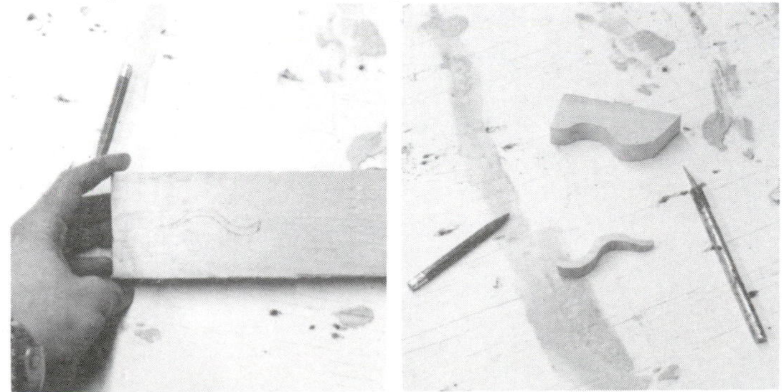

My next challenge was to create a leech that would do justice to the mount. The idea for the wooden leech came after attending Bob Berry's carving seminar at the 1985 World Taxidermy Championships. Making the leech was easy because it is a simple shape to carve. I cut the basic shape out of pine with a freehand cut on a band saw.

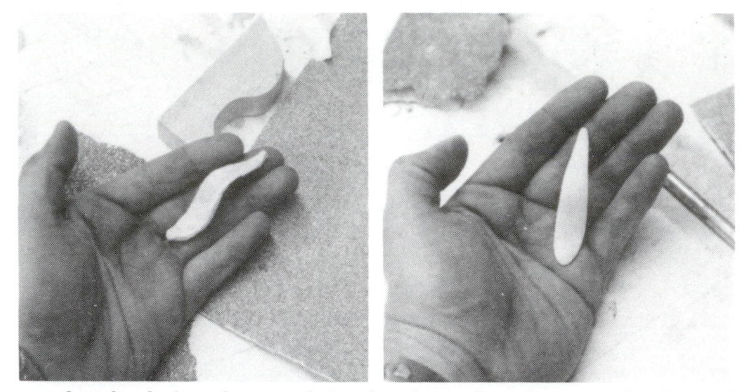

I finished the shape of the leech with an X-acto knife and sanded the wood smooth. The distinct detail of a leech is it has lines down its back. I put the lines in the back by breaking off a small section of staple gun staples that were pointed on the end and dragged them across the back. Before painting, I installed a hook by heating a T-pin and burning a hole through the sucker end of the leech. I filled around the hook with Sculpall™ The leech was now ready to paint.

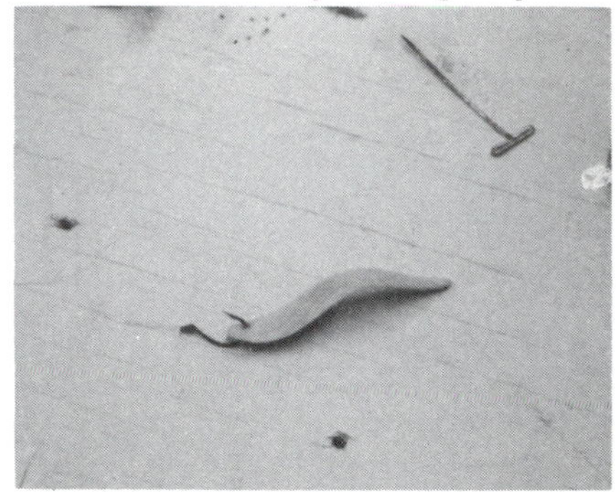

To paint the leech, I used three basic colors: *Polytranspar* Opaque Medium Green (FP61), Chocolate Brown (FP70), and Dark Brown (75% Chocolate Brown [FP70] mixed with 25% Black [FP30]). I started by sealing the wood with two coats of Fungicidal Sealer (FP220). Next, I applied a coat of Opaque Medium Green (FP61). I sprayed with an airbrush until the green on top of the white pine hit the tone I wanted. I would describe it as a medium-heavy coat. I did not cover it all the way; it was still somewhat transparent. Then

I sprayed a medium coat of Chocolate Brown (FP70) on the entire leech. Most leeches have spots so I took a small paint brush and painted spots on the entire body with Dark Brown (75% Chocolate Brown, and 25% Black. Then, I used dark brown to subdue the spots and achieve the desired color. I finished the leech with two coats of polyurethane gloss.

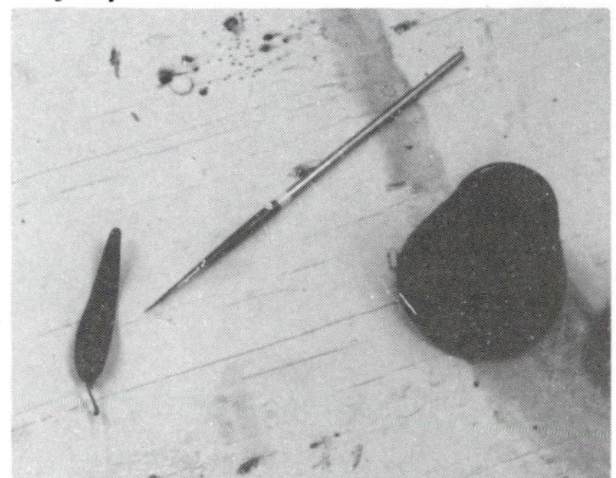

I was pleased with the first one I did, but it turned out to be a jumbo and it looked too big for a three pound fish. The second one was just right and a size that I felt more people could relate to.

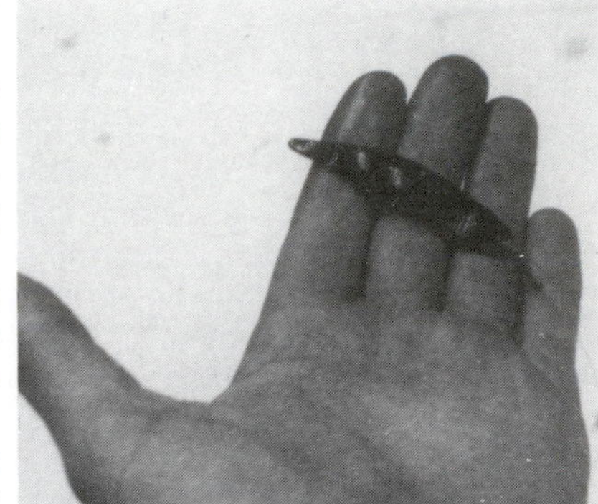

For the base, I thought I'd try something different from renderings that I had done in the past. Keeping the top view shape of a walleye in mind, I drew an extremely round teardrop-shaped base that curved with the fish. I had to leave enough room for the leech at the same time. I tried to keep the base as small as possible, but large enough to complete the picture that I was visualizing in my mind. I then cut the teardrop shape out of 3/4" walnut and routed the edge.

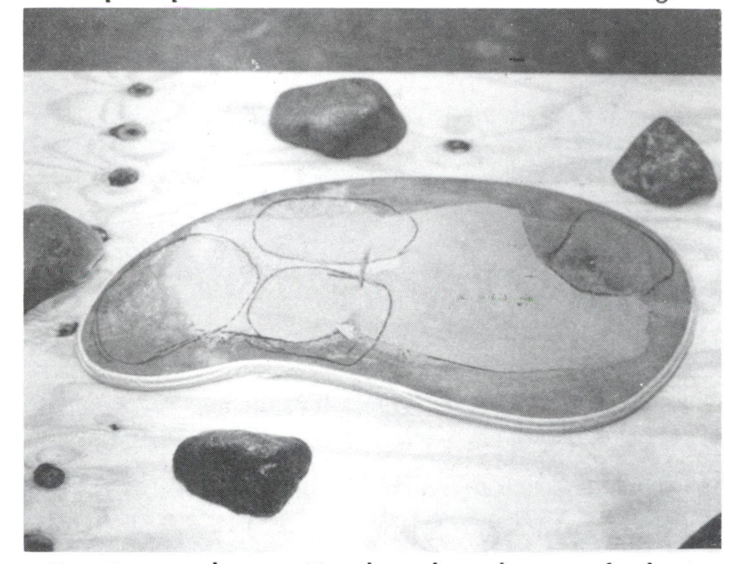

Now, I was ready to position the rocks and create a focal point. I started by taping off the routed portion of the walnut base with masking tape and positioning the rocks. When I was satisfied with the arrangement, I removed the rocks and glued a thin sheet of scrap foam to the center of the base. This was done to suggest a subtle change in the bottom.

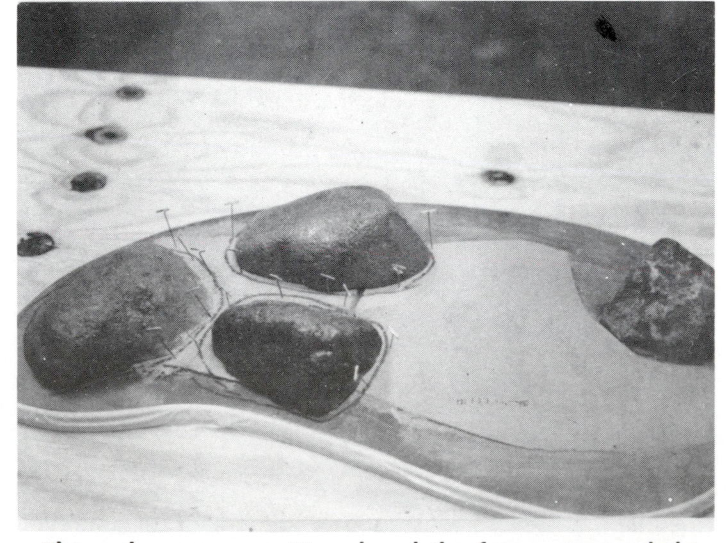

The rocks were repositioned and the foam was sanded to accept the rocks.

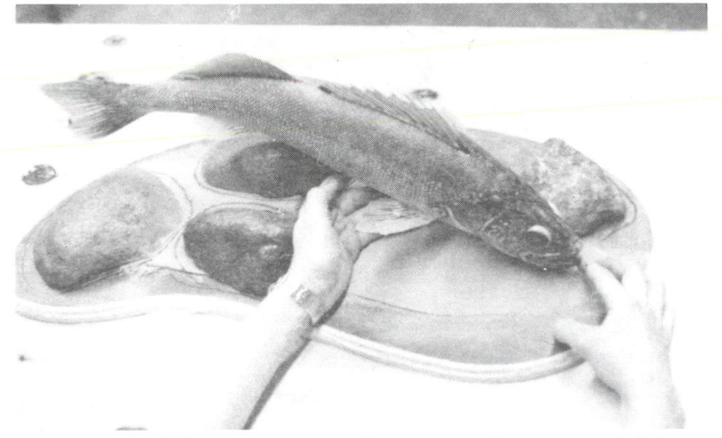

I rechecked the composition by posing the mounted fish on the base.

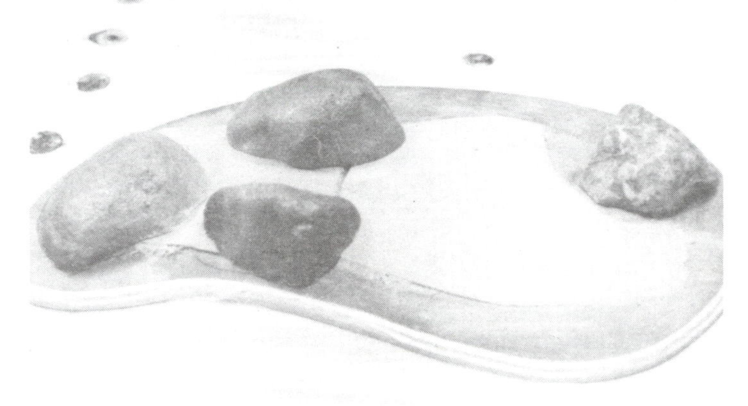

The cast rocks were placed into their original positions and marked with T-pins.

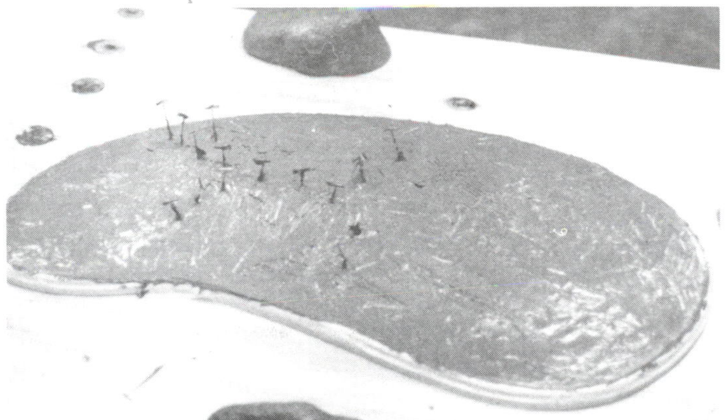

They were then removed and I applied a 1/4" thick coat of Super Paste* to the entire base, tapering to the edge.

Next, sand and gravel were added to the paste. I peeled off the tape and let it dry 24 hours. The next day I shook off the excess sand.

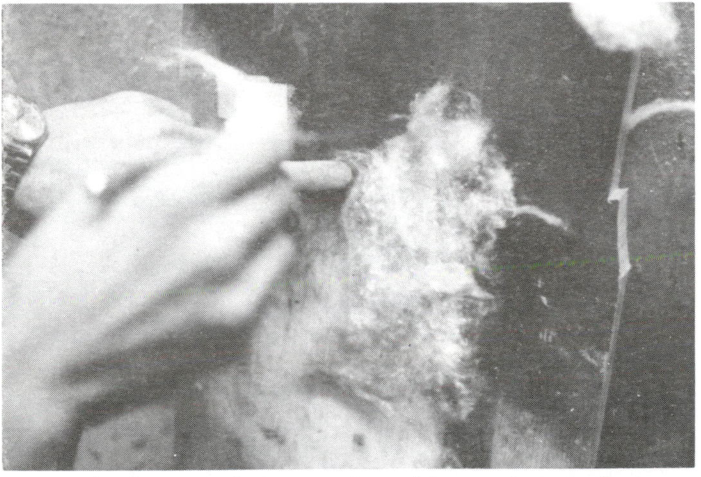

For a finishing touch, I wanted to add moss. I made the moss by tearing thin sheets of cotton bandage (bird cotton) and dowsed both sides heavily with Light Transparent Green (FP51). I quickly placed the moss on the base while the paint was still wet, which eliminated the need to glue it on later. My final step was to gloss the entire base with one coat of polyurethane gloss.

*Alan Gaston's walleye scene pictured above was constructed using the same techniques as described in this section. It won Best of Show and Best in World Fish at the 1986 World Taxidermy Championships in Lawrence, Kansas.*

*"Super Paste" is 75% linoleum paste and 25% Finglue. Finglue is H. B. Fuller Adhesive product No. S3809, 2400 Kasota, St. Paul, Minnesota 55108.

# 3. Putting Your "Shoulders" Into Habitat Bases

*by Dan Blair*

On occasion, the shoulder portions of a gamehead mannikin are not required (as in the cases of short-neck mounts, or when the head from a full shoulder mannikin is added to an altered life-size form). These unused shoulder portions should *never* be discarded for at least one good reason; they can be easily fashioned into excellent habitat bases for a variety of mounts.

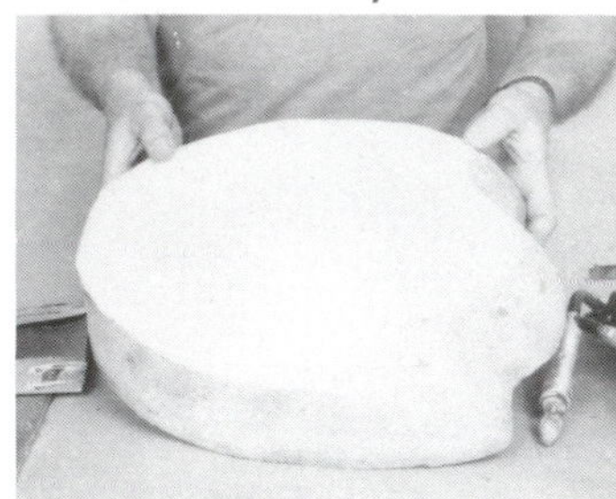

Most form manufacturers these days include a plywood backboard in their headform which, when using the shoulder portion as a base, ensures a flat even surface for the base to rest upon. The plywood offers a secure foundation from which the foam will *not* break away, and wires from mannikin feet or artificial habitat can also be securely fastened to it. When adding pieces of driftwood, etc. to the base, the plywood also offers a solid plate through which to run the long-shanked screws up into the driftwood.

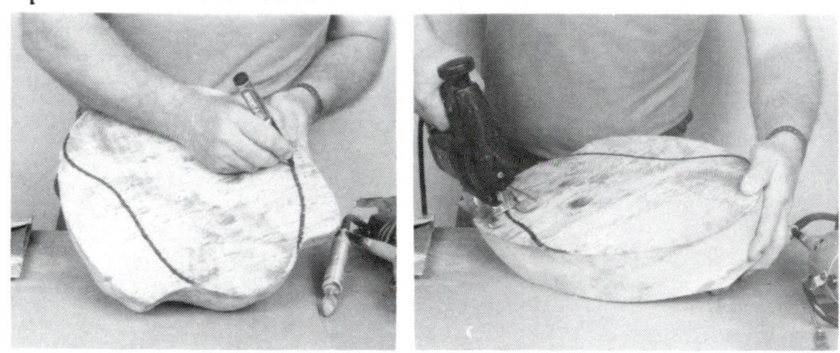

If the "shoulder" shape is unsatisfactory, or too large for the the mount, simply sketch a more preferable design on the plywood and cut it out with a saber saw or band saw. Even if the saber saw blade isn't long enough to penetrate clear through the foam, the foam will break off the rest of the way through the cut with just a little pressure.

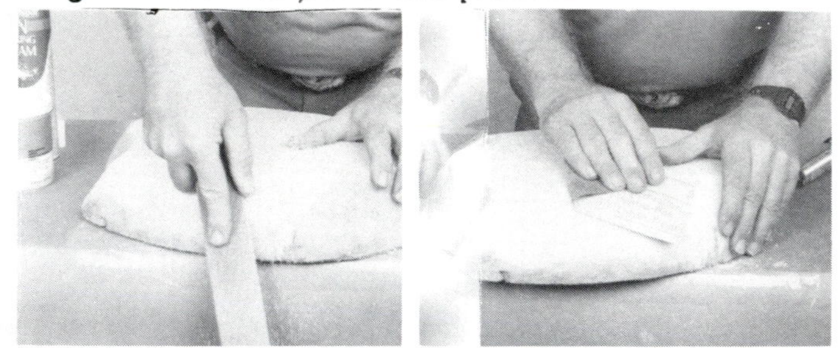

The foam can then be easily rasped, carved, or sanded into any shape to immitate rocks, earth, etc. For a rock-textured surface, the foam can be covered with number one molding plaster, or any of the maches available from WASCO. Before the mache sets up, use a modeling tool to detail the rocks with cracks and other textures (see Chapter 8 for more detailed instructions). After drying the rock, use hand or airbrush applied paints to finish the detailing.

If a preferance for a vegetation base is requested, apply a complete coat of Fin Backing Cream (FC101) and onto it, press sheet moss over the contours of the foam. Then insert the wire stems of artificial plants into the foam for artistic balance and conformation and it's ready for the addition of the mount.

An alternative to the sheet moss would be the addition of artificial snow, or sand and gravel to the wet Fin Backing Cream. Real or artificial sticks and stones can be added before or after these steps by simply gluing them in place with hot glue or more Fin Cream.

To attach the specimen to the base, be sure to keep leg wires long enough to reach all the way through the depth of the foam and plywood backboard. Predrilling the holes with a 1/4 inch drill bit long enough to pass entirely through the base will help. Rather than sharpening the ends of the leg wires, it works better sometimes to blunt them by bending the very ends back on themselves before pushing them through. Once through the bottom of the base, pull the leg wire tight enough to draw the foot of the mount firmly down onto the upper surface of the base. Do the same to each of the remaining leg wires.

Before cutting off excess wire, use a "V" shaped gouge or chisel to notch a channel into and away from each hole in the plywood. Into this groove bend the leg wire, cut to proper length and then staple or hot glue securely in place.

The addition of felt to complete the bottom of the finished piece will cover the wires and staples and also any other imperfections in the plywood surface. This *extra* step will help "finish" the work to the greater satisfaction of any customer.

## Glass Domes & Plexiglass Cubes

Carvers and taxidermists alike are faced with the task of protecting valuable and delicate works of art and one way most requested by customers is the inclusion of a glass case. But as most artists know, such cases are not commonly abundant and are rather prohibitive in cost.

An exceptionally small number of companies supplying the taxidermy and woodcarving crafts offer glass domes or cases, and those who do can almost name their own price. As a result, many artists will attempt making their own at least once before giving in and paying those prices. Glass domes (without bases) range from about $25 to $50, depending on sizes (up to 10" diameter × 12" high). Cases begin at $40 minimum and from there on, the sky's the limit. (Imagine the cost for casing a 9 ft. polar bear standing erect.)

When attempting to build a "case," its most critical (and most visible) features are the joints of the wood framing, and the unions of the glass itself. If an artist has the ability and equipment to accomplish these connections smoothly and tightly, by all means, he should do so. However, most will find the end results less than satisfactory and their customers may find it even less so. It might be best under such circumstances to request bids from local cabinet and carpenter shops. To save additional charges, offer to do all the sanding, staining, and finish work to the wood. (All one needs are the precut parts ready to assemble.)

At a recent state taxidermy competition, one master of the art was overheard telling how he hired a retired carpenter to build all his cases at very reasonable rates. The old carpenter was glad for the opportunity to work again and enjoyed the extra tax-free cash it put in his pockets. No doubt the taxidermist was pleased also because his beautiful cases were custom built to his specification and had cost him slightly more than the price of the materials.

There are also other alternatives to buying expensive glass domes. One common source, believe it or not, is K-mart. They don't usually stock the domes as a craft item, but rather as a completed novelty usually housing dried or silk flower arrangements, or exotic butterflies, etc.

CHAPTER 7

# Earth and Ground Cover

Snow and blue goose by Scott Souders of Nebraska.

Pheasant by John Lager of Colorado.

***by Jim Hall***

Almost every exhibit will have some type of dirt, mud, or sand in it. This phase of exhibit building is very fast and easy. Most artists use "actual" dirt or sand glued to a base. This requires treating it for insects and drying it out. It is a good idea to collect and have on hand a good supply of a variety of different textures and colors of dirt and sand, ie., white sand, gray sand, black dirt, red dirt, brown dirt, etc.

## Dirt Treatment

It is a good idea to "treat" dirt and sand for insects before putting it into an exhibit. This can be accomplished in a couple of different ways. One easy method is to put the dirt into a plastic garbage bag with a no-pest strip insect killer. Seal the bag with a twist tie and leave it for a couple of weeks.

Another method is to "soak" the dirt in a bacteriacide solution. Mix 3 gallons of bacteriacide solution according to the directions on the container and then add the dirt. Let it soak overnight, then pour it out into a sturdy cardboard box. Allow it to dry thoroughly, occasionally raking it in order to completely dry it. Once completely dry, place it in a sealed plastic garbage bag until it is needed.

Once the dirt has been treated, all that is required is to paint all areas of the base to be covered with dirt or sand with *Polytranspar*™ WA/FP Airbrush Paint. Use a color that will closely match the same color of the dirt or sand being used. (This base coat ensures that the base won't show through if an area is missed when applying the dirt or sand). Next, apply a liberal coat of Ultra Seal, Fin Backing Cream, or Elmer's glue to the base with a bristle paint brush. While the glue is still wet, sift or sprinkle the dirt or sand over the glue to completely cover the base. Once it has dried completely, simply blow off the excess with an air compressor.

To make mud (or wet looking sand), use calatyzed *Polytranspar* Lay-up Resin and mix dirt with it in a cup or bowl until it becomes a dough or putty-like consistency. (Don't get it too wet.) Cab-O-Sil or aerosil can be added to thicken it if necessary. Now put the mixture onto the base wherever mud is desired and allow it to cure and harden. NOTE: Footprints (tracks) should be made before it hardens if so desired. Also any weeds, grass, etc., should be added now.

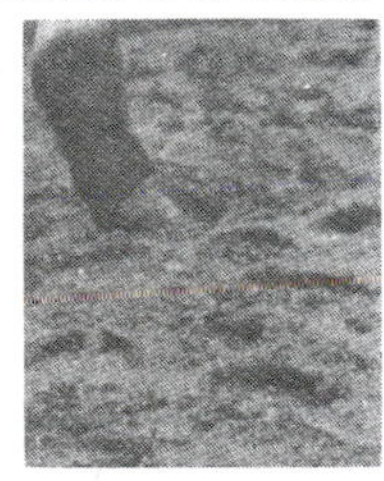

*Polytranspar* Lay up Resin will also work to "glue" on dry dirt and other habitat materials to a base. (Remember these materials must be thoroughly dry or the polyester resin will not work properly.) Simply paint a thin coat of the catalyzed resin on the base and then cover it with dry dirt. Allow it to cure and fully harden, then blow the excess off with an air hose.

## Dirt Substitute

One material that works well as a dirt substitute is sawdust. Glue the sawdust onto the base the same as would be done with actual dirt. Once it is dry, airbrush paint it with *Polytranspar* paints to achieve the desired color of dirt. To give the sawdust an interesting texture, mix vermiculite (Zonolite) to the dirt before gluing.

Peat moss is another material that looks good in an exhibit. It must first be put in the sun or under a heat lamp until thoroughly dry. Treat it for insects the same as actual dirt. Then, apply it to the base using glue in the same manner as with actual dirt or sand.

There is no use to reproduce small pebbles or gravel as they are not very heavy. If pebbles or gravel are desired in a scene, then glue actual stones to the base. The same adhesive used to glue dirt or sand will also work for small stones.

# Forest Floor Ground Cover

One of the more common types of habitat, which is useful to depict all seasons of the year, is the dirt and debris found on a forest floor. (For winter scenes, add a very light coating of artificial snow as described elsewhere in this book.)

While you may wish to gather your own materials in the summer months, an alternate material is available at your local garden shop in the form of plant soil. By purchasing different brands of soil, the artist can obtain different grades and texture of soil and debris that will produce habitat bases that are most convincing.

If you decide to try one of these packaged soil mixes, be sure to pour the soil out on a piece of newspaper so that it can dry thoroughly. Any dampness left in the soil will interfere with the proper setting of various glues and fiberglass resins.

Applying this dry material is quite simple. Merely construct the typical plain plywood, hardware cloth, and mache base (as described in Chapter 6) and permit it to dry. Then, brush on a liberal coating of catalyzed fiberglass lay-up resin and sprinkle on a layer of the potting soil. When the resin has completely cured, turn the base or habitat over and the excess soil will fall off. If a deeper appearance is desired for the surface, apply another coat of resin and soil material.

To further seal the finished surface, and prevent small pieces from breaking off during handling, we recommend spraying the potting soil with a light coat of *Polytranspar*™ Water/Acrylic Wet Look Gloss (WA240). If a light coating is used, it will be absorbed in the soil and will not show at all.

# Creating A Footprint

The resourcefulness and initiative of wildlife artists is sometimes amazing. They will go to no end nor leave any stone unturned in order to create some little detail that can be associated with a mount or a habitat base arrangement! We have heard many taxidermists claim that they spend at least 100 hours on a competition mount with habitat, much of it on some small detail.

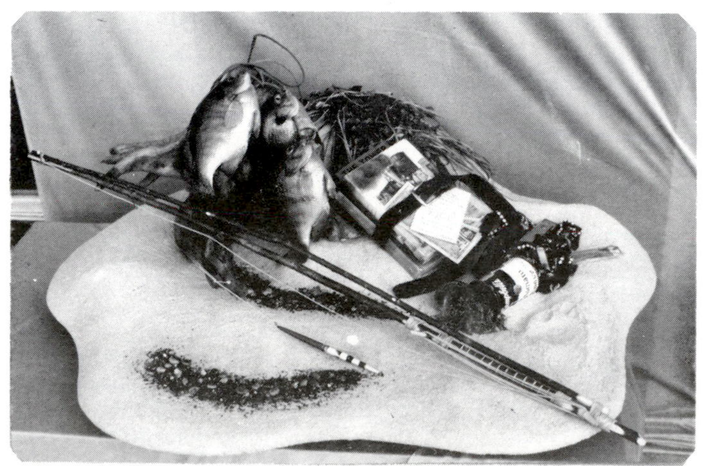

Years ago, I created a habitat setting entitled "School Days," which depicted a day in the life of a young boy who decided to play "hooky" from school and go bluegill fishing instead. This was a sand-base habitat upon which I wanted to create the bare footprints of the boy. Unfortunately, all of the young boys I could find that I could convince to press their foot into the wet sand and glue mixture had feet that measured size 12 or better. Since I had in mind a smaller foot size, I came up with the following plan.

I waited until one day my wife was in a particularly good mood, and asked her if I could make a plaster cast of her foot, to which request she more or less agreed. I mixed a quantity of casting plaster in a suitable sized container and she gently placed her foot into the clammy medium. It was quite exciting. . .

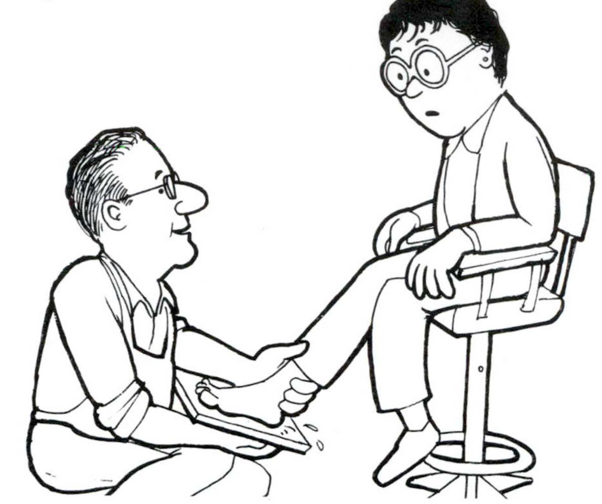

. . .When the plaster had set, we had the reverse impression of her foot, but again, the footprint was longer than I desired.

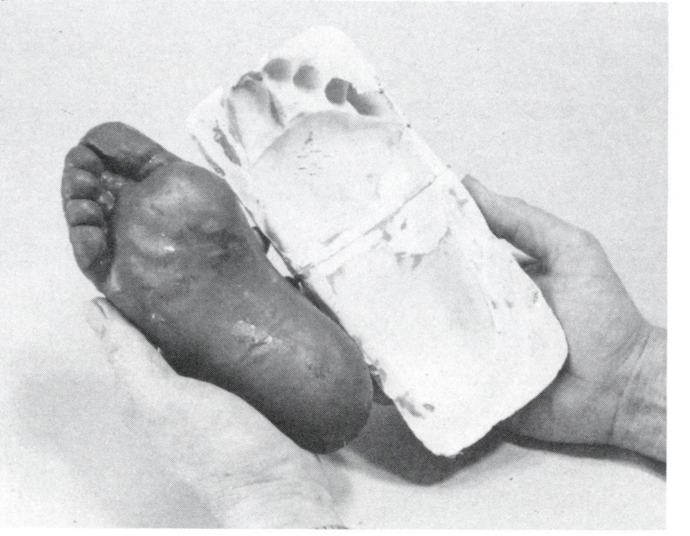

However, this was easily solved by cutting the plaster mold in half and then fastening it back together. The hairline crack can be seen at the mid-point of the mold.

After the plaster had cured and dried for about a week, the surface was waxed several times with a mold release wax such as Wildlife Artist Supply Company's Ceara Mold Release. Into this cavity was poured a quantity of RTV molding rubber (Mold Builder™ would work fine), which was allowed to cure, resulting in what I felt was the proper sized rubber "foot."

Now, I had created a foot with which I could make all the footprints I wished, and the model worked very well.

This process required quite a bit of effort and time to create a rather small detail, but it solved the problem very well, *and* I'll be able to make all the footprints I need in the future. If I had the same problem to solve today, I'd probably hand carve the footprint into the sand, or at the very least, search longer and harder for that young lad with the proper sized foot.

# Sun-Baked Earth

*by Dan Blair*

While building a habitat base for a life-sized coyote investigating a trap baited with a striped gopher, an interesting discovery was made. The base consisted of foam hot glued to a plywood platform, over which was spread a heavy coat of yellow carpenter's glue (Elmer's). Into this glue was pressed potting soil straight from the bag.

Normally the soil would be allowed to dry before applicaton, but in this case it was as damp as it had been in the bag. The glue, being water soluable, was totally compatible with this moisture.

However, as the glue and soil began to dry over the next few days, an exciting occurance took place. Just like "sun-baked earth" this imitation began to split and crack open. The realism it duplicated was as natural in appearance as if the sun itself had been the creator.

The one and only obvious flaw was the space between the cracks exposing the yellow cream color of the foam base. However, it was a simple task to "touch up" these areas with an airbrush using *Polytranspar*™ Chocolate Brown FP/WA70 and Black FP/WA30 paint mixed 50–50. (Black Umber FP/WA29 can be substituted and blends very well).

Once the soil and glue had been painted and dried completely, a light dusting of "dry" sandy soil was scattered over a spraying of Grumbacher Tuffilm. The matte finish spray helped anchor the "dust" in place but a second coat, this one on top of the dust secured it even more. (For these coats, it is recommended you use a nongloss spray like Tuffilm to avoid a wet look.)

For added realism, mix a bit of crushed leaves, or other vegeatation into the sand, and the addition of "small" twigs will help keep the habitat from appearing to sanitary. Even insects, feathers, and bone fragments can be helpful in this respect.

When recreating this scene it was also discovered that heat lamps would accelerate the drying and cracking process.

One point to keep in mind also with this particular scene is that plants, if any are used, should appear as parched and dry as the "sun-baked" earth.

# Waterhole Base

*by Wendy Christensen-Senk*

I started with a regular wooden framework for the base. In this case, I used a 2×4 framework covered with 1/4" hardware cloth and reinforced it with sheets of burlap dipped in plaster. This method allows a contour to be easily formed so a low spot was made in which the water will lie. I then nailed on the finished wood sideboards on the base and caulked the seams where the "water" would later be poured to keep it from seeping out.

The mixture I used to give the desired effects of mud is a mixture of medium grade sand colored with black dry tempera color. This colored sand is then mixed with catalized polyester (fiberglass) resin until it is of a doughy consistency. (Do not make it too wet.) I then spread approximately 1 inch of this mixture in the areas that I wanted to look like mud. Using the extra feet of Cape buffalo and various antelope, I pressed the hooves into this mixture to make it look like tracks sunken in the mud. If the mud appeared too shiny, or when I wanted it to fade out into dry ground, dry, colored sand was sprinkled over the wet fiberglass mixed sand. Any grass or plant materials were also added at this time.

The advantage of this fiberglass mud is it is extremely durable and easy to clean. An air compressor easily blows off dust without damaging base materials. Also, this fiberglass mud makes an excellent base to pour the Envirotex water over. It seals air pockets very well which avoids excessive air bubbles in the artificial water.

To make the water, I mix just enough Envirotex (available from Wildlife Artist Supply Company) to thinly cover the part of the base that I want to look like water. Envirotex is a 2 part epoxy which is mixed 50–50 "thoroughly" for 2 minutes, and then poured over the desired area. A little bit will go a long way. It is much better to pour the water in layers no deeper than 1/2." This makes the water much clearer and allows the epoxy to set better. Air bubbles from mixing will appear on the surface of the epoxy. To eliminate these, continually blow on them until they disappear. Allow each layer of epoxy to cure (at least overnight) before another layer is added. Several layers may be needed to build up to the desired depth.

The advantages of Envirotex epoxy over many other products for simulating water is that it does not shrink away from the base, one never has to worry about it cracking, it's easy to mix and pour, and it looks like water!

Another effect I tried to create on this waterhole base was the occurence of animal droppings. I simulated these by forming Sculpall™ into various sizes and shapes and rolling them in sand and finely crushed plant material. (I used dried deiffenbachia debris because it was handy.) Paint touch-up was also added. These droppings were then glued in piles at various places on the base.

Finally, shadows and colors were airbrushed over the cured soil for added results. I added some reddish tinges to match the soil conditions of the location where I collected the animal.

*Wendy Christensen-Senk's nyala antelope scene won a bronze Carl E. Akeley medallion at the 1986 WTCs.*

# Creating A Sand Base

*by Jim Hall*

One of the easiest and simplest methods to construct habitat bases that I know of is the sand surface base. These bases require a "bare-minimum" amount of time to build, and the materials needed are very inexpensive. The following procedure will provide a customer pleasing base that is also very durable during handling of the finished mount.

As a specimen to use for this sequence, we selected a small ermine mounted on a single piece of driftwood.

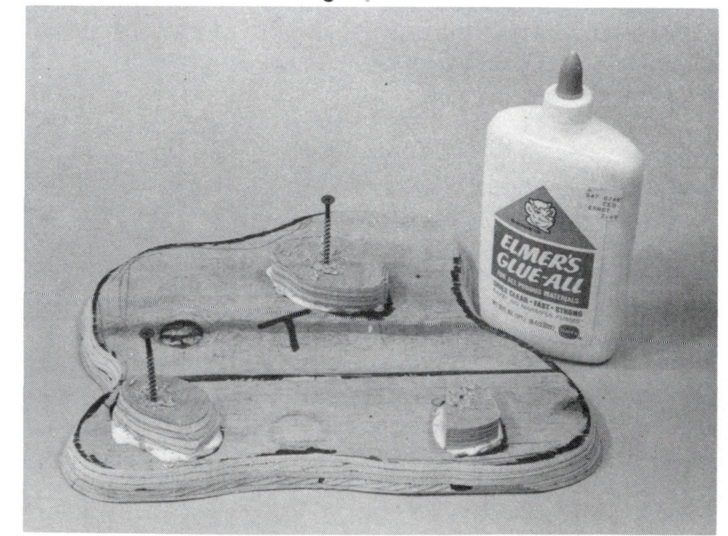


We began construction by hand-sawing an irregular outline of the base from a scrap piece of 3/4 inch rough plywood.

Three small pieces of plywood were then glued to the base to provide the screw support for the driftwood. Holes have been drilled in these spacer blocks at the points where the driftwood will ultimately touch the sand base. Two inch Grabber screws, which will be used to fasten the base to the driftwood, are shown here to show the location of the holes. (Note the beveled edges of the plywood to give a better appearance.)

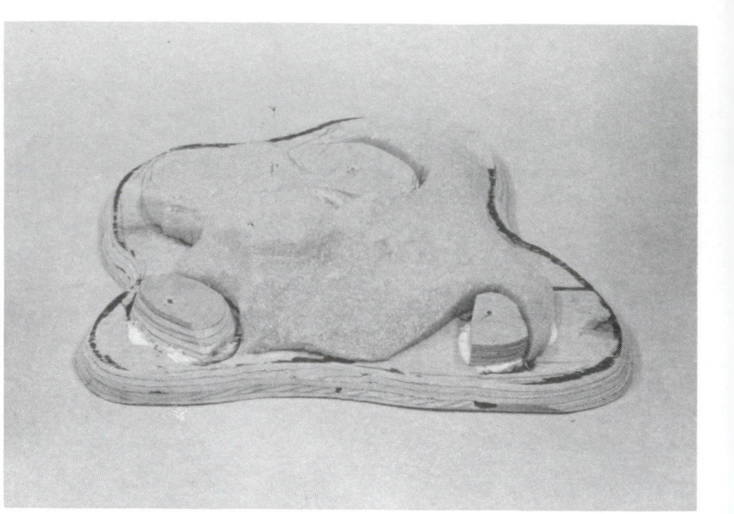

I then rough-shaped a filler piece of scrap styrofoam in-between the spacer blocks to create the shape of the base surface. I used a gouge and knife to rough shape the styrofoam. It was then glued to the plywood with white glue.

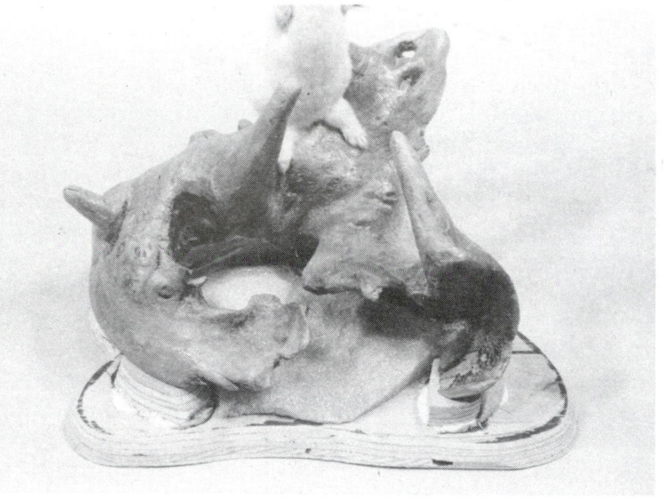

When the glue had dried, I checked the fit of the driftwood upon the base again, just to make sure that the three spacer blocks were still making contact with the driftwood.

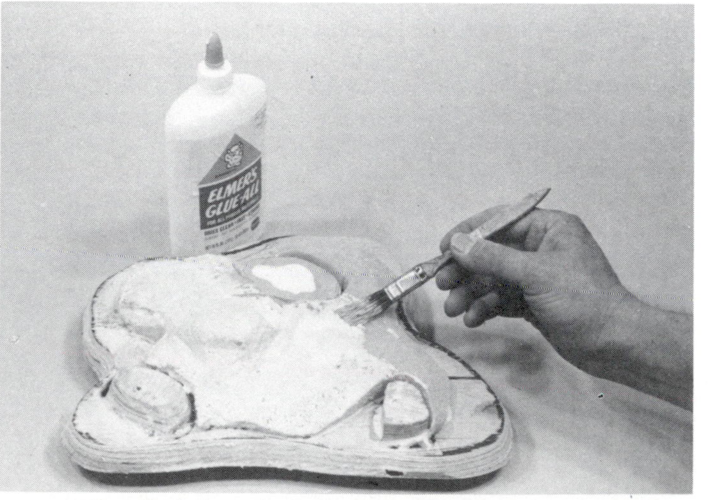


I next brushed the whole base surface with white glue (Fin Backing Cream is also good for this step) to act as a primer and sealer for the Jim Hall Mache layer to follow.

When this primer coat was semi-dry, the layer of mache was applied and smoothed with a spatula over the entire surface. I made sure that I overlapped the beveled edges of the plywood thoroughly.

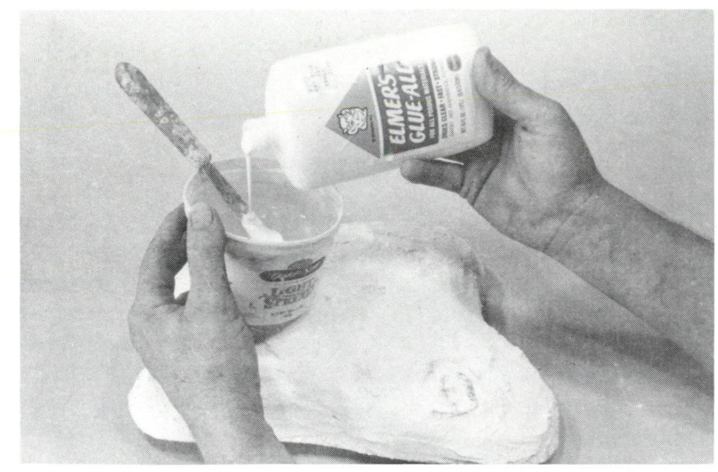


Then, I mixed more white glue with a container half full of warm water, (use about 3 parts glue to about 6 parts of water, stirring until completely dissolved). (Fin Backing Cream is ideal for this step as well.)

Sand was then added to the glue/water mix until a thick paste was achieved.

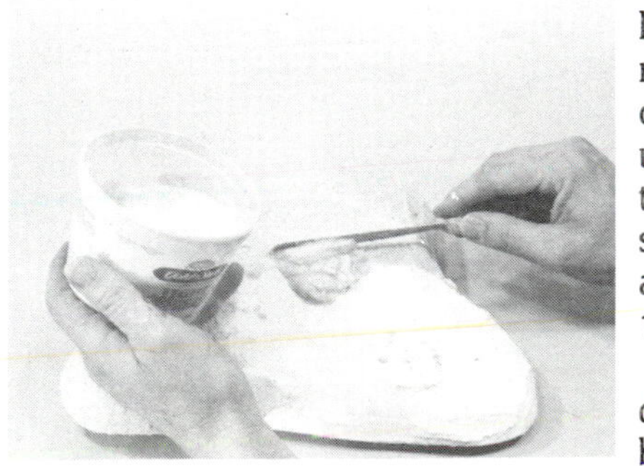

In the photo to the left, the sand and glue mix was spatulated over the mache in a uniform coating. Work the sand mix into the surface and smooth to a thickness of at least 1/8 of an inch.

If a smoother finish is desired for the sand base, brush the surface with a small bristle brush dipped into the sand/glue mixture as demonstrated in the photo below.

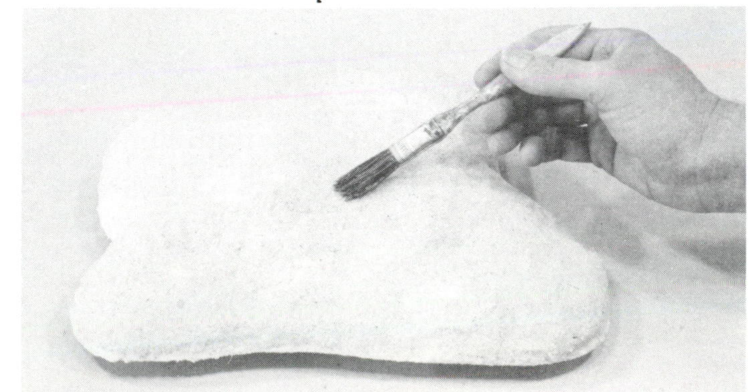

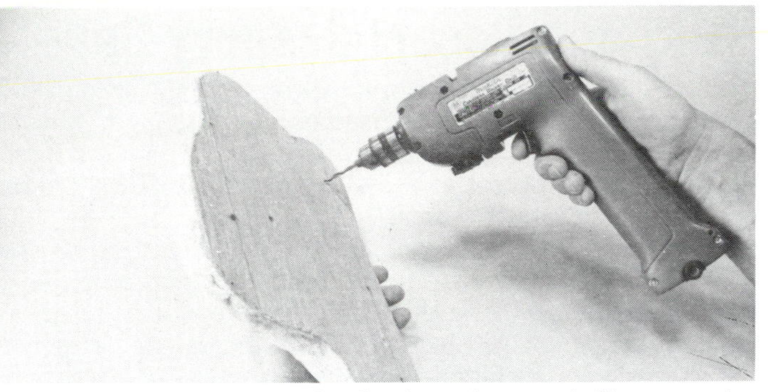

When the base is completely dry, place a drillbit into the already established screw holes and finish drilling through the mache and sand layers. Drill carefully so as not to chip the sand.

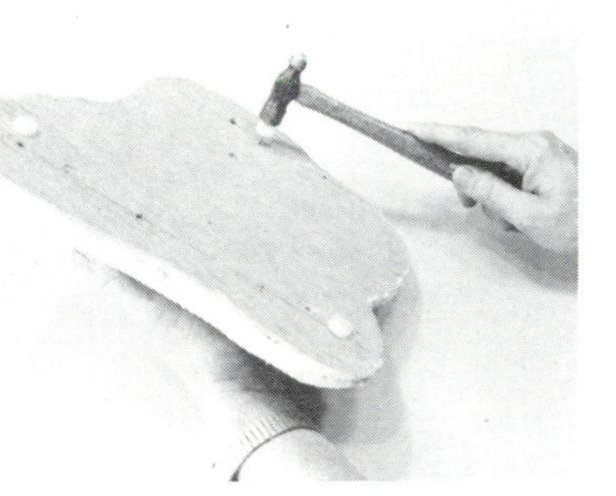

In the photo on the right, I am installing three plastic nail-on furniture glides to the bottom of the base. Three glides will ensure a steady, rock-free base. They are available at most hardware and home centers.

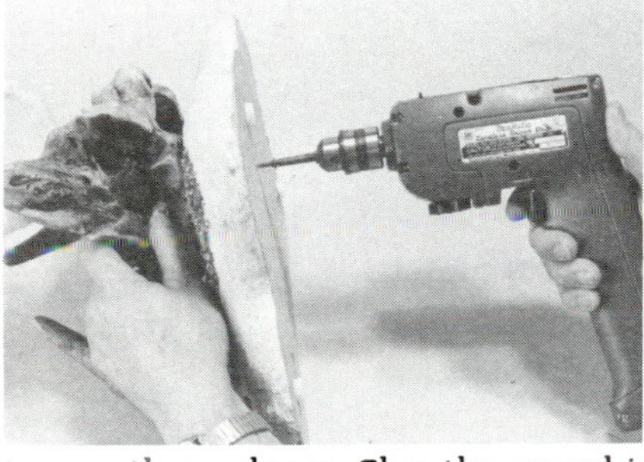

The specimen and the driftwood were then secured to the sand base with the use of Grabber screws as demonstrated in the photo to the left.

For a finishing touch to the rather plain sand base, small areas of gravel can be added in two or three places. Glue the gravel in place with Fin Backing Cream directly from the container.

# Making a Desert Scenery Base

Bob Elzner of Apache Junction, Arizona, demonstrates how to construct this simple base used in his rendering of a regal horned lizard in it's natural habitat. This scene won a 2nd place Masters Division ribbon at the 1986 World Taxidermy Championships in Lawrence, Kansas.

***by Bob Elzner***

This habitat is very easy to create. It consisted of a desert landscape and a regal horned lizard. The regal lizard lives in the desert and looks like a creature from prehistoric times. The lizard's diet consists of ants and that's primarily all they they eat. The habitat incorporated the lizard in it's natural enviroment seeking out it's favorite food at the base of an ant hill.

To begin, an oak oval base in which to create the scene was selected.

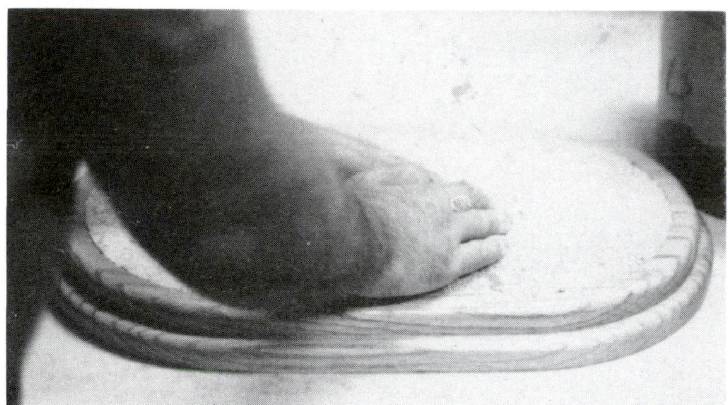

Tom Sexton Fish Filler available from Wildife Artist Supply Company, is mixed to a creamy consistency and applied to make a base for the desert floor and to build an ant hill.

Small rocks, natural to a desert setting, were selected and embedded into the fish filler to create a realistic setting.

Using Elmer's wood glue, I brushed a coat over the entire surface of the fish filler. The glue seals the rocks and the fish filler to the oak base. (Note: Ultra Seal or Fin Backing Cream works equally well.)

Actual dirt and gravel were added over the wood glue. I collected this from the desert. (Note: White sand works well also instead of desert dirt.) The dirt was sifted through screened wire and thinly sprinkled over the surface of the wet glue. At this time a small amount of gravel was added and also embedded into the scene.

Natural grass was used which I had collected from the desert. It is easy to collect at certain times of the year when grass covers the desert floor. I used an ice pick to make a hole in the fish filler and inserted the grass roots into the hole and "tamped" it solidly in place. Next, the oak base was wiped down with a damp towel to remove any excess glue from the wood. At this time, I allowed the fish filler and the wood glue to dry overnight.

Once the filler and glue were completely dry, I shook off all of the excess sand and gravel. Wood stain was then applied to the oak base and allowed to dry.

Finally, the lizard was placed in position and holes were drilled to accept the wires. The holes were filled with Sculpall™ epoxy and the wires were inserted. I allowed it to completely dry. The lizard had been located directly in front of it's favorite place, the artificial ant hill. All the scene needed was a few ants, which were glued into place with Instant Bonding Glue and the scene was complete. The entire desert scene took about 35 minutes to create.

# CHAPTER 8

*There are many methods for producing realistic artificial rocks for habitats. This chapter explains these methods as used by some of the top wildlife artists in the country.*

# Creating Artificial Rocks

It is desirable to be able to produce realistic looking artificial rocks for a couple of reasons. One is readily obvious, "weight." Natural rocks are quite heavy and adding them to an exhibit will make the exhibit undesirably heavy also. Another reason is, that often a scene requires a certain size and color rock that may not be readily available in the immediate area or the artist simply may not have the time to go collect the size, shape, and color rock needed. Having the knowledge of how to create various artificial rocks enables the artist to produce anything from Florida sandstone to Wyoming lava stone without ever leaving their studio. Rocks of all sizes can be custom built to the exact size, shape, color, and specifications required for each scene.

There are a number of different methods of recreating natural rocks. We have outlined a variety of methods of producing them. Which is best? That's hard to say because so many artists have such varied needs. We recommend that serious artists try "all" the methods at some time or other, as they will surely need experience working with different techniques and materials. If just "dabbling," then refer to the following chart to aide you in choosing a suitable method and give it a try.

| METHOD | COST | HOW FAST CAN THEY BE MADE | DETAIL ACCURACY | REPRODUCING THE SAME ROCK |
|---|---|---|---|---|
| **1. Fish Filler/Mache** | Very Economical | Fast | Good | One Time |
| **2. Foam Carvings** | Economical | Medium | Good | One Time |
| **3. Cotton/Urethane** | Economical | Medium | Excellent | One Time |
| **4. Fiberglass/Wire Mesh** | Moderate | Slow | Good | One Time |
| **5. Alginate Mold** | Moderate | Medium | Excellent | Pour Multiple Molds 2 to 3 Per Mold |
| **6. Latex Mold** | Moderate | Slow | Excellent | Fair Multiple Molds |
| **7. Silicone R.T.V. Mold** | Expensive | Slow | Outstanding | Best Multiple Molds |
| **8. Fiberglass Direct Cast** | Moderate | Slow | Excellent | One Time |

*Well executed artificial rocks are difficult to discern from the real things.*

As a general rule, most any of these methods will yield a lightweight rock or stone that is accurate enough for commercial purposes. Museum or competition grade rocks will require more time, effort, and as a rule, more expensive materials, therefore, one should choose a method to build them accordingly. As with anything, it is much easier to obtain "top" results if reference is used.

# 1. Fish Filler/Mache: Fast, Easy, Economical Rocks

## STONE CLONES—The Fake Rocks That You Can Take for Granite

***by Dan Blair***

Craft catalogs, trade publications, carving schools, taxidermy schools, seminars, and video tapes, to name a few instructional sources, have gone to some length at instructing wildlife artists in a variety of methods of creating realistic artificial rocks for habitats and exhibits.

Some artists go to even greater lengths and expense by constructing latex and silicone RTV molds for making castings of rocks. Such molds, though effective, have a somewhat limited value in many commercial studios simply because it takes several molds (and a considerable amount of time) to produce a large enough *variety* of rocks to prevent obvious duplications in a scene.

As you know, certain species of fish such as walleye, smallmouth bass, etc., prefer rock bottom habitats. To accurately portray such a habitat could require as many as a dozen or more "different" rocks in sizes and shapes as well as complexion and color. Birds and mammals displayed in such a scene present the same problem.

When using a large number of rocks, or when using a few *very large* rocks on a base, one extremely important concern is, and should be, *weight*. (Otherwise we'd simply use actual rocks, right?) But, there is a quick, simple, and inexpensive method of making lightweight and realistic looking rocks.

If you're one of the many taxidermists who uses the fish-fill method of mounting, you most likely have all the materials that you need to make rocks on hand. And if you, as many taxidermists, allow the leftover fish filler to harden in the bowl or throw it away after filling a fish skin, you're wasting fish filler *and* potential habitat material. Likewise, if mache is used around your studio and extra mache is mixed, it could be converted into artificial rocks with just a few minutes work.

By using the following methods, you will save time, money, *and* fish filler/mache and be able to produce top-quality rocks for top-quality habitats.

*These realistic rocks were hand made from Tom Sexton Fish Filler.*

The Tom Sexton Fish Filler we refer to is sold by Wildlife Artist Supply Company of Loganville, Georgia. Many other supply companies have fish fillers but this is the best that we have tested. It is a white, granular powder to which one must add water prior to making a mache-like filling for half cast and sack-type fish mounts. Tom Sexton Fish Filler is very lightweight making it very desirable for this project. Alternatives would be Jim Hall or Sallie Dahmes Mache. These offer a stronger, more durable rock (with Sallie Dahmes Mache being the stronger but would also add slightly to the weight).

When using the half cast method for mounting fish, once the fish skin is filled and shaped, it is usually allowed to "set" or harden before carding fins, etc. During this time, (or anytime that you have leftover mache or filler), it is a quick and simple task to take the remaining filler or mache and hand form it into one or more rocks, depending on the shapes and sizes that one requires.

It is easiest to mold the shapes in the palm of the hands to obtain a nice rounded stone like those usually seen on river and lake shores. By simply adding dry tempera colors, available from Wildlife Artist Supply Company, to the mixture, a very natural and realistic color may be obtained. This "base" color can later be shaded and other colors added to create a variety of effects by painting the dried rocks with an airbrush.

Keep in mind that most rocks "as seen" are set firmly into gravel or dirt and therefore the artificial rocks should be shaped with flat bottoms rather than like watermelons and footballs.

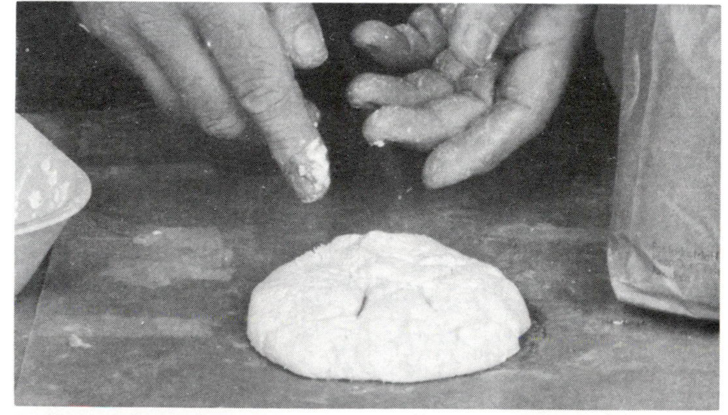

To flatten them, first shape the rock to your preference and then firmly "plop" it bottom down on a flat surface. Newspapers make a good surface protector and the paper helps to siphon off moisture from the rock which accelerates the drying time.

Cracks which form when the rock flattens out and expands may be left for effect or smoothed over with damp fingers. Or, for a different effect, two rocks can be merged together while wet creating a crack between them. Once dry, they will almost always remain bonded together. If not, a touch of hot glue or most any other adhesive will quickly reattach them.

An interesting "fossil-type" effect easily can be made by forming the rock twice the size preferred, and then once dry, saw ing it in two on a band saw. In this case, don't bother flattening the bottom as the band saw cut takes care of that.

If larger rocks are required, they can be made in much the same manner, except that they should be formed around a "core" of foam, or a sawdust sock. (The bigger the rock, the bigger the core.) Always remember to keep the outer surface of these type rocks at least one inch thick for sufficient bonding strength.

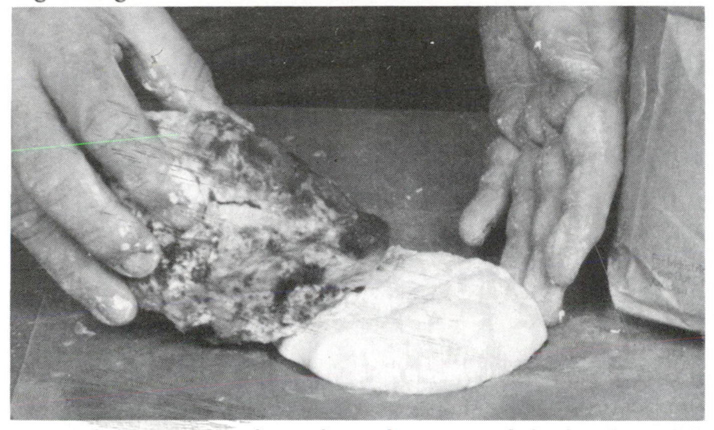

Pitting or texturing the rock can be accomplished with modeling tools, or by using a dry and very porous sponge to roughen up the surface, or a natural rock with a good texture may be used by pressing the rock into the surface. This will leave a "negative" impression but will still look very natural. The drier the fish filler mix, the harder it will be to texture the rock. So, if a coarsely-pitted surface is desired, it may be advantageous to make a batch of fish filler or mache strictly with this purpose in mind.

Another method of texturing this type of artificial rock is by simply forming the rock and then pouring sand or gravel over it. By tapping the surface lightly, one can "set" the sand or gravel into the fish filler more securely. Once it has dried, brush, or blow away the excess. Then, seal the remaining sand onto the rock with Wet Look Gloss FP-240 or Fungicidal Sealer FP-220. It is also easy to create a very porous effect by using a heavy bristle brush to remove the gravel. The latter method requires a bit more effort, but will produce a rock of somewhat volcanic appearance which is ideal for desert or mountain scenes. Some artists allow the filler/mache to thoroughly dry, then paint with a coat of Ultra Seal, *Polytranspar*™ Fin Backing Cream, or Elmer's white glue onto the surface and sprinkle the sand, etc., onto it while it is wet. When these products dry, they hold tenaciously.

Once "completely" dry, one can go to any extreme to paint the rocks. The quickest way I've found is to simply spray a heavy and complete covering of *Polytranspar*™ Airbrush Paint using the the color of my choice and then wipe it back off by hand.

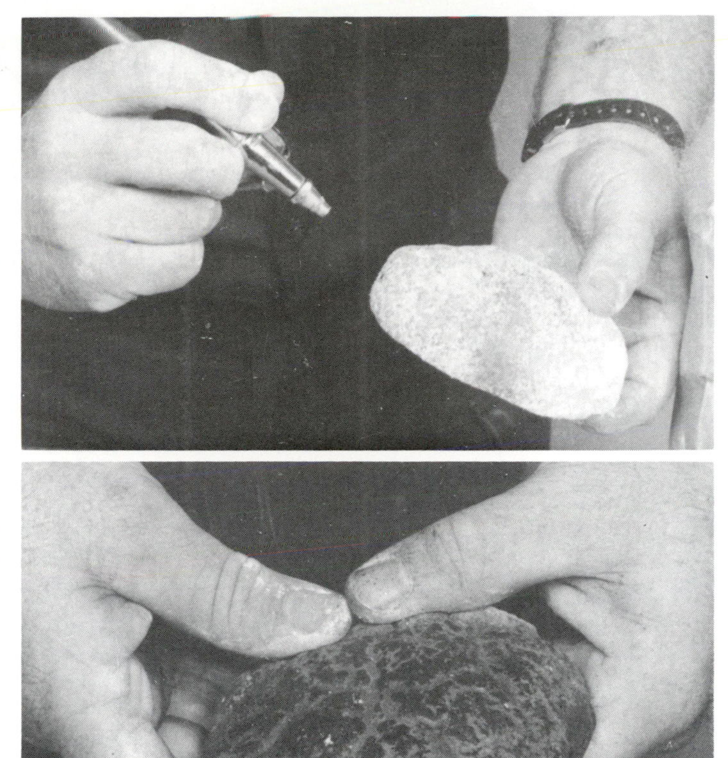

Black, when wiped off, creates a dark grey, granite-like appearance which I like. And the cracks and pores retain the high contrast black which helps define details. To give the rocks an underwater look, Medium Bass Green (WA/FP 51) can be airbrushed over the upper surfaces and lightly glossed with Wet Look Gloss to highlight its transparency. This coating gives the rock a somewhat mossy green appearance and a slightly reflective sheen as well.

By "lightly" depressing the airbrush and lowering the air pressure one can make it "spit" or "splatter" drops of paint which gives the false impression of a porous surface. Also, a toothbrush with paint on it may be "flicked" with your thumb to "speckle" a rock. Additionally, by painting different colors of paint, one can give the suggestion of different types of rock or sand within the same rock. One advantage of these mache rocks is that lacquer paints don't melt away the rock as with many types of foam rocks.

One significant factor about the fish-filler/mache rocks which you're bound to appreciate is the fact that these rocks will readily accept pins to hold toes of birds in place and can be drilled to accept the foot and leg wires of bird mounts or carvings. If the bird is sizeable and there is a need to reinforce the rock to prevent it from chipping away at the wire, simply drill the hole out to 3/8 of an inch and insert wooden doweling of the same size to which an Ultra Seal, Fin Backing Cream, or Elmer's Glue type adhesive has been applied. Then drill the wire holes into the dowels and position the mount as usual. This type of base is ideal for birds of almost every type, but especially for Chukar partridge, bobwhite, pheasants, and most ducks.

As with most types of learning, "experience" can be one of the best teachers. So don't hesitate to experiment with different ways to display your mounts in habitats built on and around a good rock foundation. Try the method of making fish-filler/mache rocks and just see if your customers don't take your rocks for "granite."

# 2. Carved Foam Rocks: Light and Easy to Make

*by Bob Williamson*

This method of creating artificial rocks is economical and requires little skill. The items shown in the photograph are typical materials that are used to create a carved foam rock.

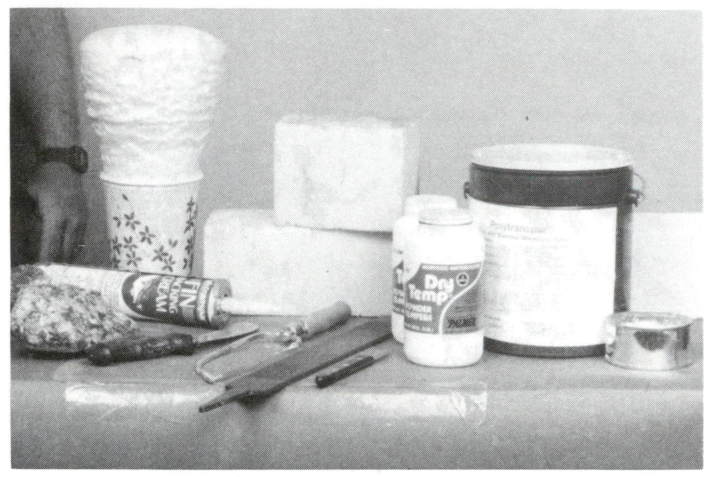

*Materials to construct foam rocks are found in most taxidermy studio*

1. Natural rocks for reference and/or texturing
2. Fillet knife or any sharp knife
3. Hacksaw
4. Rasp
5. Sanding wire
6. *Polytranspar*™ Hide Paste, Fin Cream, Ultra Seal
7. Tempera, glitter, artificial snow, sand, gravel, pumice
8. Mache and sponge
9. Blue foam, urethane foam, beaded foam
10. *Polytranspar*™ Airbrush Paint

One foam that is commonly used for carving rocks is called Type 1B Fabrication billet foam and is available from Dow Corning. Many boat dock builders and insulation companies supply this material. Usually it is available in 7" high, 14" wide, and

*Type 1B Fabrication billet foam*

108" long blocks and is a light blue color. Many taxidermy supply companies sell this type of foam for carving fish bodies. One problem with using this material is that lacquer paint, thinners, polyester resins, and other oil-based products will soften and decompose it if they come in contact with it. Hot-melt glue doesn't work with this type of foam either as the hot glue will melt it readily. The best glue that we have ever found for gluing pieces of this foam together is *Polytranspar*™ Hide Paste. This glue has tremendous strength for gluing foam pieces. Water based products such as Ultra Seal and Fin Cream can also be used to "seal" the foam and once sealed with the correct sealer (or layered with mache), paints such as lacquers, etc., can be used without fear of melting the foam.

Another common foam used for carving rocks is a very light density (1 to 1½ lb.) urethane foam. This may be purchased in logs also and is usually a light tan color making it an ideal base color. More dense urethane such as scraps from mannikins (which range from 3 to 6 lb. densities) that have been altered will work but are tougher to carve, especially the hard crust "skin." Urethanes can withstand lacquers, polyester resins, and the like and require no sealer coat. They also may be hot-melt glued with no ill effects. Additionally, they lend themselves very well to water based paints, adhesives and maches. The main problem with this type of foam is that it is more easily dented than billet foam. This problem is generally offset by painting, adding sealers, or adding a thin layer of mache to toughen the outer surface.

Both types of foam are easily carved. Since rocks don't have any particular size or shape, basically all that is required is to rough carve with a fillet knife and pocket hacksaw the approximate shape desired.

Then, use a rasp and sandpaper to round contours and final shape it.

To add realism, press a natural stone with good detail into the foam to imprint the texture.

Another type of foam is "beaded" styrofoam (closed cell), this is available in billet logs also. This foam is generally white in color. It is the least desirable for this method of recreating rocks. The biggest problem with it is that it's hard to carve, sand, and shape. It will hold wires fairly well but this is really not much of a problem with the other types of foam as anchor blocks and plywood bases can easily be glued into them for this purpose.

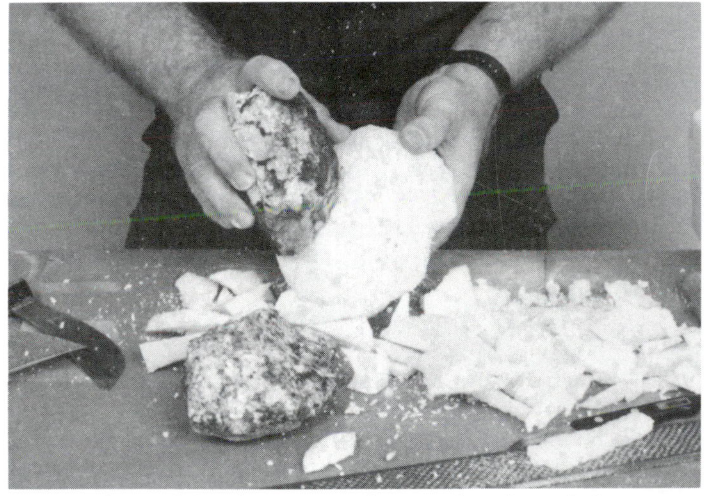

Once the rock has been carved, rasped and sanded it is time to seal the rock. Several products may be used to seal the rock are: clay, mache, *Polytranspar*™ Hide Paste, Ultra Seal, Fin Backing Cream, and Elmer's Glue, to name a few. Basically, covering the rock with clay or mache will require a sealer coat after it dries and hardens. Mache can be tinted with dry tempera colors prior to mixing to obtain a good base color. Clay will generally already have an adequate color base coat.

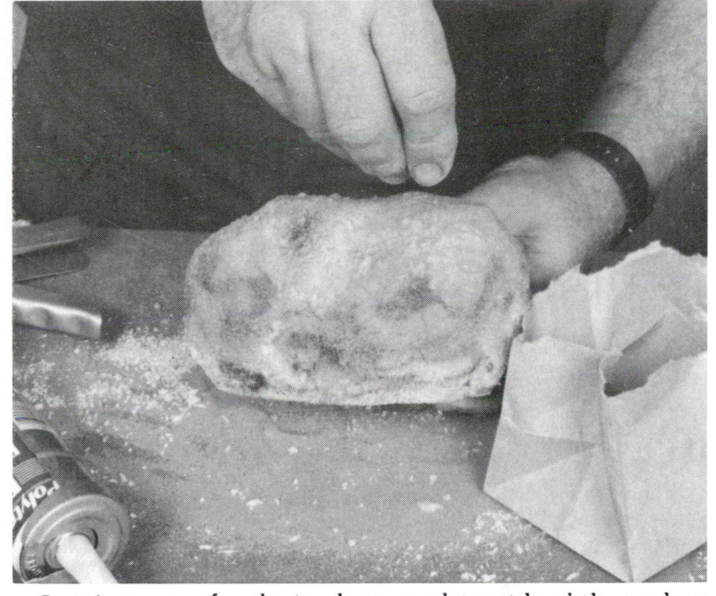

Certain types of rocks (such as sandstone) lend themselves better to coating the foam with a good adhesive such as *Polytranspar* Hide Paste, Fin Backing Cream, Ultra Seal, or Elmer's Glue, and then before it dries sprinkling sand, pumice, etc., over the glued surface. Reapply as necessary. For small areas, use spray adhesive and sprinkle sand onto the sprayed area.

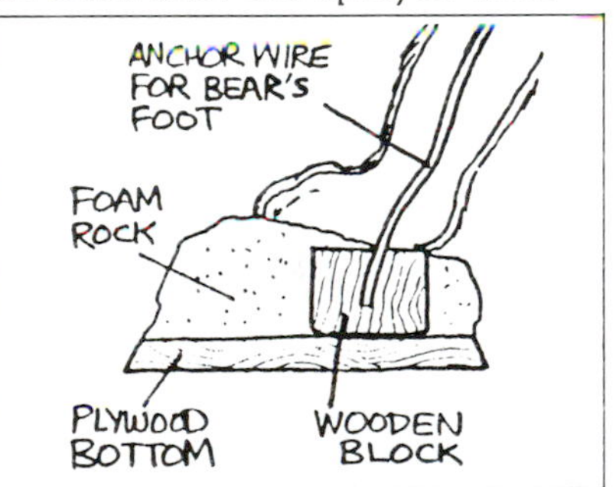


If wires are to be inserted, they may be glued with *Polytranspar* Hide Paste. If a heavy object is to be glued to the rock, wooden blocks or dowels may be glued into place to accept the wires. Also, it may be wise to glue a plywood base on the rock too.

Refer to page 114 for exact step-by-step techniques for carving and texturing artificial rocks by Kelly Seibels. Kelly took the coveted Best in World honors at the 1986 World Taxidermy Championships with a harlequin duck mounted on an artificial waterfall using carved styrofoam rocks.

# 2. Free-Formed Urethane Foam & Cotton Rocks

***by Ken Edwards***

A method for creating individual "free-form" shaped rocks was developed by Randy Nelson of St. James, Minnesota. The method uses two-part urethane foam, ordinary cotton, and hot water. The rocks created with this method are surprisingly lifelike, lightweight, easy-to-make, and no two rocks are ever identical. The process *does* take some getting used to, and your first efforts will probably not be your best, but after a few trials, you will get the hang of it.

The actual working time of the foam/cotton method of creating artificial rocks is very quick, so you must have all the necessary supplies laid out in front of you before beginning the project. To get started, you should have the following items ready at your work table:

1. Two-component urethane foam (14 lb. density)
2. Measuring spoon.
3. Coloring agents (water soluble dies, paint, tempra colors, etc.)
4. Paper cups and tongue depressors for mixing the foam.
5. Ordinary cotton (nonsterilized) available at any pharmacy.
6. An empty bucket for mixing the foam with the cotton.
7. A large bucket of hot water (hot tap water is sufficent).
8. A working area covered with plastic wrap.

9. Silicone spray (mold release spray).

10. A selection of latex texturing molds for shaping the rocks (see box on next page). The texturing molds used to make these rocks were obtained from tree bark.

11. Good ventilation and respirator. Read the safety precautions on the urethane foam label.

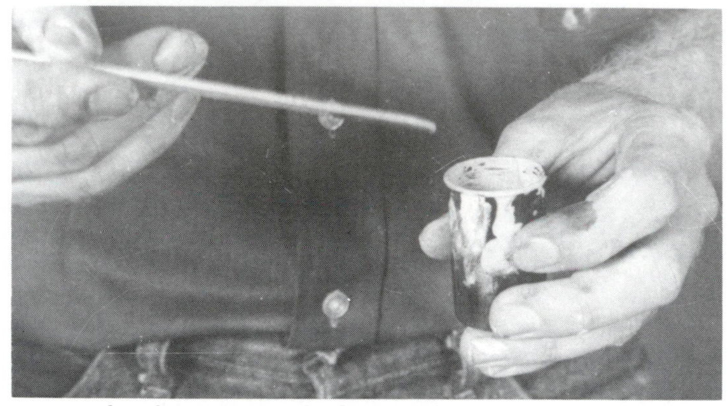

Begin by determining the basic color of the rocks that you wish to create and mix small portions of water soluble dyes or tempera paints until this color is obtained. Randy uses water-base coloring pigments from Pittsburgh Paints (these shading pastes are the ones used to shade paints), but most water soluble coloring agents (including dry tempera pigments) will work. The water-soluble coloring will react with the foam in the hot water to give varied surface colors on the final rock. Only a minute amount of coloring agents will need to be used in the foam; therefore, the initial color of the dye will appear much darker than the finished rock.

The natural color of the foam is tan, so to create a cool-gray rock, it must be colored by using black coloring with some blue added. For warmer, or sandy-colored rocks, add red, yellow or brown colorings.

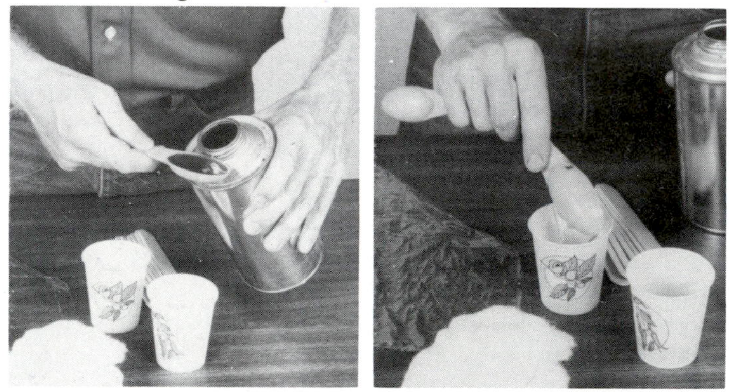

Pour measured amounts of urethane foam part A and part B into two separate cups. Add the coloring agent to part A and stir thoroughly.

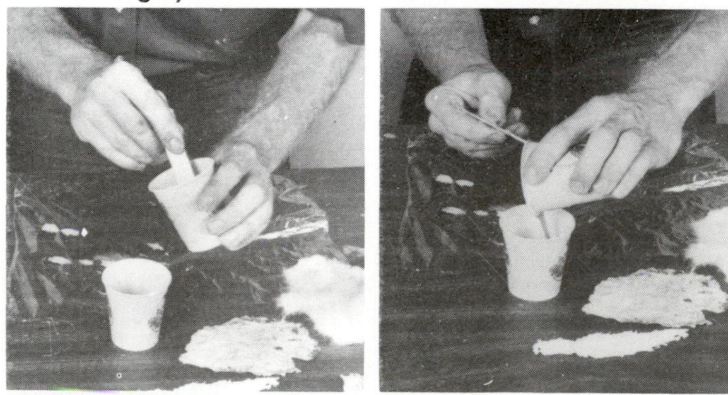

Mix part A and part B together and stir thoroughly with a tongue depressor or a drill with a paint stirrer attachment on it.

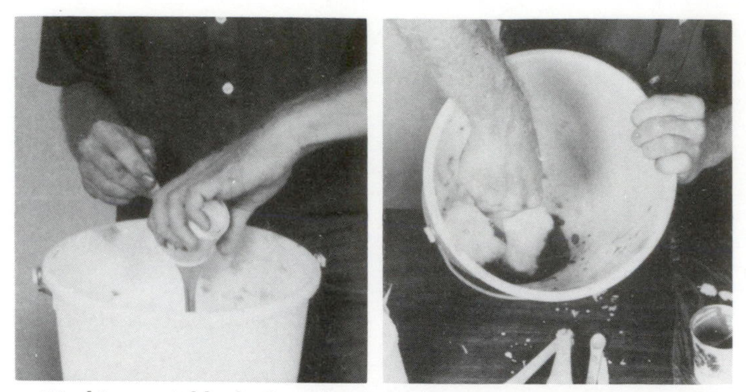

Working quickly, before the foam begins to expand, pour the foam mixture into an empty bucket and add a few pieces of cotton.

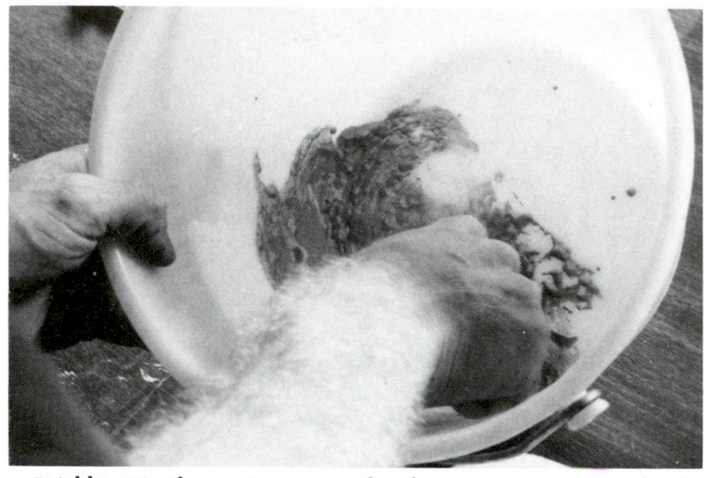

Briskly stir the cotton into the foam mixture, completely saturating the cotton until no white can be seen.

Let the foam begin to rise slightly (about 3 or 4 seconds), and without stopping, pick up the foam/cotton mixture using two tongue depressors and drop it into a second bucket of hot water. The water should be as hot as possible (very hot tap water directly out of a faucet). The hot water will immediately cause the foam to expand further and the cotton will keep the foam together and create interesting shapes. Using the tongue depressors, turn the mixture over in the water so all sides will rise equally. When the foam has reached it's full expansion and stopped bubbling in the hot water, test the mixture by touching the surface of the foam. If the surface doesn't collapse under the pressure of touching it, then it is ready to remove. If the surface is extremely soft and collapses under pressure, wait a few seconds and test again. (Remember, hot water and foam can burn you also, so take care.)

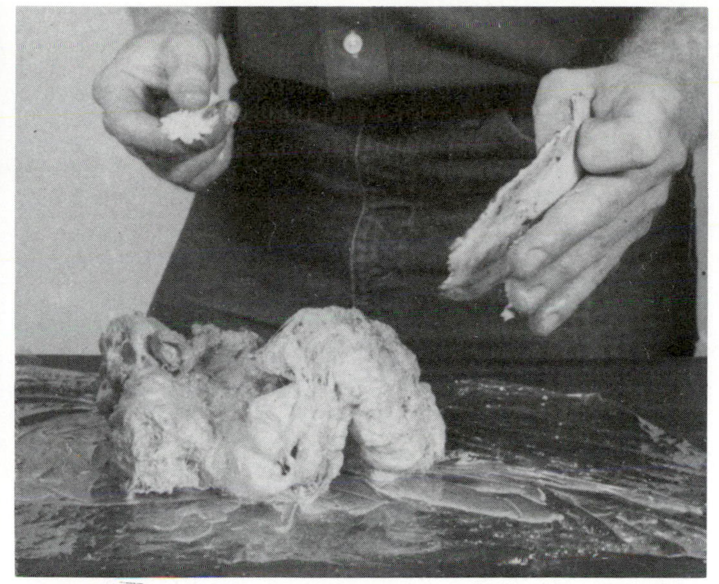

Once firm, remove the mixture from the bucket and place it on a plastic-covered work surface. At this point the foam will still be soft and pliable.

Before using the latex texture molds, coat with silicone spray. Holding a texture mold in each hand, begin custom shaping the rock surface. As the foam begins to harden, detail can be pressed into the surface and the general shape can be manipulated. Go with the natural shape as much as possible—too much squeezing will collapse the cotton and ruin the interesting natural shapes, crevices, and lumps that have been produced.

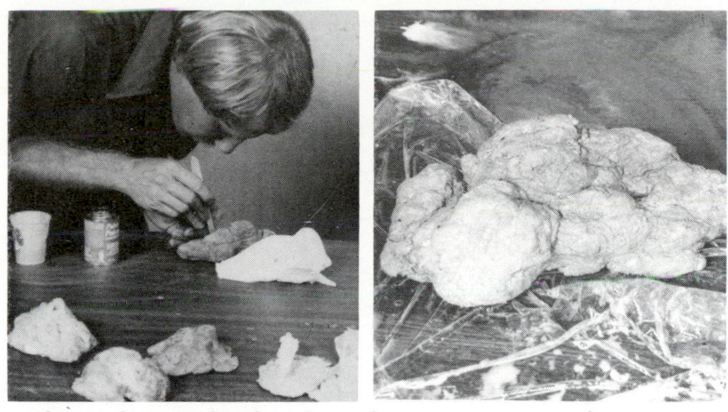

The rock can also be shaped using a tongue depressor to accentuate crevices or depressions in the surface. You will notice at this point that the coloring agent in combination with the cotton and foam creates interesting variations in the surface color; much like a natural rock.

Before the foam has completely cured, it may be placed on a base or added to other rocks to create groups of rocks, ledges, or cliffs. If the rock is still slightly soft, it will adhere to other rocks and bases as it dries. By placing smaller rocks together as they dry, one can "build" custom shapes and designs.

By varying the coloring material, amount of cotton, size of cotton pieces, and amount of texturing manipulation, many different types of rocks can be easily imitated with this method. One word of advice; don't become discouraged if your first attempt is disappointing. The first few rocks that are attempted should be approached with an attitude of experimentation and learning. After you have made two or three rocks, you will more clearly understand the concept of this procedure as well as develop the correct "timing." Then you will begin to create better and better artificial rocks with each try.

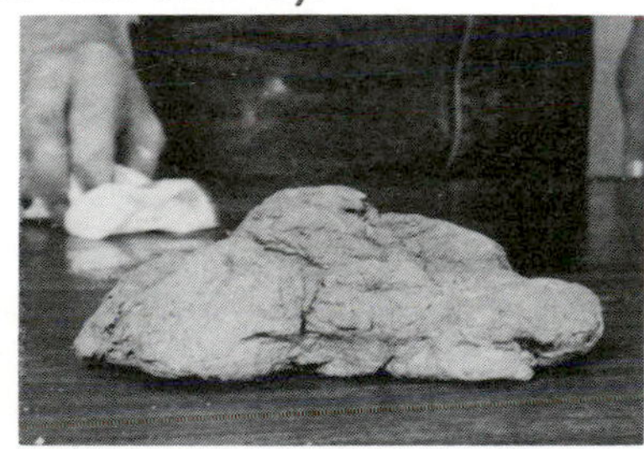

Once satisfied with the shape and texture, allow the rock to harden. It can then be highlighted and detailed (if necessary) with airbrush paint, tempera paints, acrylics, or oil colors.

Randy Nelson likes to use *Polytranspar*™ Airbrush Paints straight out of the container and apply the paint with

## Randy Nelson's Versatile Texture Molds

When it comes to reproducing natural textures, Randy Nelson has some very strong opinions on the subject. "Don't try to sculpt or create something yourself," Randy says, "when you can find something already made by God."

Randy has a small variety of latex texture molds that he uses for almost all his natural ice, water, and rock textures. These molds were made by simply pouring catalyzed RTV rubber over natural textures and letting it harden.

His favorite texture molds are pictured at right. They are made from tree bark, rocks, and wood grain from post ends (and burned wood). He made some of these molds in his studio and some in the field. (The tree bark molds were made by pouring RTV rubber on live trees in a forest.

Randy uses the same molds over and over on many different projects to reproduce many different effects. He feels that the basic shapes and textures in nature are recreated in many other different forms. For example, the wood grain of a fence post makes good ripples for artificial water surfaces. "Why should I spend hours sculpting individual water ripples," Randy says, "when the same shapes are already available in other forms?" He uses the same textures to shape a variety of elements for totally different results. In creating a winter scene, Randy used the tree-bark texture mold to detail urethane-cotton rocks, artificial water ripples, and sheet ice.

Randy finds that his small collection of

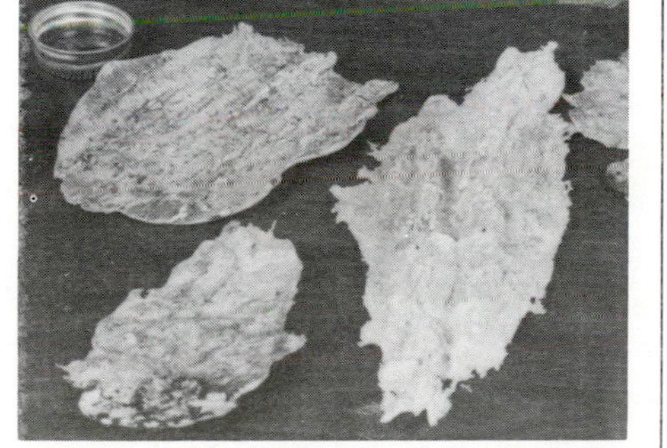

texture molds is all he needs for most instances. However, he is always on the lookout for new and interesting textures appearing in nature. If he sees an interesting fence post end grain, rock, or tree bark, all he has to do is coat the surface with RTV rubber, allow to harden, and peel off. Presto!—Another mold is added to his collection. (Note: Mold Builder™ will also work as a molding material for textures.)

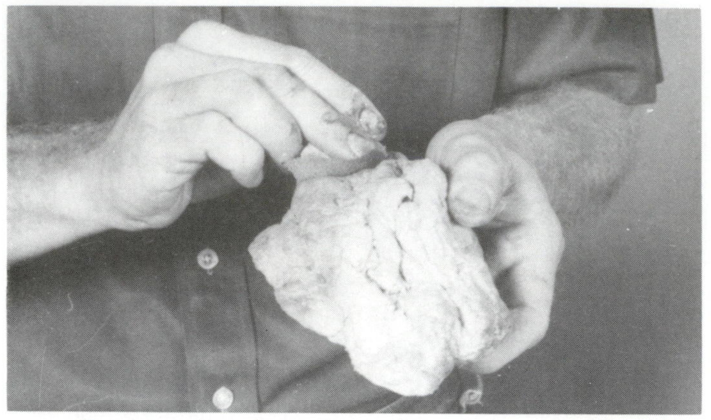

a small sponge directly to the surface of the rock. The rocks can also be "spatter-painted" with tempera paints and a toothbrush as described in the previous section. If you wish for the rock to appear glossy, seal it with an acrylic spray or a *Polytranspar*™ gloss coat. For a flat finish, no sealer is necessary.

Note: If the rocks are dropped in the water too quickly (before the foam has begun to expand enough), the cotton will soak up too much water which will arrest the rising action of the foam. This technique can be used to make a very realistic looking artificial mud so don't discard it—simply add it to the rest of the habitat materials on hand.

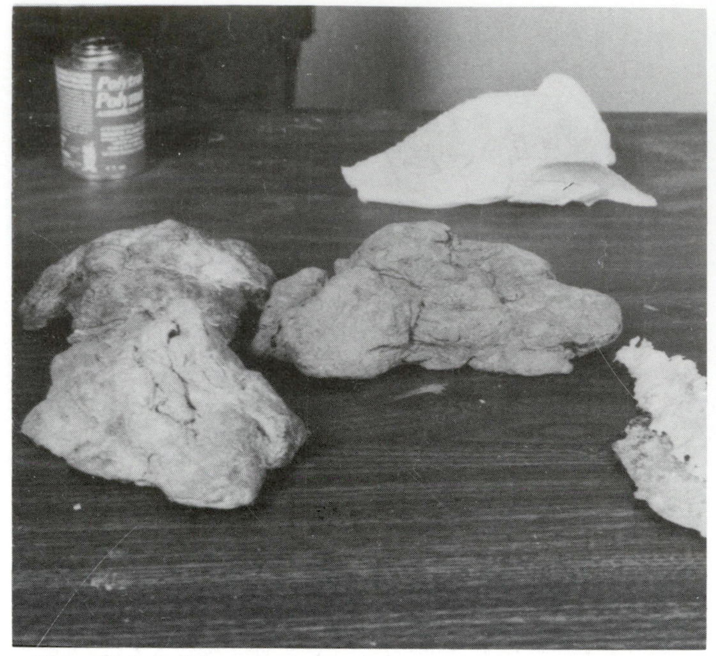

These lightweight, convincing foam and cotton rocks can now be used in any number of habitat projects where rocks are needed. They may be hot melt glued or regular adhesives will work fine to adhere them to bases.

# 4. A Wire Mesh And Fiberglass Rock Construction

***by Jim Hall***

Once in a while, a customer will request a base that incorporates special requirements. The lifesize mountain lion mount, sitting on a rock ledge base used for documenting this section was just such a case. To make a fiberglass rock from scratch (without a mold) is not difficult. Generally fiberglass reinforced artificial rocks are used for large heavy mounts that require a strong support. A western hunting outfitter wished to have a lifesize mount that could be readily transported to and from various sportsmen shows where the outfitter advertised his hunting services. The mount and base needed to be as light and compact as possible to enable one man to easily carry it and set it up. Also, the base had to be very durable and easy to grasp to ensure safe handling. Naturally, the design of the base had to compliment the over-all piece. Another consideration was that the piece could only take up a minimal amount of room in the booth and be easily displayed by setting it on a table or hanging it on a pegboard if so desired.

A natural rock base was decided upon for this mount as this type of base can be most durable especially if it is sealed with fiberglass resin as this one was. The finished base and prepared mannikin are shown in the photo to the left. As can be seen, the piece takes a minimal amount of room to display, yet looks quite natural.

As with most habitats, "design" can make (or break) the rendering. Forethought can and will save both time and materials. It was decided that three of the feet of the mountain lion would not be directly fastened to the base in order to present a more natural pose. Since this mannikin was manufactured with metal rods protruding from the feet, three of the original metal rods or bolts were cut off flush with the feet.

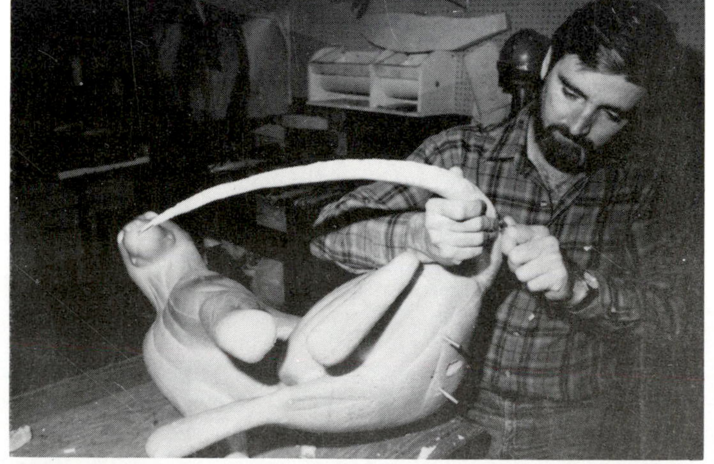

Instead of using the preinstalled bolts, two pieces of 1/4" "all-thread" bolt material were strategically placed within the hip section of the right rear leg. These metal rods are approximately 10 inches long and were securely fastened into the form with Smooth-Out™ epoxy. The third metal rod used was one of the original ones which was located in the paw of the right front leg (not shown). Always use at least three solid points of contact on a lifesize mannikin to provide a stable attachment.

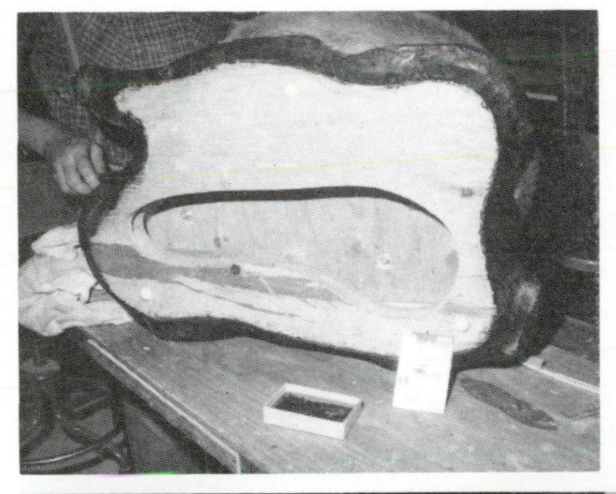

The photo to the left shows the underside of the base and the location of the three mounting bolts. Always remember when designing your base, to keep in mind that you need access to these bolts and nuts. Note the open access to the bolts in this design.

Once the preliminary design work is done, we can begin construction of the base. Begin by positioning the mannikin on a sheet of plywood and sketching the basic shape that you have in mind.

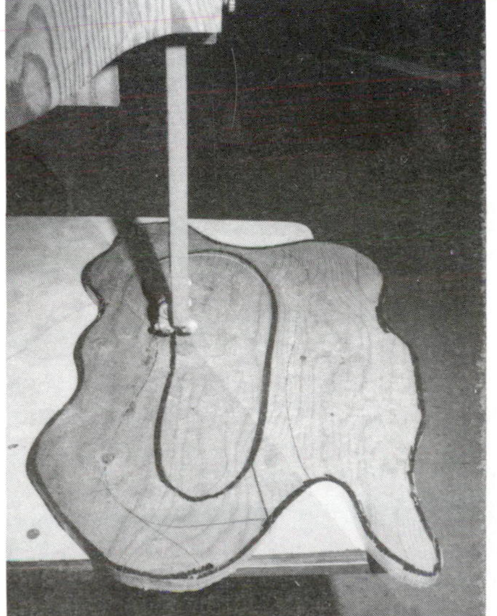

This type of base utilized three horizontal pieces of 3/4" plywood; the top piece to which the mannikin is bolted, the intermediate piece which will create the "ledge" effect, and the bottom which formed the base.

These various shapes were each cut out with a bandsaw (sabre saw can be used).

Now, cut various spacers from scrap pieces of 2" × 4" pine to establish the heights of the three pieces of plywood. Place the subject on the base and prop everything up, as shown in the following photograph, building block fashion, until you are satisfied with the design.

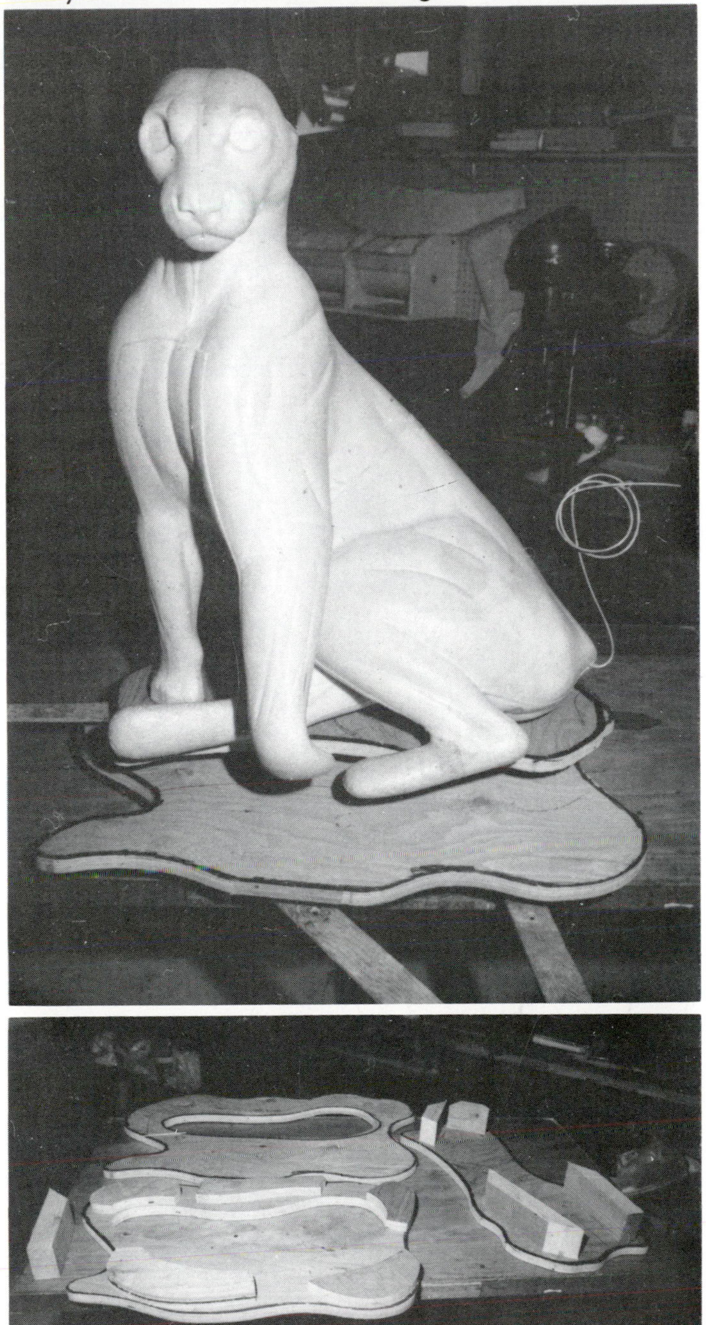

Above demonstrates the three individual tier pieces with blocks in position before final gluing and nailing.

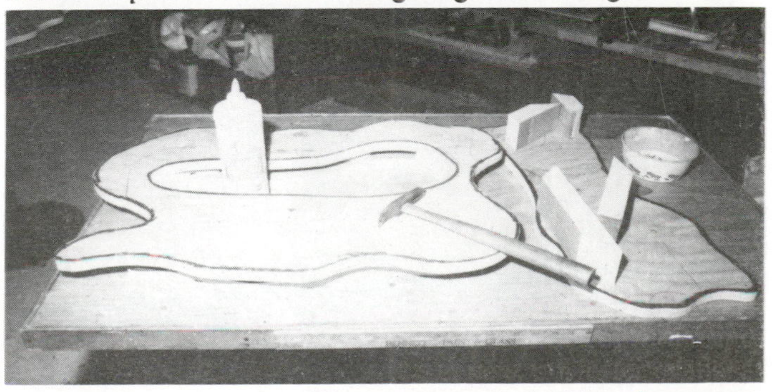

Assembly was next and was accomplished by gluing with Elmer's wood glue and then nailing the three sections together.

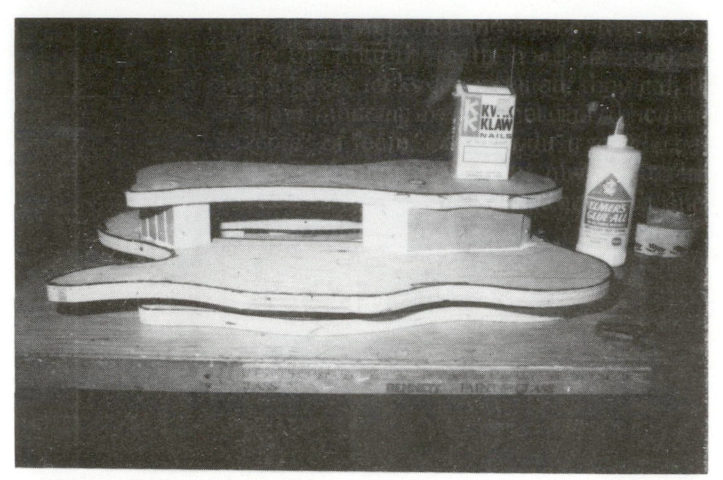

Above shows the final framework completed.

The next step was to locate the mounting holes for the mannikin.

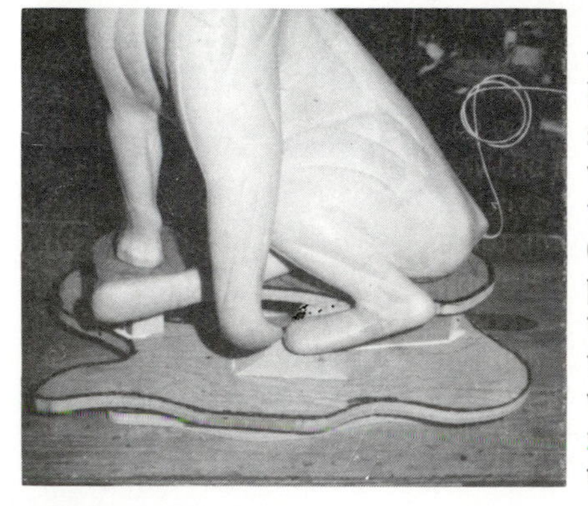

In the final position of the lion form on the base an addition of small blocks of wood were placed beneath the left hind foot (demonstrated in the photo to the left). Once the best pose and location was chosen, holes were drilled to accommodate the metal threaded rods.

The mannikin was then removed and the actual preparation of the base was the next step. I began by stapling narrow strips of hardware cloth or screen to the wooden framework. This provided the necessary support for the mache layer and was the base for the sides of the rock ledge. The base was completely covered except for the bottom, with the wire mesh.

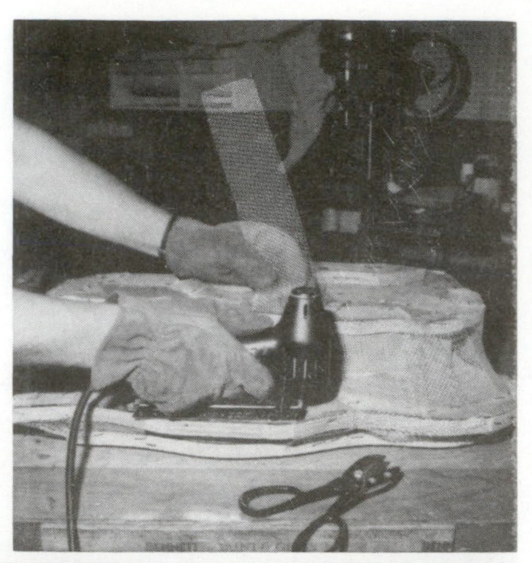

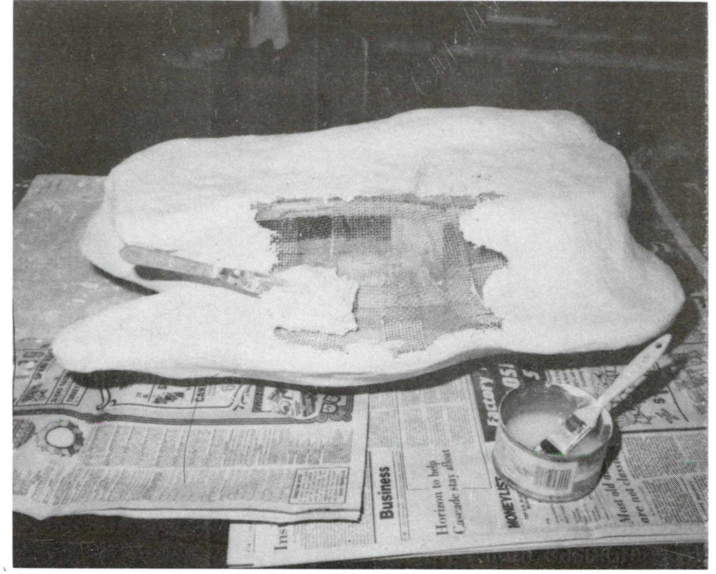


A layer of Jim Hall Mache was mixed to a creamy consistency and then spatulated over the mesh. (Press the mache quite hard so that it will form a good bond with the wire.)

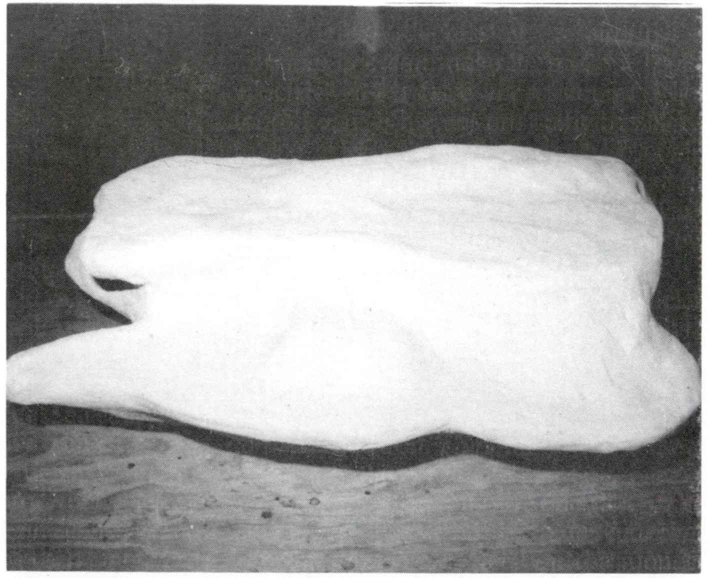

I completely covered the base with mache. It is important to concentrate on the folds and edges of the base and sculpture detail that will eliminate any appearance of the plywood. Then the mache-covered base must be set aside for drying, which should take 5 or 6 days. At this point the mache could be sealed, painted, and finished to look like a rock ledge. However, since this exhibit would be handled and moved to various sportsmen shows we chose the more durable fiberglass to coat it.

The texturing of the rock for this base was accomplished by first applying a sealer coat of catalyzed *Polytranspar*™ Fiberglass Lay-Up Resin LR101 directly to the dried mache base. This resin was tinted with the Wildlife Artist Supply Company Dye Kit and fiberglass chop was added for strength. A light green tint was used to provide a base color. (Remember when using dyes that very little is needed.) I allowed this first sealer coat to cure for approximately 3 or 4 hours.

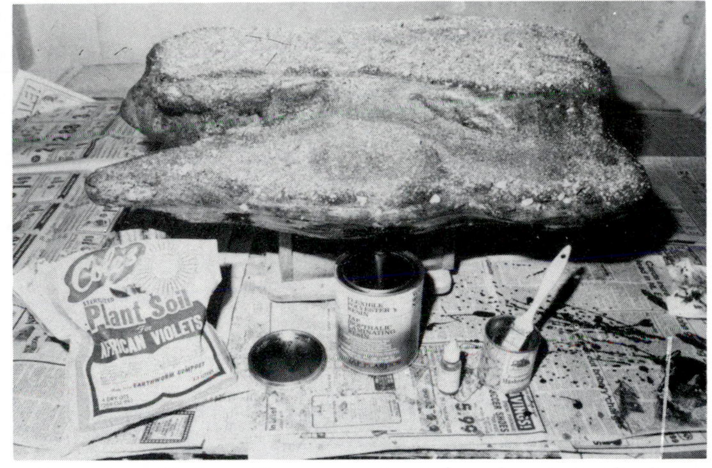

Next, I applied another coat of catalyzed resin to the base. This time I tinted the resin lightly with black to offer a more mottled look to the rock. Now with the resin still wet, the final texture was given to the base by sprinkling a mixture of "dry" potting soil and zonolite insulation material over the wet resin and then allowing it to dry. The texture was enhanced by adding small gravel, twigs, leaves, debris, etc., (literally anything that comes to mind can be added at this point.) Important: If the potting soil that you are using feels damp at all, spread it out on a piece of newspaper and dry it completely for one or two days before adding it to the base as it can prevent the resin from curing if wet or damp. When the resin dried and cured, all loose debris that had not adhered was blown off with an air gun.

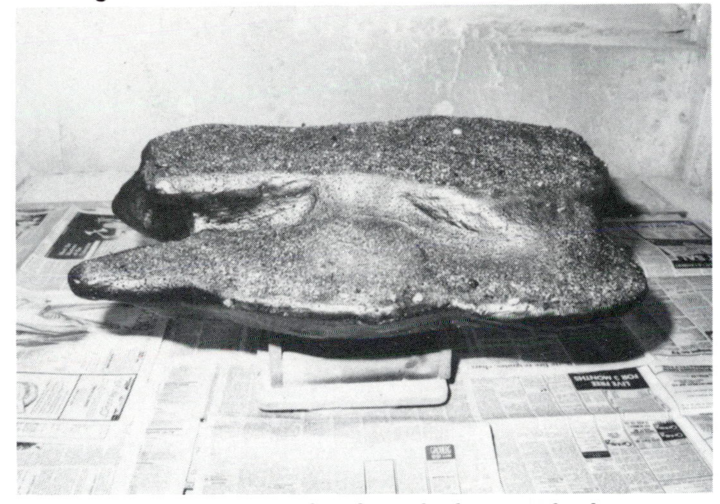

Above shows the completed "rock" base. A final spray application of *Polytranspar* Fungicidal Sealer WA/FP220 was given to the texture to further protect it against handling and to give it a slightly higher sheen. The mountain lion was then mounted and bolted securely to the base. It was light, portable, durable, and displayed the lion quite naturally. A special purpose base that was quick and easy!

# Molding Rocks: 5. Alginate 6. Latex 7. RTV Silicone

***by Bob Williamson***

There will be many times when the artist is constructing various habitat bases that they will wish to incorporate one or more small stones. If the stones that one decides to use are about the size of gravel, then we recommend actually using gravel, as it will be the most convincing and will not add an excessive amount of weight.

If stones about the size of a baseball are desired, then it would be wise to consider molding and reproducing the stones. There are three advantages to using reproduced stones; namely, the great reduction in weight of the habitat; the fact that these artificial "rocks" can be cut in half, drilled for specimen mounting screws, or otherwise easily modified, and lastly; they are really fun to make.

The most accurate method of recreating rocks is making a mold of an actual rock and then casting an artificial rock from the mold. There are a vareity of materials that can be used for molding rocks. The three most popular methods are:

1. Alginate method
2. Latex method
3. RTV Silicone method

Primarily the "procedure" of making the mold is the same with all methods. The big difference lies in the "longevity" of the mold and the speed in which the mold can be made. Alginate molds, for example, can be made very quickly but the mold deteriorates rapidly allowing the artist to make two or three parts at the most. Latex and RTV Silicone molds take much more time and expense to make but allow for multiple rocks to be made from the mold and will last indefinitely.

Which is best? It really depends upon "demand." If the artist intends to make innumerable rocks of essentially the same size and shape, then a Latex or Silicone RTV mold would be the best bet. On the other hand, if it's a "one shot" deal and the artist will not need several rocks of a certain size on a regular basis, then an alginate mold might be a better choice.

## Alginate Mold

In this first example, an alginate mold is being made of a rock. A bed consisting of high fiber mixed with water has been placed midway on the rock.

Alginate sets within 2 to 3 minutes after adding water and stirring: (faster if hot or warm water is used) so the artist must work fast. Mix enough Alginate to cover the entire rock with one pour.

In order to apply reinforcement coats, one must remember to always pour them while the first coat is still tacky or it will not stick. It's best to have several cups of Alginate, water, and stirring tools handy.

Immediately after applying the first coat, add water and stir another cup and apply it before the previous coat completely cures.

After building 2 to 3 coats of Alginate, a "mother" mold made of plaster or bondo may be used to support the flimsy Alginate mold; however, since the Alginate mold cannot be used repeatedly due to the deterioration of the mold, a bed of high fiber

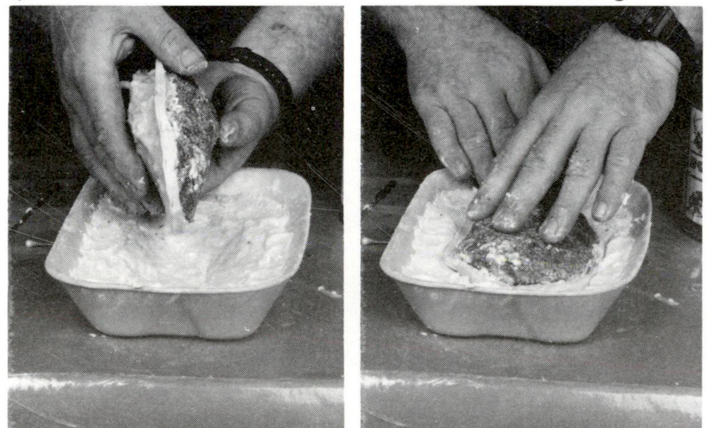

or sand can be made to support the mold instead of making a mother mold while casting the artificial rock if so desired. (Always posiiton the mold with the rock still in it, into the bed of high fiber or sand prior to removing the rock from the Alginate mold. Once the depression is made, remove the rock from the mold and place the mold into the depression in the bed.)

Alginate molds must be used to make the artificial rocks as soon as the Alginate molds are made because they deteriorate after just a few hours. Now the artificial rock may be produced.

Molding plaster is a good medium to use to make the rock when using an alginate mold. Mix a splash coat of plaster mixed with tempera pigments to the desired color and pour it into the mold. Swirl it around until an even coating completely covers the mold and then pour out the excess and let the splash coat harden. Once it has hardened, a reinforcement coat consisting of a thicker coat of plaster or fish filler, auto body filler, etc., can be used to reinforce the splash coat. It is not necessary to fill the mold entirely with reinforcement coats. The idea is to make the artificial rock as light as possible. Once the reinforcement coat has hardened, the Alginate mold can be removed from the artificial rock. If more than one artificial rock is desired, make it immediately as the Alginate mold will deteriorate and crack quite rapidly.

An alternative to using plaster as the medium would be to mix a small quantity of A and B urethane foam and pour this into the mold. The excess will expand out of the mold. Once it has cured the mold may be peeled off and the excess foam may be trimmed. Fiberglass resin is not a good medium to use in conjunction with Alginate molds because it does not react properly when exposed to the water contained within the Alginate mold.

To use fiberglass resin to produce rocks, use the following technique:

# Latex & RTV Silicone Molds

***by Jim Hall***

The very first step is to select the desired size and shape of a small rock. Try to pick stones that have many irregular surfaces to create character in the reproduction.

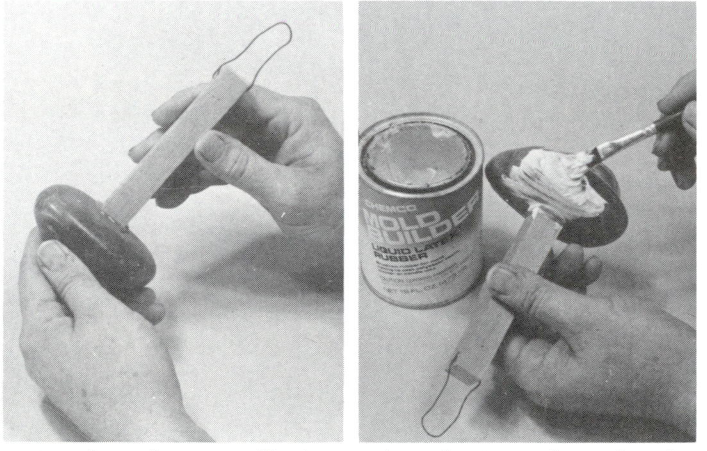

Next, hot glue a small 1/2 × 1/2 inch piece of wood to the bottom of the rock to serve as a handle as shown in the above left photo.

Apply the first layer of mold builder latex to the stone with a small brush (right photo above).

NOTE: The more expensive RTV Silicone may be substituted for Mold Builder if so desired. Simply catalyze it according to the directions on the can. One thick coat will usually be sufficient. If two or more coats are desired, add before the previous coat dries as in the Alginate method. Producing a fiberglass rock from an RTV Silicone mold is the same as Mold Builder. Use the same directions as Mold Builder with the exception that RTV Silicone can be used after allowing it to cure for 24 hours.

Brush the first layer of Mold Builder on smoothly and evenly, as it is the layer that will show the detail in the finished reproduction. Allow this layer to dry according to the instructions as outlined on the manufacturer's can. Additional coats of Mold Builder can be applied after about 4 to 6 hours drying time. The total number of coats will depend on how much "stretchability" will be needed to remove the mold from the stone. The molds used in this system contained 8 to 10 layers of Mold

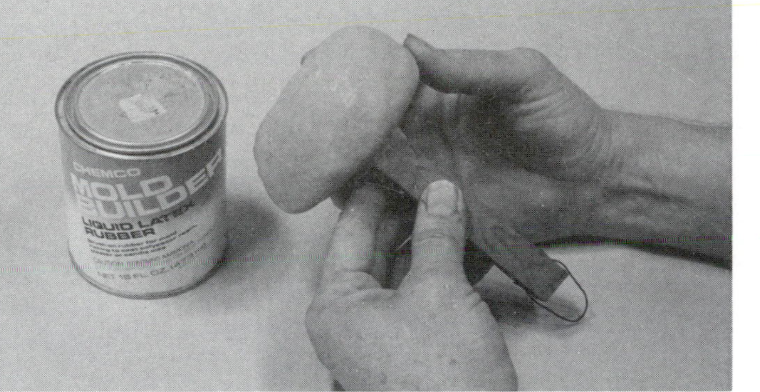


Builder. After the layers have been brushed on, permit the Mold Builder to cure for 3 or 4 days before removal.

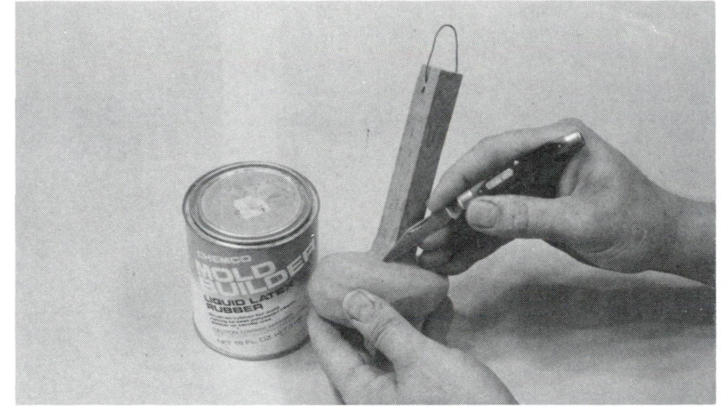


When the rubber is completely cured, cut around the wood handle with a sharp pocket knife or scalpel.

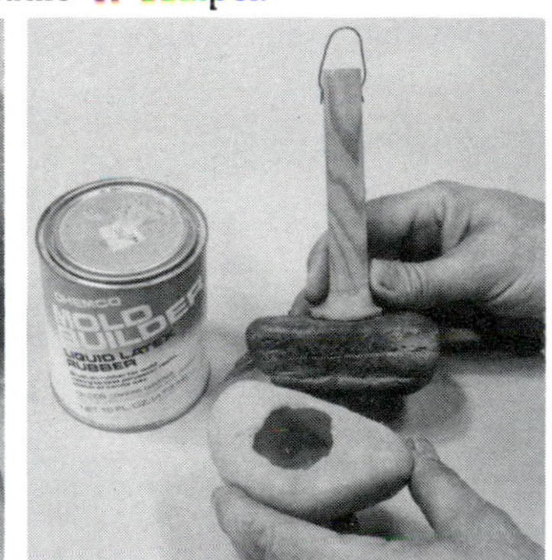


Then, the mold can be gently "peeled" from the actual stone leaving a complete stone mold containing all of the actual stone characteristic on the inside.

This type of rubber mold can be used without a second supporting or mother mold very easily. The above photo on the left shows a mold made by this process and several fiberglass "stones" cast from it. Note how varying the position of the fake stones seemed to change their shape. Because of this, only 10 or 12 molds (photo above on right) need to be made to create just about any effect desired.

# Casting The Artificial Stone In Fiberglass

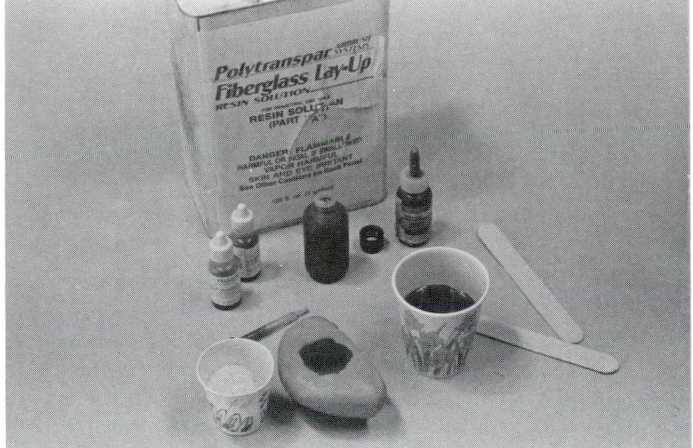


The basic materials required to cast a fiberglass stone are shown above, and include; *Polytranspar*™ Fiberglass Lay-Up Resin, catalyst, mixing containers, and stiring tools, fiberglass Wildlife Artist Supply Company polyester dyes, various fillers such as sand, vermiculite, etc., an eye dropper or syringe for adding catalyst, and a rubber mold.

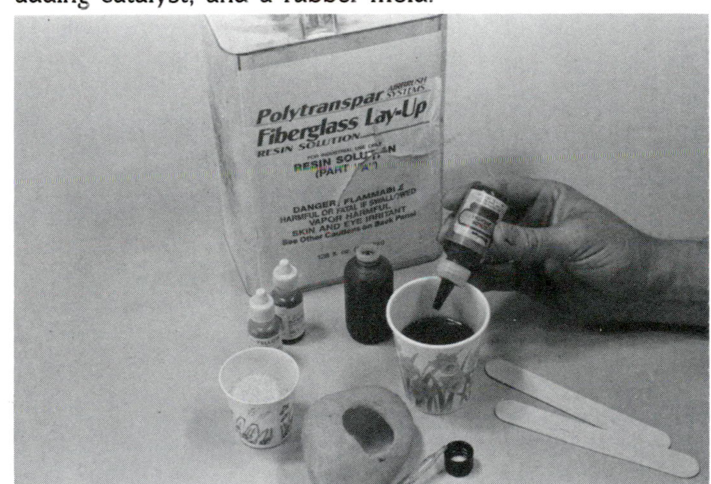


The first step in making a stone is to color the resin. Various colors or combination of colors can be used, depending on the desired final color. We recommend having black, brown, red, green, and yellow on hand.

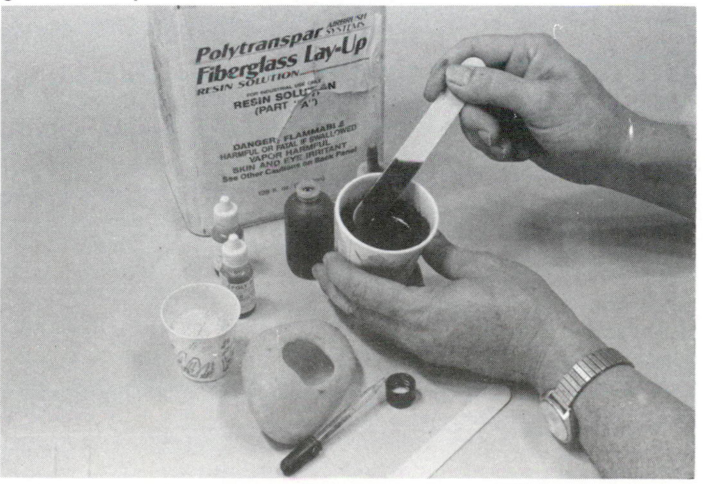


The next step is to catalyze the resin. About 10 drops of catalyst for each ounce of resin will be sufficient. Thoroughly

mix the catalyst into the resin. Be sure to scrape the sides and bottom of the container while you are mixing.

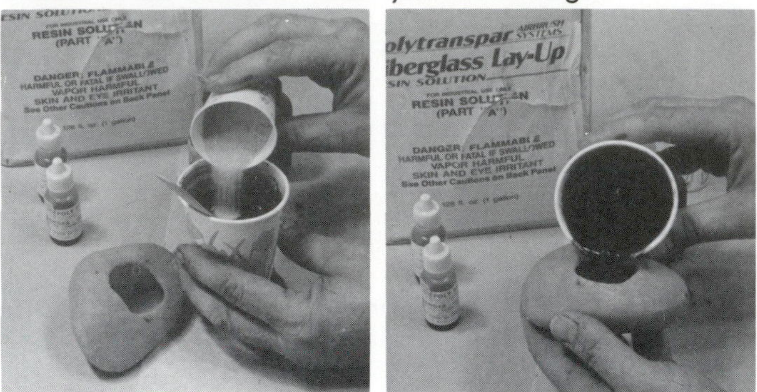


Once catalyzed and thoroughly stirred, additional fillers can be added, in this case about two ounces of sand was added to the resin. Then the combined mix is then poured into the rubber mold.

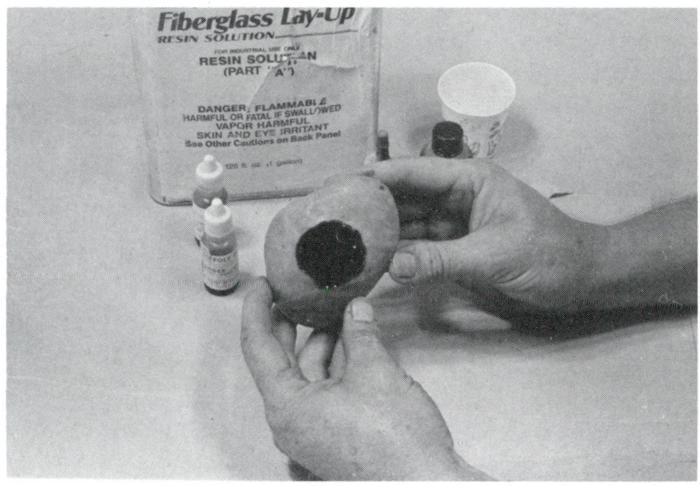


Since this stone was to be hollow, only about three ounces of resin and filler were poured into the mold. Then, the mold was rotated to make sure that all surfaces were covered.

Rotate the mold until the resin begins to set, and it can then be set aside to cure.

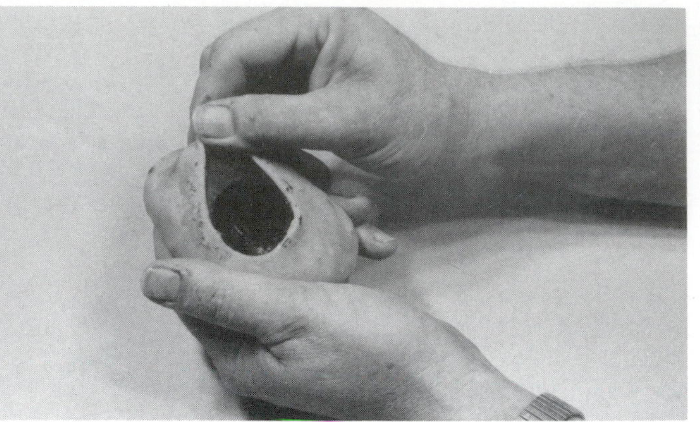

When the resin is completely cured and cool to the touch, the reproduced stone can be removed from the mold by gently stripping it over one end and peeling it off.

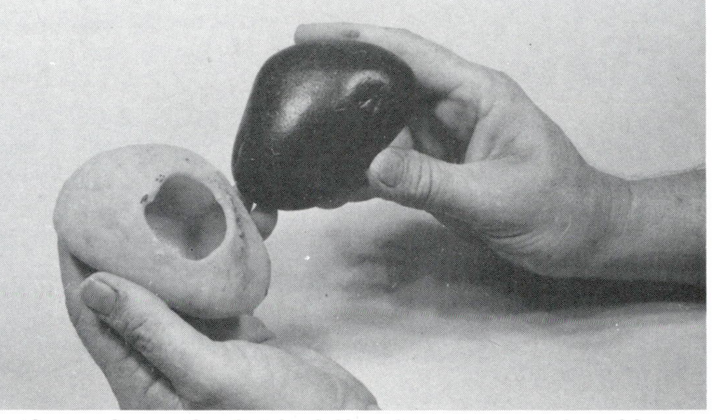

Above shows the finished fiberglass stone and mold.

NOTE: Refer to Page 53 for World Champion Alan Gaston's technique for casting artificial stones using latex molds. His method was convincing enough to win Best in World as well as Best of Show at the 1986 World Taxidermy Championships.

# 8. The Quick & Easy Direct Cast Fiberglass Rock

*by Jim Hall*

The design concept for this type of habitat base originated with Master Taxidermist Bruce Babcock of Kalispell, Montana. We ran across Bruce and this unique idea at the 1985 Taxidermy Review Competition held in Twin Bridges, Montana, where Bruce won 1st place in the Big Game category and Best of Show for his female mountain lion and kitten display.

The slab-rock reproduction used beneath the cats was created using the same technique described here. The method is so simple that it lends itself very well to all types of commercial work, and yet is convincing enough for the most demanding competition.

We utlilized Bruce's technique and created a "rock ledge" for a lifesize mountain lion.

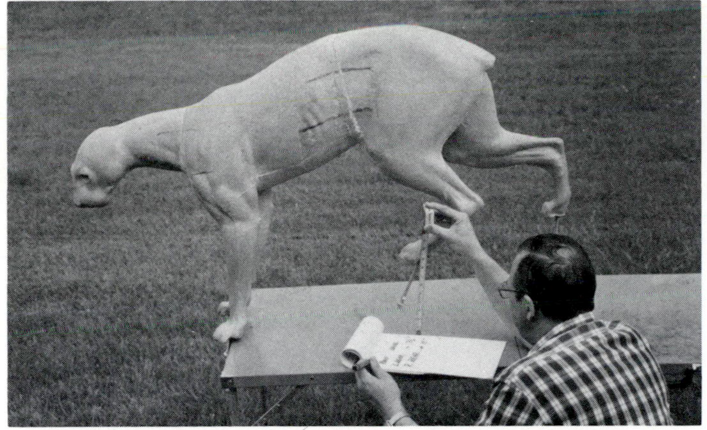

As with any project, we began by first planning the project ahead of time. First, we maneuvered the lion form around until the position appeared natural, and then made a few measurements of the height of the feet, the spread of the front and rear legs, and the distance between the front and rear legs. These measurements were noted in a notebook for future reference. (Remember to make these measurements as accurately as possible, so that the mannikin will "fit" the finished base nicely.)

# Constructing The Framework

One of the best attributes of this method of base making is that the framework is so simple and fast to construct. No contour outline is needed at all, and the only actual support needed is a base outline and whatever structure is required to hold the mannikin in the correct position.

In the photo above we have laid out the desired width and length of the base on a sheet of 3/4" plywood. We selected the rectangular shape, as the finished "rock" was to be trimmed with a 4" deep box of black walnut.

The layout for the leg bolt positions of the mannikin were redrawn, based on the measurements taken previously.

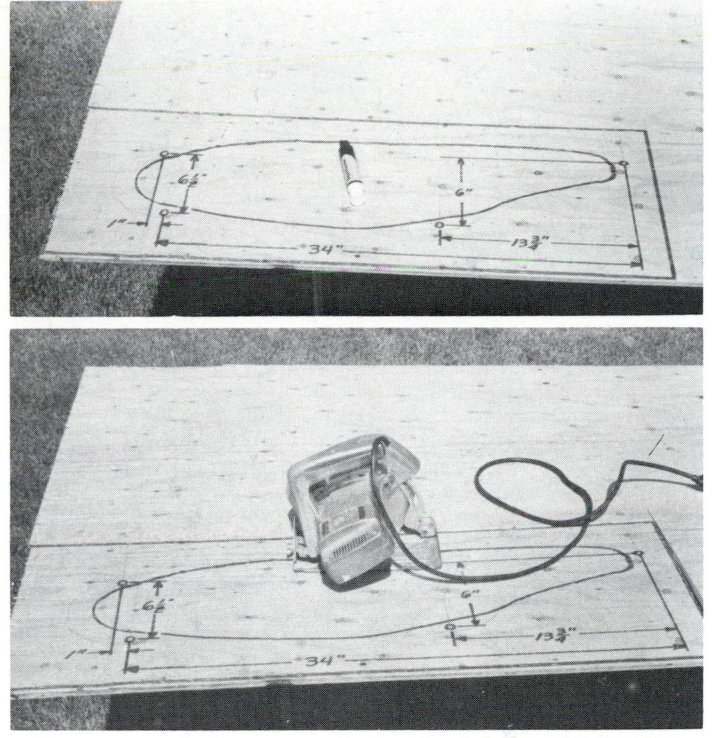


Since we needed access to the inside of the "rock" to tighten the nuts on the leg bolts on the mannikin, a cut-out was made. Notice how the cut-out coincides closely with the leg bolt hole pattern. The rectangular outline was then cut with the aid of a skill saw.

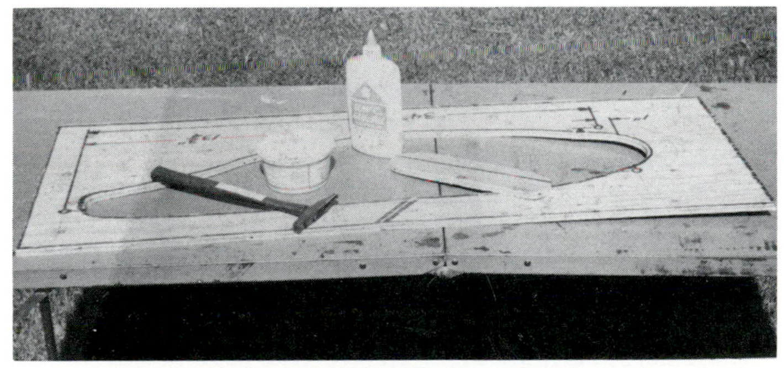

The cut-out can be sawed with the use of a sabre saw or a bandsaw. The sabre saw has the advantage that a cut need NOT be made from the outside of the rectangular base, but also tends to be somewhat slow.

We used a bandsaw to make the cut-out, which required an additional "patch" to be glued and nailed over the bandsaw cut. The patch will strengthen the base and will not be seen after it is completed.

To obtain more height for the rock, we raised the position of the front feet of the mannikin off of the base by about 3", and fabricated the simple platform. This platform was then glued and nailed in place.

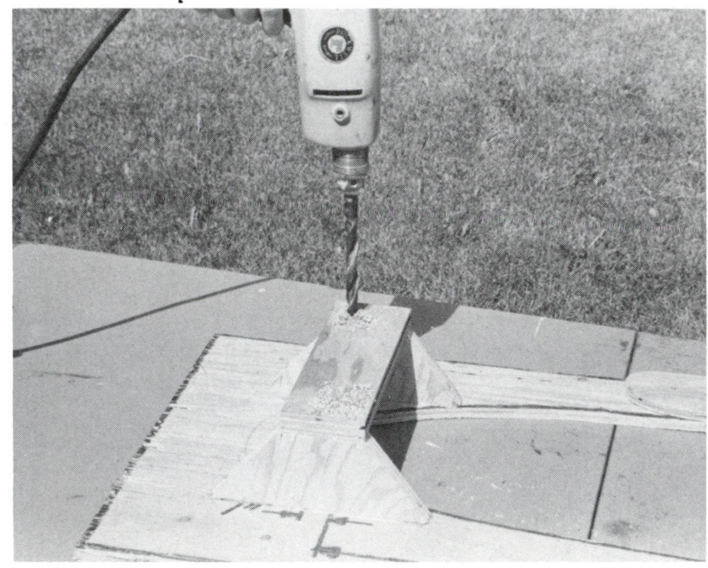

The hole pattern for the front feet was then drawn on the platform to conform with our dimensions, and 1/2" holes were drilled for the leg rods on the mannikin.

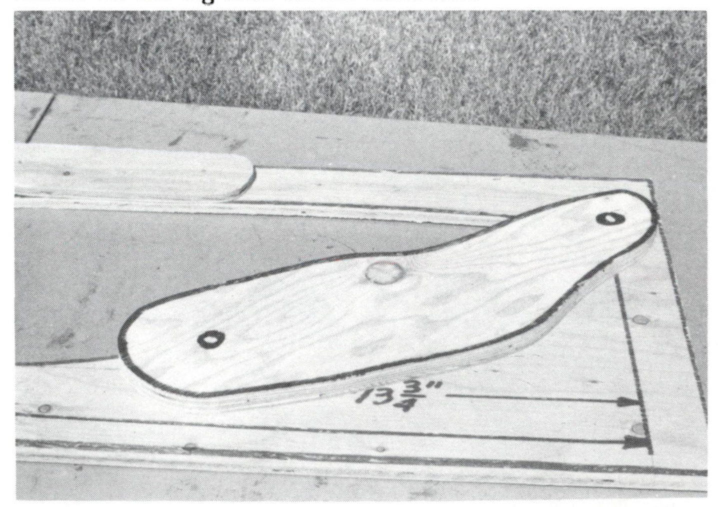


A second platform was then constructed for the rear feet in much the same manner, as can be seen in the Photo above. Our dimensions reflected a difference in height between the front and rear legs of 11." We cut a riser block of 1 × 4 pine to raise the rear platform to the correct height.

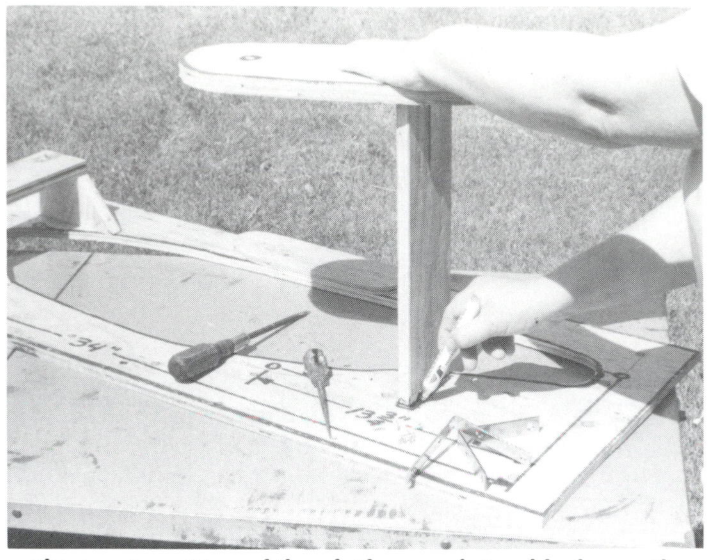

The exact position of the platform and riser block was then determined and outlined on the base.

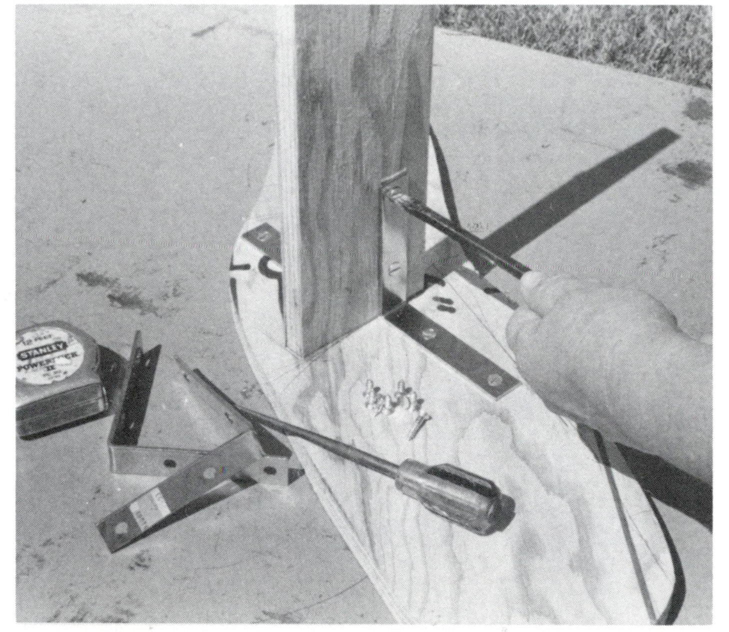


The platform was then fastened to the riser block with two 4" corner brackets and screws.

The bracket was then inverted and placed into position on the base. Two more 4" brackets were then installed to hold it in place. While this may seem like a weak support, it must be remembered that the fiberglass covering over the base framework will strengthen it immensely.

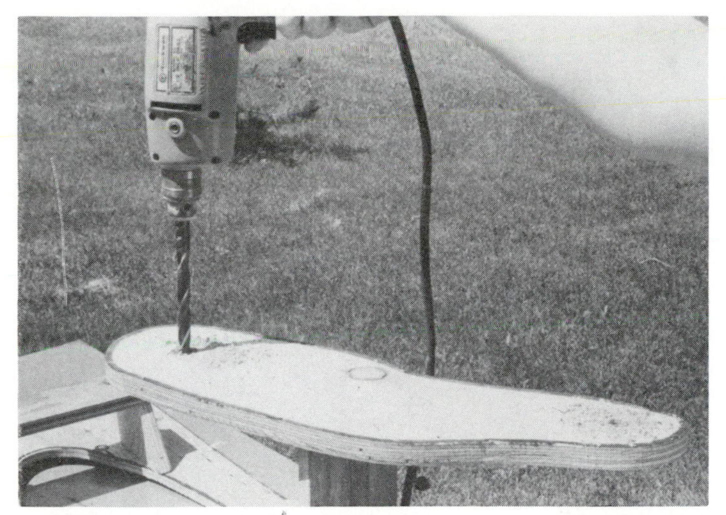

Referring to our initial measurements, we then drilled the 1/2" holes for the rear leg rods.

Since the right rear leg position was 2 ½" higher than the left leg, a block containing a 1/2" hole was glued and nailed into position over the right rear leg rod hole.

As a final double-check, we placed our mannikin on the framework to make sure that everything lined up correctly before the fiberglass covering was applied.

# Creating The "Rock" Covering

Since our purpose now was to cover our simplified framework with a fiberglass "rock" surface, we selected four rather large slab rocks that displayed good surface character to cast. The larger rocks of course yield a larger piece of fiberglass, which gives the artist more latitude of position and adjustment than would a smaller rock. This rock casting process could be performed in the field if a particular rock shape was located that was too large or heavy to transport to the studio. If it is decided to cast a rock in the field, plan on covering the framework for the rock there also, as the fiberglass impression MUST be used while it is still SEMI-CURED and in a very soft, flexible

condition. For the sake of convenience, we simply purchased four slab rocks with good texture at a local masonry shop, and set them up outside the shop for casting.

We arranged the stones that we wished to cast across some planking at a convenient work height. After thoroughly brushing any loose debris from the rocks, we gave them two complete waxings with Ceara Mold Release Wax.

While the wax was drying, we cut several strips of 3/4 ounce fiberglass mat from a roll and set them aside.

The photo above demonstrates how to rip the strips of mat into smaller pieces over the edge of a piece of board or plywood. Ripping the mat will produce a ragged edge that is preferable to a cut edge, as the fibers will lay together much nicer and form a tighter bond during the lay-up process. Rip enough pieces of the mat for one layer that will completely cover the surface that is to be cast.

The materials used for casting the rocks were *Polytranspar*™ Fiberglass Lay-Up Resin, Catalyst, and the 3/4 ounce fiberglass mat.

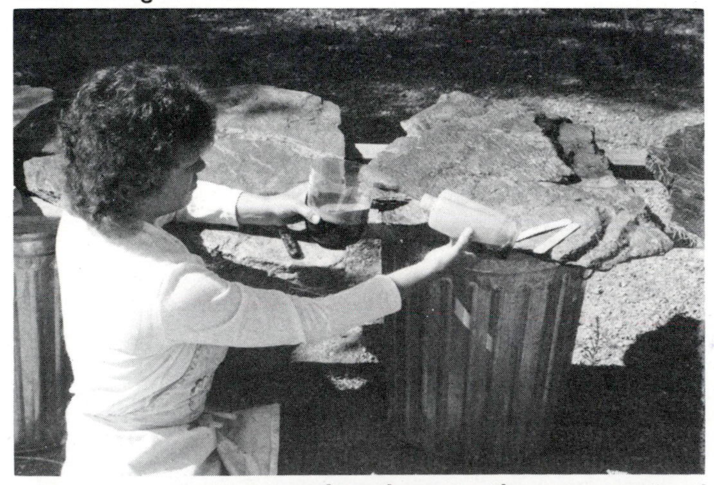

The next step was to catalyze the resin. The correct ratio of catalyst to resin for this type of lay-up operation is as follows: (These ratios are all "liquid" measure).

2 ounces catalyst per gallon resin
1/2 ounce catalyst per quart of resin
1/4 ounce catalyst per pint of resin
1/8 ounce catalyst per 8 ounces of resin

The catalyzed resin was then thoroughly mixed and then brushed over the surface of the waxed rock.

When the resin began to set up, a second batch of resin and catalyst was mixed. This resin was used to completely wet-out or saturate a "single" layer of the 3/4 ounce mat as demonstrated in the photo above. Make sure that the surface is evenly covered with a uniform layer of mat and resin.

When the resin started to set, it was gently lifted off of the surface of the rock. This is the most critical step of the whole process.

The sheet of resin needs to come off of the rock without tearing, but still remain soft and flexible enough that it can be easily formed. Note the flexibility of the resin sheet in the above photo. As can be seen, the rock surface has been duplicated "in reverse." This sheet is now ready to apply to the base framework.

We recommend that you work with only one sheet at a time, as the more the resin cures the more difficult or even impossible it will be to bend and shape. Notice how nicely the detail of the surface of the rock has been captured in the fiberglass sheet.

# Applying The Fiberglass Sheet To The Base

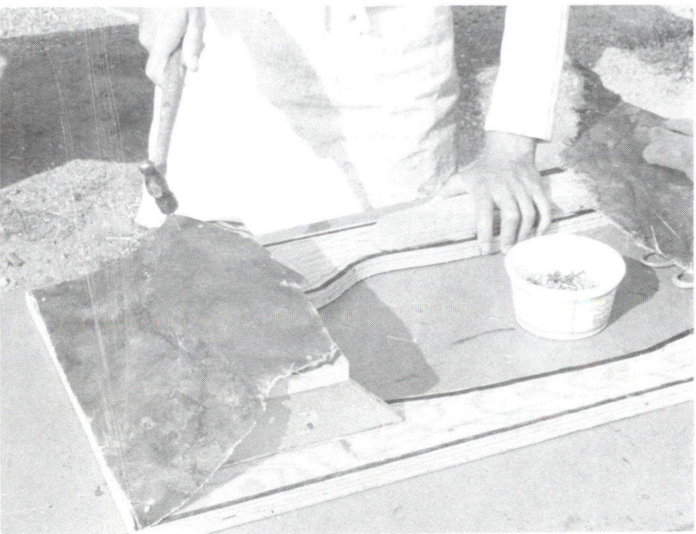

While the resin sheet was still soft and pliable, we positioned it in various ways on our base framework to take best advantage of the detail of the rock surface. When it was determined which area we wished to cover, a section was cut to the proper shape with metal-cutting scissors. This piece was then pressed into place and tacked down with small brads.

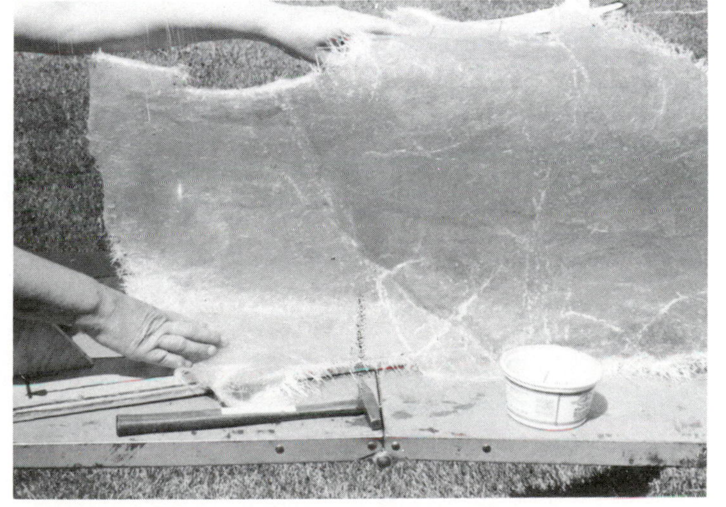

A second sheet was then cast from another one of our sample rocks and formed into position.

We continued to apply sheets of the uncured fiberglass to the base framework, pushing, cutting, and forming as we proceeded. A staple gun proved useful in many places. Note how the crushing and forming of the fiberglass sheet easily recaptures the rock-like appearance, even without being painted.

In areas where the fiberglass sheeting overlapped, we bonded the pieces together with a hot melt glue gun.

In places where we encountered a rather large gap, we used Ultra Lite body filler and chopped fiberglass as a filler. This plastic filler was first catalyzed, then spatulated into place, and brushed with a small brush and lacquer thinner.

Smaller areas and seam lines were also covered and brushed with Ultra-Lite. The artist should leave rather coarse brush marks in the filler to make it blend better with the "rock" texture.

A sanding "flap wheel" was then used to remove some of the excess filler material where needed. Always use safety glasses when using this type of equipment.

# Painting The Fiberglass Base

While all aspects of habitat base making are interesting, it is a welcome break in the monotony of everyday work to put color and the finishing touch to the base. It is here that we finally bring realism and a sense of artistry to our work. (This is the fun part.)

Since the base that was constructed for this section was for a light-colored mountain lion, we decided to color the base rather dark to provide good color contrast.

We began the finishing process by wiping the base down with a paper towel dipped in lacquer thinner. This removed any trace of wax left remaining from the casting process.

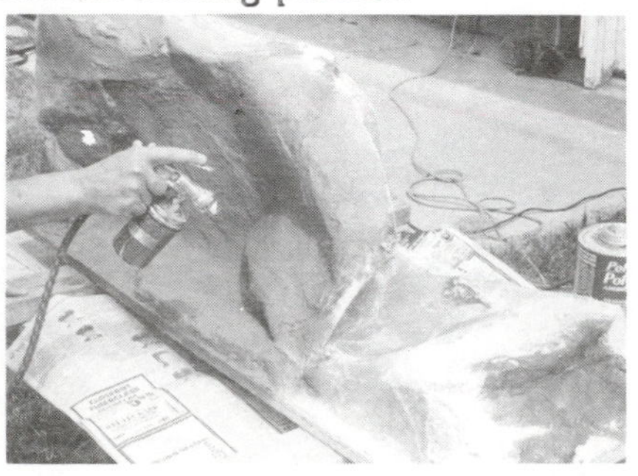

We then sprayed the rock base with a very light coating of *Polytranspar*™ Fiberglass Gray Primer (FP 195G) to prepare the surface for the subsequent colors and to give it a more uniform base color. We felt that after spraying the base gray that a silver primer (FP195S) might have given us a nicer effect. Whichever color that is chosen, do not omit the primer step as other paints will not adhere to fiberglass.

To tint the base to a desired reddish-brown colored effect, we selected only earth colors such as Yellow Ochre WA 141, Sienna WA 200, Chocolate Brown WA 70, White WA 10, and Black WA 30. *Polytranspar*™ Water/Acrylic airbrush paint proved to be an ideal paint because it had a longer drying time which permitted us to stipple (jab) the various earth colors together to obtain the final blended effect that we wanted.

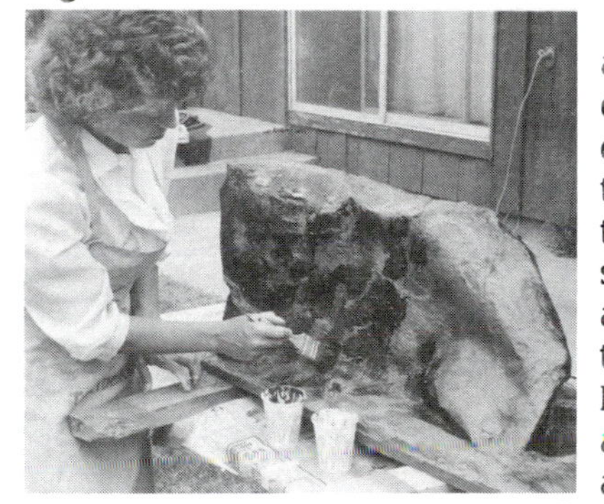

Stippling may seem like a somewhat crude method of painting but it produces effects that cannot be obtained any other way, particularly on a rough object such as a rock base. (It is also just about impossible to fail using this technique!) If too much color has been applied; yellow, for example, simply stipple or jab another color into the yellow until the color blend is satisfactory. The above photo illustrates how the various colors are applied in an irregular, "splotchy" pattern. By continuing to stipple the surface with the brush tip, the colors will all begin to blend together in a nice mottled pattern.

Additional texturing can be added at anytime. At one point we noticed a spot on our "rock" where the fiberglass mat pattern was showing through. We lightly dusted this area with sand, and merely stippled paint right over it. The sand camouflaged the defect and the paint colored and held the sand in place. Small pieces of gravel or imitation lichen can be applied in the same manner if so desired.

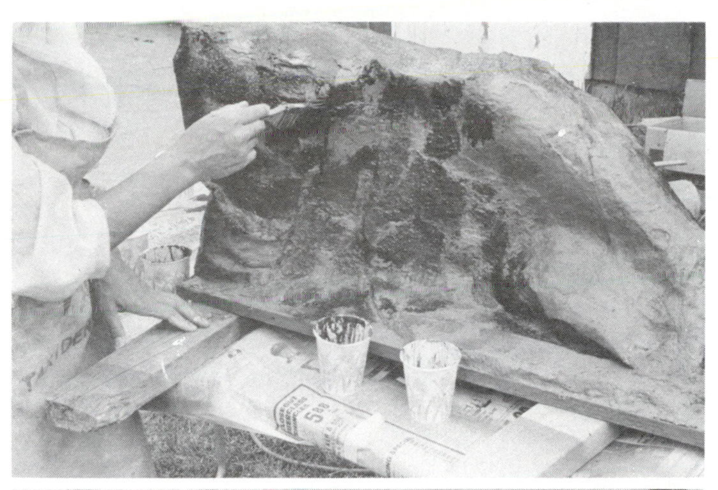

The entire rock was stippled with our basic colors as shown in the two previous photos. The entire painting process took about 20 minutes, and when we were satisfied with the result, set the base aside to completely dry. *Polytranspar*™ Water/Acrylic paint dried to a very dull gloss and no other finish was necessary.

In the photos above, both sides of the completed base are shown.

This unit was designed to fit into a 4 inch deep walnut finished trim border, which can be seen in the photo of the finished mount, below.

A punch was then used to pinpoint the leg bolt locations (above left photo) and the fiberglass was drilled for the bolts from the show side (above right photo).

The last step before setting the base aside was to recheck the fit of the mannikin.

*The photo above shows the completed mountain lion by Master Taxidermist Jim Hall. One of the definitions of "aesthetic" in Webster's 3rd New International Dictionary describes it as "having a sense, real or affected, of beauty. . ." Few, if any medias are more suited to using "aesthitic" in self-descriptive narration than taxidermy and woodcarving arts. And few methods of recreating rocks can be any more aesthetically pleasing and beautiful than that just descibed by Jim Hall in the scene above.*

# CHAPTER 9

*This chapter is a combined effort of some of the best wildlife artists in the industry using live and reproduced vegetation, with a variety of methods and styles.*

# Plants Real & Reproductions

## Cattails Made Easy

***by Jim Hall***

If any single item could be used to typify a marsh habitat scene, it would have to be a cattail. They are instantly recognizable by anyone as "belonging" in this type of environment. Many wildlife artists use cattails in their background habitats or paintings very successfully. The below photo is of live cattails in their natural setting.

There are really only two disadvantages to using live cattails; one is that late in the season, they are very tall and don't blend well into small habitat situations; the other is that cattail blossoms can "go to seed" long after they have been installed, creating a very large mess.

The size problem can be solved somewhat by gathering cattail items and leaves earlier in the year, when they are smaller. The blossoming problem can be solved by simply replacing the seed pod with a reproduction.

To create this reproduction, we began with scrap pieces of urethane foam that were approximately the size of a natural cattail blossom. Urethane foam was used because it is easily carvable and will accept any kind of paint.

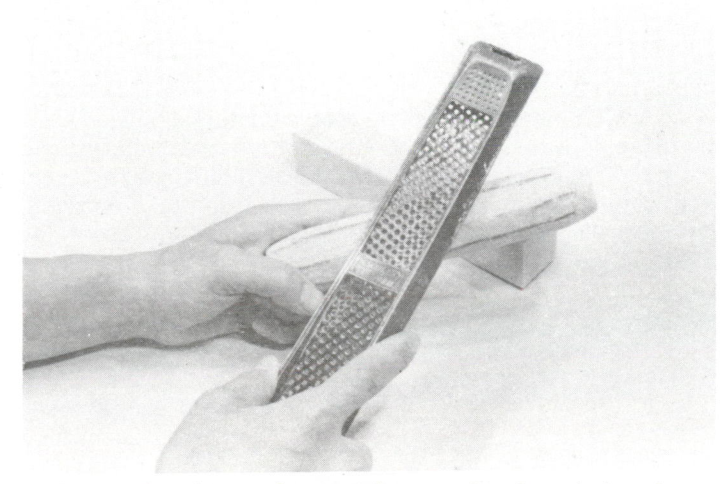

The "seed pod" was then gently rasped and sanded to shape.

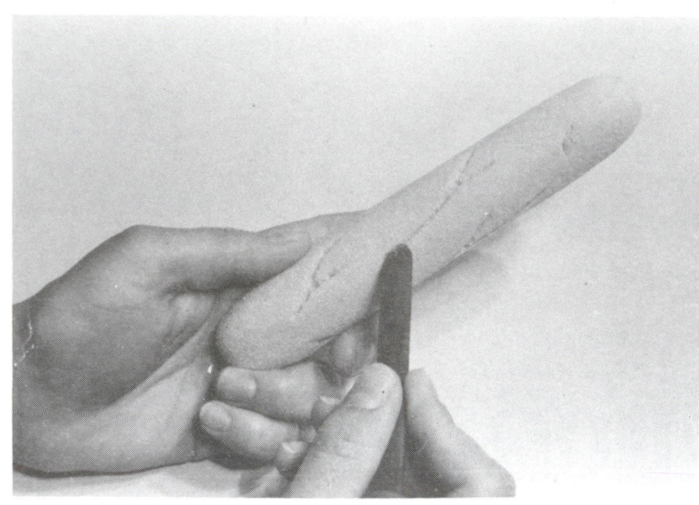

The individual texturing was carved into the cattail with a dull, blunt edge of an old table knife, using a natural cattail as a reference.

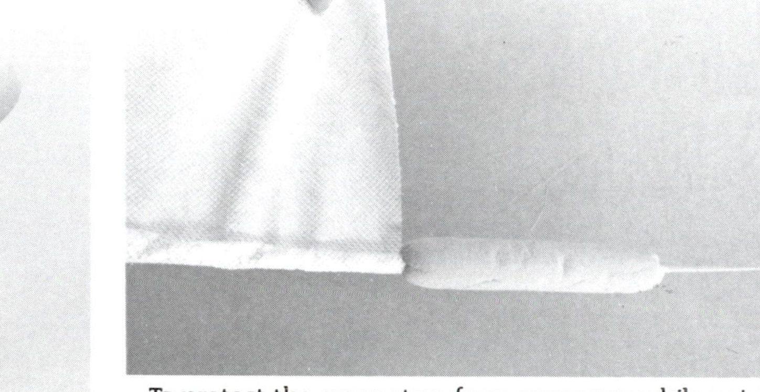

A hole was then drilled lengthwise through the cattail to make access for the stem.

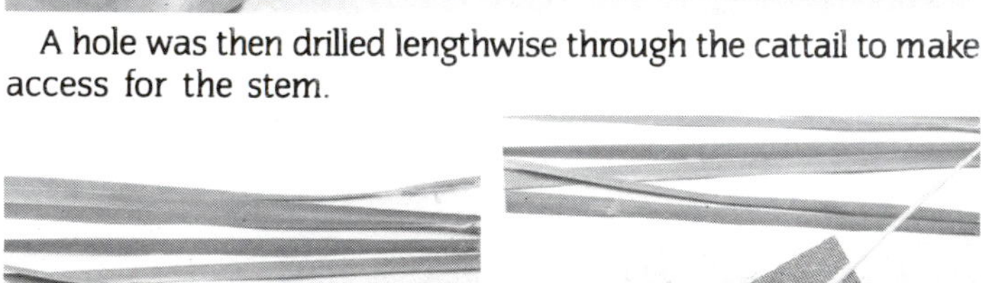

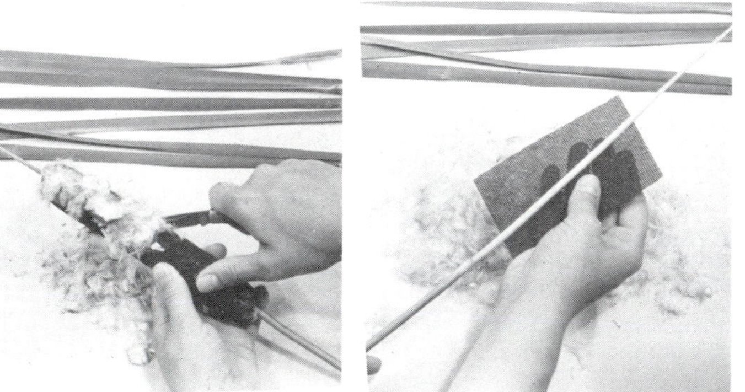

We solved the problem of creating a uniformly tapered stem by using an actual stem. The above photo on left shows how we cut the natural blossom from the stem, followed by some light sanding to true everything up nice (top photo on right).

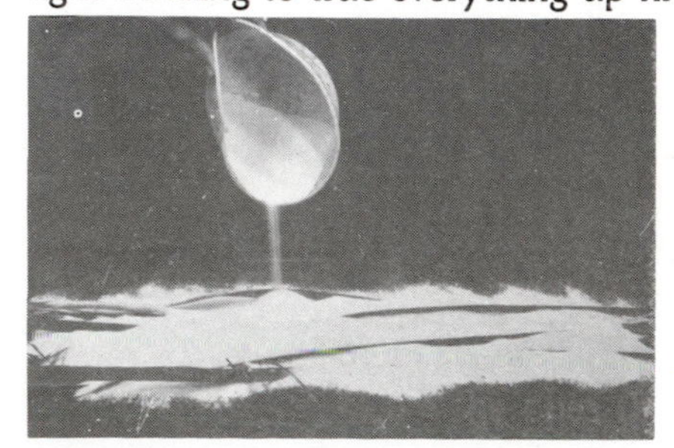

This stem, as well as the associated leaves we intended to use, was then buried for several days under a layer of powdered borax, which removed excess moisture while permitting the natural color to remain.

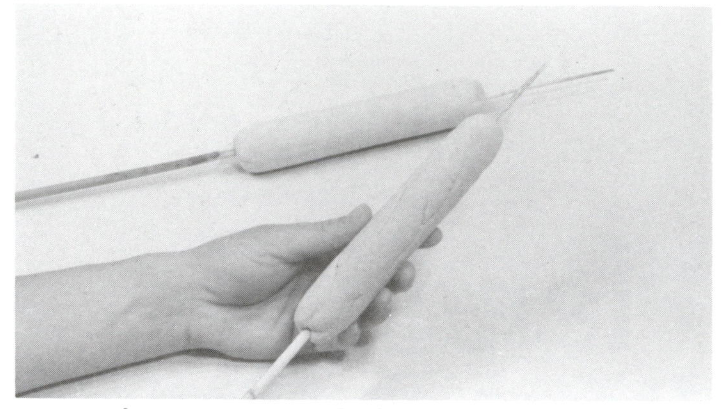

When the stem was completely dry, we placed a few drops of Ultra Seal on the bare spot (where the original cattail blossom was) and pushed the carved blossom into position. The cattail was now ready to paint.

To protect the green stem from overspray while painting, we wrapped a piece of paper towel around it close to the blossom.

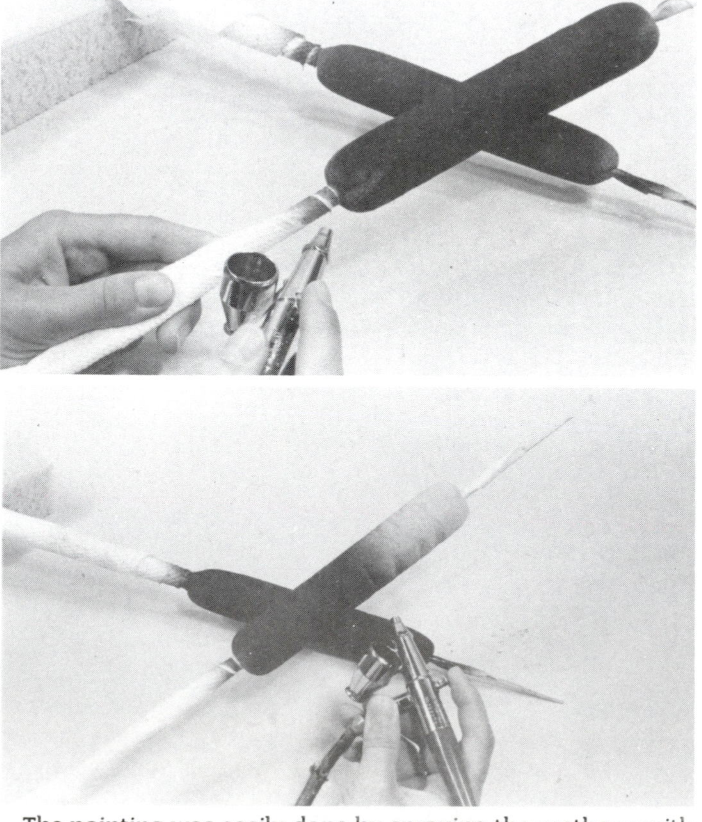

The painting was easily done by spraying the urethane with *Polytranspar*™ Chocolate Brown (WA or FP70). This color was airbrushed over the urethane until the desired color was achieved. No other finish was applied.

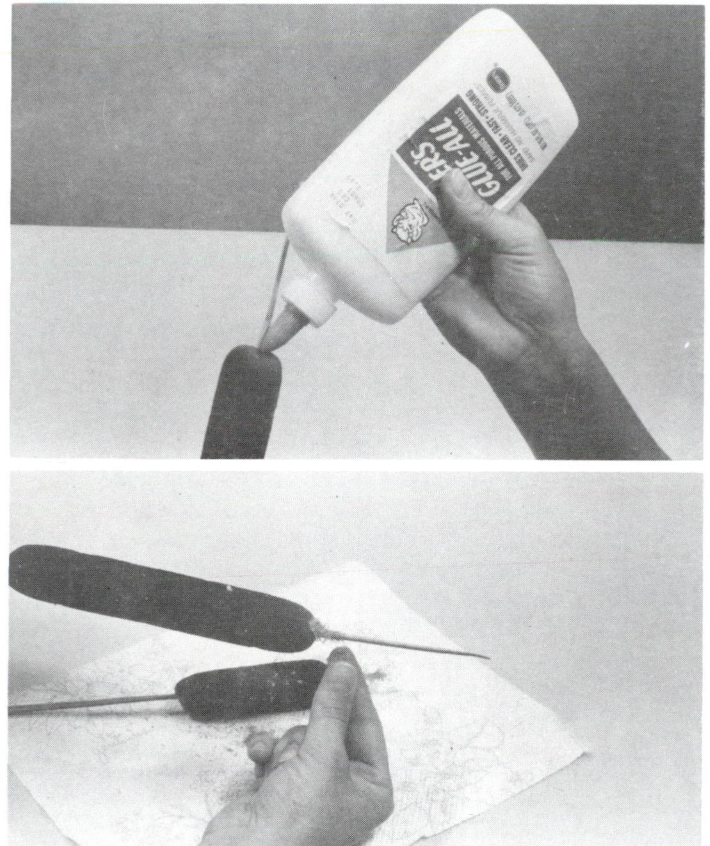


As a final touch, a spot of white glue (Ultra Seal would work fine) was placed near the upper tip of the cattail, and some small pieces of natural cattail were glued in place.

Above is an arangement of cattails. The two in the center are the ones that were reproduced.

# Instant Mushrooms

One of the nicer additions to any spring, summer, or fall habitat base or scene is the mushroom. Once a good mushroom mold has been made, one can easily produce as many fake but very convincing mushrooms as is desired, and making a mushroom mold is not difficult.

The mushrooms for this project were purchased at a local market.

We then selected two of the more firm mushrooms that had nice long stems for our subjects. Next, we found two containers to assist us in the molding operation. These containers should

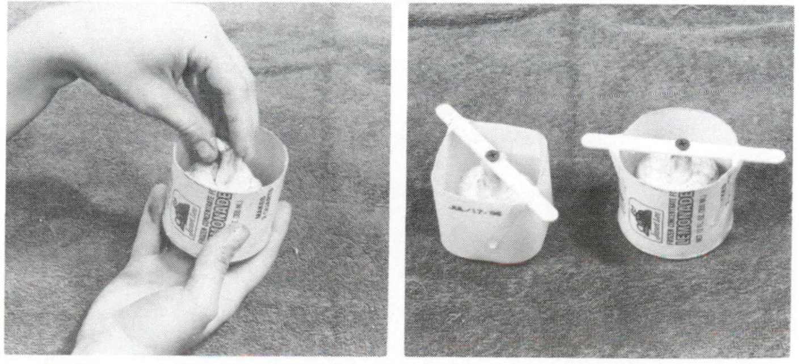

be large enough to provide about 1/4 to 3/8 inch clearance all around the mushroom. A small ice cream stick was then gently fastened to the mushroom with a small screw. This should leave the mushroom completely suspended within the container.

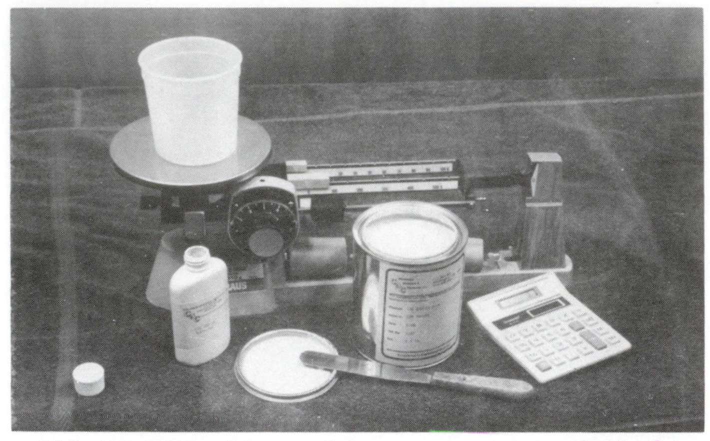

CAC 100 RTV rubber molding compound (available from WASCO) was selected for this mold making process. It must be weighed and catalyzed very accurately for best results. Complete mixing instructions accompany each supply of RTV rubber.

When thoroughly catalyzed and mixed, the RTV rubber is poured into the mold containers surrounding the mushrooms. This RTV rubber cures overnight at room temperature.

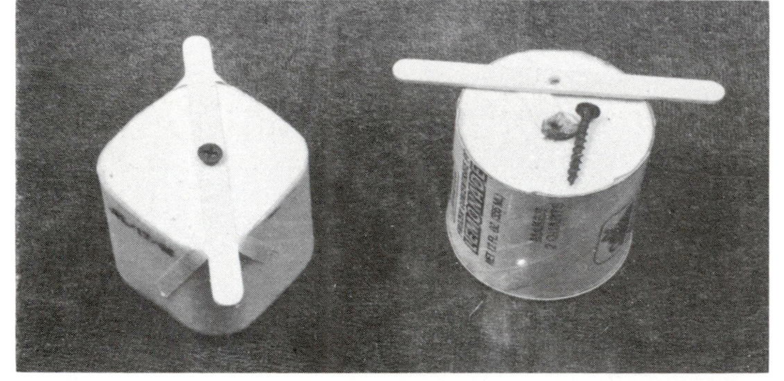

When the RTV rubber was completely cured, the temporary screws and sticks were removed, and the molds were taken from the mold containers.

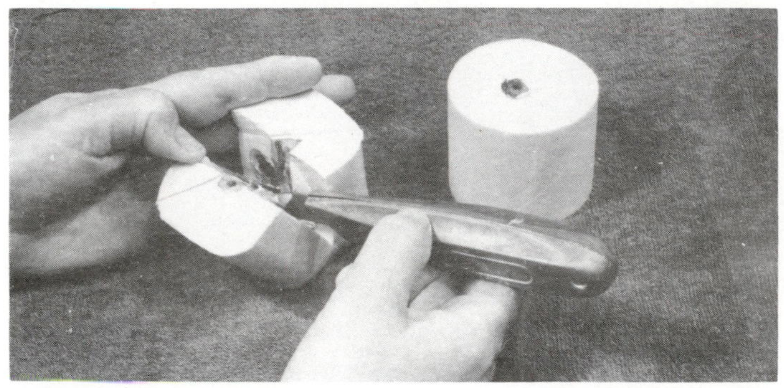

To remove the mushroom from the mold, the mold was split about two thirds down its length with a sharp blade.

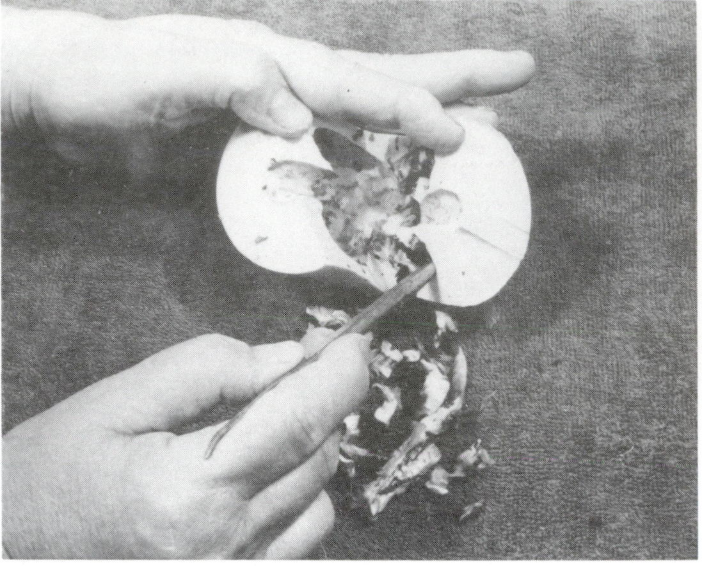

The original mushroom was dug out of the cavity, and when as much of the mushroom as possible was removed we then washed the cavity in hot water.

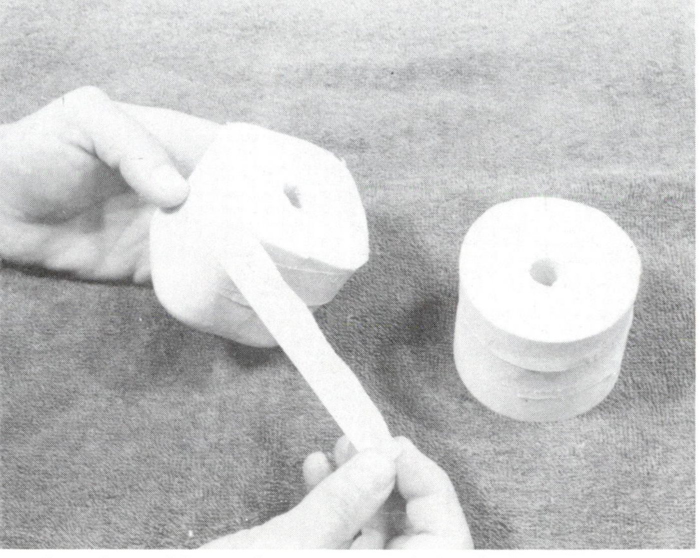

When the mold was dry, we proceeded with the casting operation. First, we taped the mold shut with masking tape.

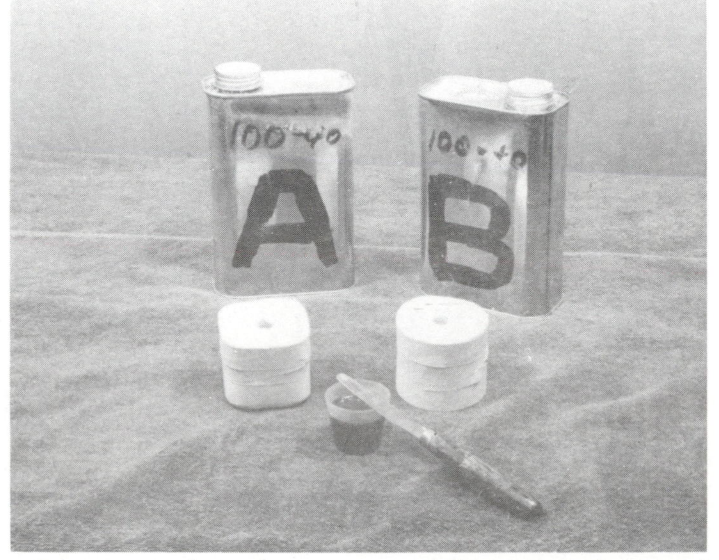


Two-part urethane foam was used for casting the actual mushroom.

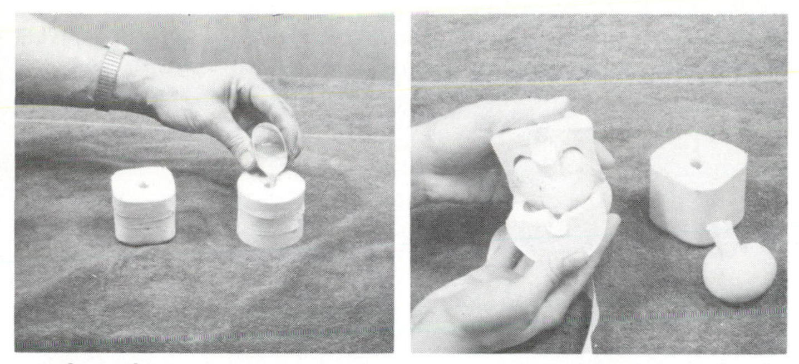

When the two components were measured and completely mixed, they were simply poured into the molds. We waited approximately 20 minutes for the foam to cure and cool, and removed the "new" mushrooms by carefully prying the mold halves apart and pulled the mushroom out.

The last operation was the painting of the mushroom, which was easily accomplished with the use of *Polytranspar*™ Water/Acrylic airbrush paint and a fine brush. An actual mushroom was kept handy for color reference.

The above photo shows a comparison of a real mushroom on the right, with the one that was reproduced in urethane foam. Or, was it the one on the left?

# A Wall Of Vegetation

***by Dan Blair***

Certain small game animals, such as rabbits, squirrels, muskrats, etc., (as well as most upland game birds and waterfowl) can be displayed appropriately against a solid "wall" of vegetation. The challenge is in creating the "wall," whether it be cattails, corn stalks, seed-type grasses, or plain old weeds. But I've found one simple solution to the problem which has several variations and applications while keeping it quick to create and inexpensive to make.

First, determine the type of habitat materials that the specimen being portrayed would most likely be found in, and choose an appropriate sized base. Now, cut the base out of 3/4" plywood according to the contours selected in the design.

Next, using a suitable sized piece of panelling or 1/4" plywood, trace the complete perimeter of the base on one end of the potential backboard. The height of the backboard panel should be determined by the length of the vegetation that is used. The backboard panel, in this instance, is shorter by 4 or 5 inches to keep it from being exposed as the grass in this habitat fans out toward the top and sides. Once the height has been determined, cut a half circle around the top end, equal in diameter to the width of your base. (To better understand this, see the diagram below).

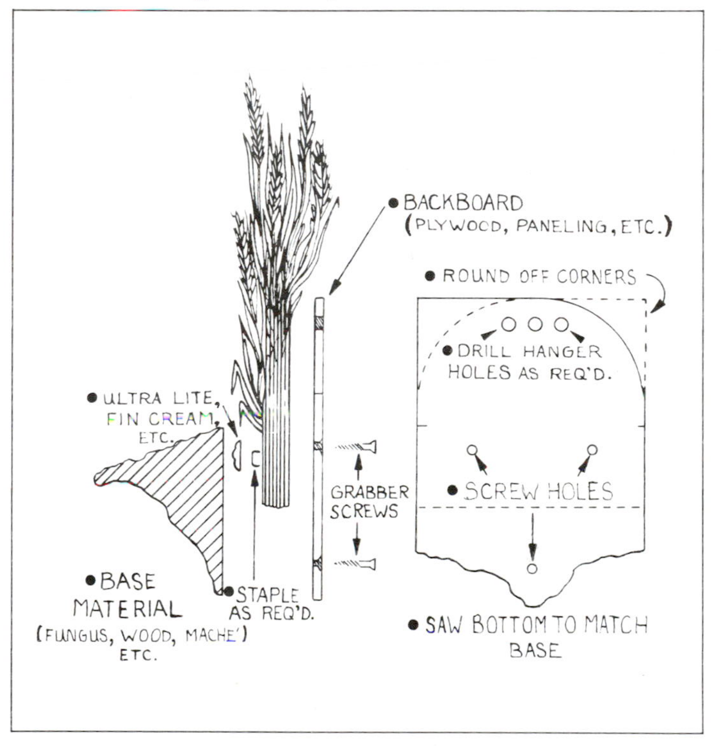


After cutting off both ends of the backboard panel with a band or sabre saw, sand off the rough edges lightly to avoid splinters. (The artist *and* the customer will be grateful for the brief moment that it takes to do this.) Next take *Polytranspar* airbrush paints in compatible colors and spray the "show" side (roughest side) of the panel with a base coat that will closely match the colors of the foliage and habitat materials being used. An exact match is not necessary. For fall colored vegetation; greys, tans, browns, or blacks may be splattered and blended together in a camouflage-type manner. Greens, greys, some blues, and black will work nicely for green vegetations. These randomly placed and combined colors add depth and help to disguise the backboard by breaking up the outlines which is the simplest rule of camouflage. Once the paint has completely dried, the exhibit will be ready for attaching the plants.

Vegetation comes in countless forms and from equally countless sources. The artist can buy numerous types of preserved plants in an array of colors to match any season from floral supply companies. These companies carry an extensive assortment of dried flowers and plants in addition to synthetic plastic and silk plants. However, there is a ready-made source of natural

dried weeds, cattails, wild grasses, etc., "free for the taking," growing along side most any roadside.

Once the proper vegetation has been decided upon and collected for the exhibit, lay it out flat across the backboard so that the bottom ends extend 2 or 3 inches below the top edge of the base as in the diagram on the previous page. This is the time to trim the length of the plants to the proper size. I recommend starting in the center with the longest plants, working right or left using the shorter stems on the outside edge. It is a good idea to "fan" the vegetation somewhat (similar to a turkey's tail) to create the final effect and make it look natural. If a plant needs to be trimmed, always trim it from the bottom end of the plants to keep them from appearing pruned.

After prearranging the vegetation across the face of the panel, staple it two or three times per stem or cluster of stems to the backboard being certain the staples do *not* protrude out the back. (The staples are only a partial measure of attaching the vegetation and will hold the plants during the next step.)

Now apply a strip of *Polytranspar*™ Fin Backing Cream, Ultra Lite filler, or Sculpall™ sufficient in length to reach from side to side. Cover the stems just *below* the top surface of the base. (Refer to the diagram again on the previous page.) Before the Ultra Lite or Fin Cream sets up, place the base material into proper alignment and press it down firmly, bonding the backboard and base together while also securing the plants between them. Be careful not to use so much adhesive that it oozes out and onto the base as it will be difficult to remove.

Once the backboard and base material have been glued together, recheck the alignment. If it is straight, then use a sufficient number of Grabber screws to secure the two even more tightly. This final step is important as the screws will pull the backboard tightly against the base thoroughly securing and supporting the vegetation. And once the adhesive between them dries, even the smaller diameter stems should be firmly fastened in place. A natural snag of driftwood, a log end, or similar item can now be secured to the base if so desired.

One realistic scene that is easy to construct is a "muskrat push-up." To build one, use sheet moss obtained from a floral supply house or collected in the field to cover a wooden bowl that has been cut in half and turned upside down. Use an imitation rock, a shelf fungus, and the exhibit will be ready to add the bird or animal.

But to professionally finish off a natural weed or grass habitat such as this, consider going a step or two farther. It takes only a few more minutes to "lodge" an autumn leaf or two between the stems of cattails. A moment longer to glue a wooley bear caterpillar to a leaf or stem and as little time to place a bone or shell or even a small section of a snake's shedded skin at the edge of the grass. These added items of "habitat" give a more natural and complete look to the finished habitat, increasing the value without a significant increase in cost. (Every kid in the neighborhood for the price of a soft drink can be a potential supplier of bugs, bones, and other habitat goodies.)

By following the few steps described in this method of habitat base construction, one can create simple and inexpensive bases which appear complicated and expensive. Instead of offering the antiquated "stuffed fish or bird on a board exhibit," offer the customer a realistic habitat scene. Most will gladly pay extra. Additionally, it offers the artist an opportunity to be creative, not to mention that they are "fun" to build.

"Small" efforts made by the artist make "big" impressions on customers. As one Iowa taxidermy studio puts it, "Our customers go those few extra miles because they know we go those few extra steps."

Trying this method of habitat base construction is a step in the right direction.

# A Matter of Life Or Death

This title sounds like just one more in a long line of environmental oriented articles about rapidly declining and deteriorating living conditions. However, the subject in this case is one which rarely, if ever (until now), has been approached regarding the field of taxidermy. As a matter of fact, when you think about it, it reminds one of the old adage about "not being able to see the forest for the trees."

When we *do* think about creating life-like habitats for displaying our finished mounts, we immediately consider a long list of sources for an even longer list of supplies providing us with plastic, silk, foam, freeze dried, naturally dried, artificial or naturally colored plants of every description from flowers to ferns, grasses to grains, mosses to mushrooms, and twigs to trees. And from this veritable jungle of vegetation we propose to primp, prune, and prepare an example of our craft depicting "life." But there is an alternative, rarely, if ever considered; "Living" plants!

With a multitude of house plants available from an even greater number of sources (on the assumption that greenhouses, garden centers, and florists still outnumber taxidermy supply companies), you can create unlimited varieties of "living" habitats for virtually any kind of mount, from mink to moose and birds to bass. Your possibilities are as endless as your selection.

The dollar value spent each year on house plants and related paraphernalia far exceeds those spent by your taxidermy customers, and indicates an area of tremendous popularity which could easily be combined with taxidermy services.

Granted, maintenance is greater with a living habitat than otherwise, but the *involvement* can be more pleasureable for "some" (though not all) customers. You have simply to make the offering available and the customers will usually let you know how receptive they are to the idea. Which brings us to a point: "What exactly is the idea?"

It could be difficult to mount game of any species on a *totally* living habitat. Most people have enough lawn to mow or hedges to trim and don't relish the thought of adding more. A black bear standing on a 'lawn" in the living room which requires a "Bush Hawg," a "Weedeater," and a "Dust Buster" to keep under control is bound to make anyone *lose* control.

On the other hand, consider the same bear mounted on a normally constructed base from which a living ivy twines from its pot concealed within the hollowed stump beside which ferns and violets emerge from their hidden containers recessed into the foam base material.

Picture the grassy stems of the hardy Spider Plant or other plants similar to lake-bottom vegetation which come in a variety of sizes and containers and can be easily concealed in or behind the wooden snag which supports a mounted fish or two.

A variety of ferns and ivy's could be combined for concealing the base and the support wire to the wing or leg of a flying ringneck pheasant or other game birds.

A javelina, or a variety of other desert dwellers like quail, rattlesnake, etc., could be very attractively displayed on a habitat containing containers of large or small living cactus and other desert type plants.

No doubt the examples could go on as endlessly as the variety of plants and flowers available to you in the nearest greenhouse or garden supply center. And *there* is perhaps the best place for you to look for examples. Take your inquiries to either business and explain what your plans and needs are. Most often you'll find the staff eager to develop a new potential market for their products. They'll gladly give you advice on which plants require least or most maintenance, which require little, if any, sunlight (some don't), how much watering is required, proper container sizes, estimated rate of growth and size potential, preferred temperatures, etc. (Be certain if you buy plants to get this information as you'll want to be able to pass it along to your customer with the plants.)

For the sake of experience, try creating one or two examples to have on hand for customers to consider and criticize, and encourage them to do both. Look over the photo examples most garden centers keep which shows the adult plants and which give a brief synopsis on the care and maintenance requirements. After selecting a few "easy keepers" build them into and around your habitat from the very beginning. Premeditate the location of the containers to keep them concealed from view as much as possible. But try not to complicate the necessary task of feeding and watering them by making them inaccessible to a watering spout.

No doubt you're bound to be asked and may even be thinking so yourself, "But what if the plant dies?" The answer is obvious! "Sooner or later, everything that lives dies." That in itself is insufficient response and would be considered poor business etiquette. But dying is a fact of life and, much like the subject of "tanning," it's something few of us want to take responsibility for. So use a little discretion here.

If you've kept the plant growing healthy for a month or more before the customer picks it up, and then 3 or 4 months later he brings it back with the complaint that he did everything you told him to do, but it died, it is probably safe to assume that he *didn't!* Be sure your customer is aware that *he* has a responsibility to the habitat also, and just like having a pet, he has to feed and water it on a regular basis.

If he asks about a warranty, offer him the *same* warranty on the plant that the greenhouse offered you—but nothing more! If (and when) the plant dies, assure him that you will be glad to replace and restore the habitat "with a new plant" at a cost slightly more than materials. Such a restoration may be as simple as lifting out one potted plant and dropping in another.

You can almost hear the most common comment already, "Now that *really* looks life-like!" Right? So consider adding a little "life" to your wildlife habitat with living plants and convince the customer who prefers the unique and unusual over the typical that with you — habitat is "a matter of life or death."

# Disposing Of A Bi-Product

Sculpall,™ one of the new versatile resin-type plastics (sculpting epoxies) on the market, is constantly filling voids in more ways than one. As innovative taxidermists discover the multitude of uses for which it works so well, and continue to create even newer techniques with it, the consumption of the product is bound to increase. Unfortunately, so will its waste.

Since measuring the amount needed for any particular task is left entirely to guesswork, it is frequently the case that the artist will mix a bit more than actually required. The resulting leftover Sculpall is, more often than not, left to harden and is eventually discarded as an unusable bi-product.

However, in one midwestern taxidermy school, the subject of "waste and how to prevent it" was taken up as a project and the favorite discovery for "saving" the leftover Sculpall has become an important addition to the habitat class. Consider this: Unless the two components of Sculpall are preheated, their curing time will vary from about one to three hours. Experimentation and experience will be the best advisor here. But, once the original task has been completed for which the Sculpall was mixed, the leftover portion will still be workable.

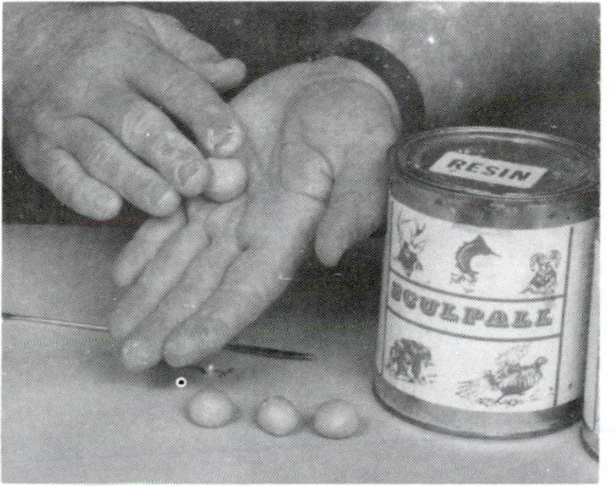


Take the remaining piece and make as many marble-sized balls as possible, but preferably in even numbers (2, 4, 6 . . .). Then roll one ball between wet fingers into a cylinder shaped to the length and circumference of a stem for a mushroom. Insert a straight or T-pin into it so that at least 1/4 to 1/2 of an inch protrudes from the bottom.

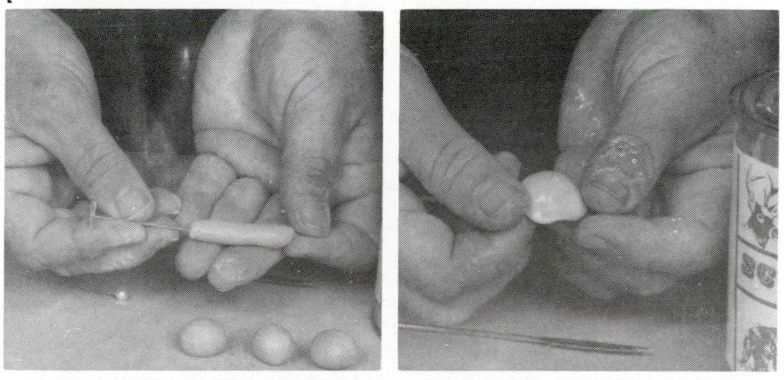

Take another ball and shape an appropriate sized cap to cover the pin head. By tapering the edges with wet fingers, one can make remarkably realistic toadstools and mushrooms as simply as just described. (A modeling tool can sculpt additional details on stems and caps if desired.)

The natural tan color of the Sculpall™ makes a good base color. It is very similar to actual mushrooms and little painting is required. An airbrush can be used to accent details, or create an aged appearance, as well as duplicate the yellows, browns, or even reds of certain fungus as preferred.

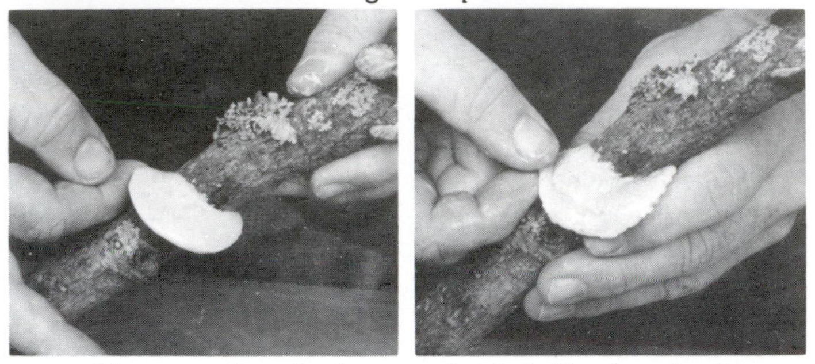

In addition, realistic-looking, artificial shelf or bracket fungus can be created by taking a portion of Sculpall and simply attaching it at the desired place on a branch and tapering the edges out and away accordingly. Then, the undersides can be textured with modeling tools, palette knives, or just about any sharp instrument. A "wet" coarse terry towel will also provide a satisfactory texture if the Sculpall hasn't hardened too much.

Toadstool type fungus can be "pinned" into a piece of foam or cardboard and allowed to stand until completely hardened. It is advisable to check the mushrooms from time to time as they cure to prevent them from drooping over. Long, thin-stemmed mushrooms should be pinned and suspended upside down in order to keep them straight until hardened. For slightly curved mushrooms, it is best to bend them slightly—just before they have completely hardened.

Once the mushrooms have cured and been highlighted with an airbrush, they can be pinned like a push pin into the habitat. The pins make it easy to adjust the location of the mushrooms to obtain the most desirable and balanced arrangement. (Even environmentalists are apt to approve of *this* method of disposing of a bi-product into the habitat.)

(Note: It is always a good idea to work Sculpall with wet fingers. Keep a water bowl handy for fingers and modeling tools. Finger prints in the Sculpall can be "polished" away before it hardens by simply dipping a cotton swab in water and buffing them away. Modeling tools will also work smoother if first dipped in water.)

# The Green, Green Grass of Home

A variation in materials, and simple creative procedures can provide a habitat artist with a vast assortment of hardy, insect proof, and realistic looking substitutes for habitat grass and weeds.

One of the quickest methods of making fake grass is using long pine needles which have been dried and then coated a few times (heavily) with clear acrylic spray. This reinforces the pine needles, making them less fragile and susceptible to breakage. It also makes them insect proof. (An alternative to spraying the needles is to dip them directly into clear lacquer or *Polytranspar*™ PlastiCoat habitat paint.) After the primer coat has dried, airbrush the needles to colors compatible with the season of the year being depicted in the habitat.

The pine needles usually grow in clusters on the branch and, if at all possible, should be kept in the cluster. This makes handling, painting, and "planting" the fake grass much easier. The clumps look very much like natural bunches of grass when finished.

Hot gluing is the quickest and surest method of attaching the clusters of fake grass to a base. But in the event a hot-melt glue gun is not available, the application of *Polytranspar* Fin Backing Cream (FC101 from WASCO) will work equally well, but with a slower drying time.

The creation of the newest paint addition to the *Polytranspar* line, PlastiCoat habitat paint has done the most for those who have tried creating their grass stems and leaves from acetate, vinyls, plastics, etc. Until now, most paints just wouldn't adhere well, making the synthetic grasses practically "untouchable" for fear of cracking or otherwise damaging the finish of the painted surface. Now, thanks to PlastiCoat, we can develop complete scenes of plastic grass without fear of its quality deteriorating with time or handling.

A very quick method of reproducing quantities of grass is first using an airbrush to color acetate, vinyl, or plastic sheets using one of the *Polytranspar* PlastiCoat paints available in four different shades of green.

HP10 Palmetto Green is a darker green suitable for grass blades, the leaves and stems of most plants, and of course, palmettos. HP20 Summer Green is a lighter, more yellow green with much the same uses. HP30 Spring Green is a bit darker green, and HP40 Lily Pad Green, all though created specifically for painting lily pads, is equally compatible for other types of vegetation. A fifth color, HP50, is Cattail Brown which can be used to shade plants down at the ground level, or to suggest wilting, or insect damage around edges or tips of grass and leaves. These paints can either be used as a primer (to which all other *Polytranspar* paints, as well as tube oil colors and acrylics will adhere) or PlastiCoat can be used as the actual finish paint.

After spraying a sheet of acetate with PlastiCoat and allowing the paint to thoroughly dry, lay the acetate on a smooth surface. Using a metal ruler as a straight edge, cut with a scalpel, razor blade, X-acto, or other sharp knife (or use a paper cutter). Alternate from top to bottom when cutting the *widest* part of the grass to keep working with a basically square sheet (see diagram).

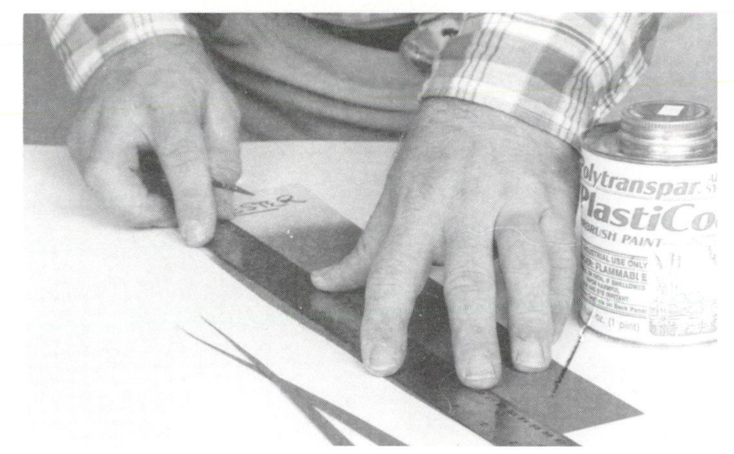


It is a simple step to add a crease in the center of the blade of grass for extra realism. If the base of the grass stem is 1/2 inch wide, *before* making the cut, place the ruler 1/4 inch from the edge of the blade of grass and crease the sheet up and over the ruler with the back side of your thumbnail. The crease will fold open again somewhat. Then move the ruler away from the crease at the bottom of the stem 1/4 inch and make the second cut. Once this system is tried a time or two, it becomes very easy to "cut the grass" with much haste and no waste.

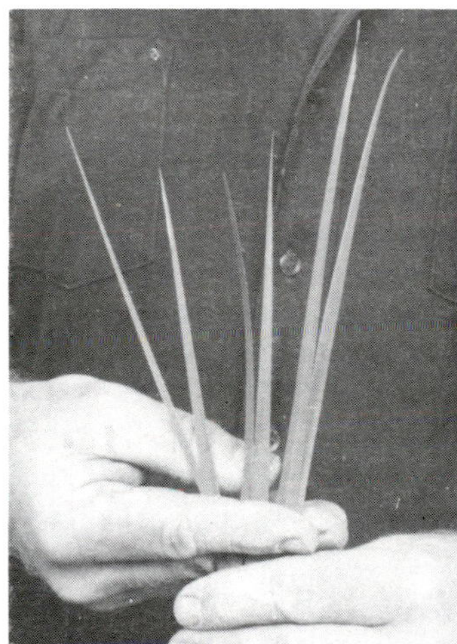

Grass blades can be made long or short, depending on the need, and the shorter pieces can also be used as leaves by adding them to stems of artificial plants with hot glue. As an example, a wire inserted into plastic (or surgical) tubing makes a very sturdy stem onto which artificial lily pads, cattail tails, etc. are secured. But by making the tubing and wire stems shorter, a variety of weedy grasses can be suggested using the blades of artificial grass as leaves, hot glued to the stems.

Any glue joints can be airbrushed with the PlastiCoat paints and then blended to match the finished colors. Remember that plants vary somewhat in color at the ground level and at the outer tips of their leaves and stems, so paint them accordingly. Also varying colors slightly will give an impression of plants being of different maturitites since the colors darken or fade, depending on the plant's age. Dying or wilting plants usually begin doing so at the farthest points from their heart or root system, at the tips of leaves and stems.

With only slight variations in the shape of the leaf, an accurate facsimile can be made of a palmetto fan. Once shaped, cut to size, and painted, the leaves can be fanned out and hot glued or even stapled together at the bottom and attached to a stem. Touch up the union of the stem and the fan with wax or hot glue and then airbrush with HP10 Palmetto Green to further blend the attachment.

With these procedures, it's a simple matter to make your competition believe the grass is greener on *your* side of the fence.

# Using Artificial Plants

Many commercial artificial or synthetic plants such as the so-called silk flowers may lend themselves well to use in habitats, but should be "fine tuned" by trimming frayed or fringed leaves and dipping them individually in melted wax. Rarely are they good enough to be used "as is" in a quality arrangement.

Knowing your reference material well will also help you to modify petals or leaves from certain plants and accurately adapt them to others. As an example, the white petals from a silk tulip can be reshaped with a sharp scissors to the configuration of a water lily petal, then dipped in white wax, and then touched up or tinted with an airbrush. These petals can then be arranged and shaped in the natural circle of the flower and secured with melted wax applied like solder with a hot tool such as a wood burner or soldering iron. (The tool should have a *very* sharp point for fine detailing.)

Other plastic plants purchased "off-the-rack" can be used in similar ways. The plastic leaves of aquarium plants can be clipped, trimmed to different shapes, painted, and otherwise modified to use in habitat construction. With a little effort, even low-quality plastic plants can be improved enough to use in artistic habitats. Don't overlook these plants as a valuable time-saving alternative to constructing plants from scratch.

# Flower Construction

Throughout the Science and Biological Centers, Natural History Museums and Arboretums of the Southcentral United States, one name is synonomous with 'top quality' museum habitats. Mozelle Funderburk is a Georgia based artist whose experience and ability for recreating accurate, breathtakingly realistic habitat environments for museum settings, verges on the legendary.

Mozelle's range of knowledge and skill using a multitude of methods for reproducing extremely life-like examples of flora and fauna is without measure, and ranks internationally among the best. We are delighted to offer her techniques to our readers.

Working in a variety of media, Mozelle provides us with the following examples of her craft and the procedures for accomplishing their production.

"The most important consideration when recreating any specimen, be it plant or animal, is *always* using top quality reference materials." No artist should ever try working from memory. It just can't be done accurately enough.

Prior to actually starting a project, collect and view all reference materials. Make notes and sketches highlighting the most pertinent features. If working with subjects accessable to photography, record your subject from all angles required to keep it in proper perspective. Be sure to use a flash to keep out shadows and to keep details sharp and well defined. (The use of a white poster board backdrop may be helpful to take out unwanted background details and focus attentions on the immediate subject.)

The *best* reference is most often the living specimen itself. Zoos, game farms, arboretums, etc. may be the best sources for life studies. But when not available, a good public library becomes your next best source of excellent reference materials. A multitude of books and magazines can be found there describing in detail most any subject you care to create.

Once all reference materials have been collected, the next step is to determine which materials are best suited to create exact duplicate molds, castings, and constructions of virtually anything visible to the naked eye.

Some of the methods still used and prefered today have been in use since the turn of the century. One of Mozelle's favorite materials to work with is wax. Although it can be the most delicate and fragile to work with, little if anything can beat it for the quality of its texture and luminescence. Wax, when mixed properly and melted to a liquid state, will easily conform to any configuration within the mold. The natural transparency of the wax is also more like leaves and petals of living plants than other casting materials.

The wax used by Mozelle is a custom melting of "about" 5 parts beeswax and 4 parts carnauba to which is added about 2 parts parrafin wax. Once melted, the wax can be tinted to the desired colors by adding Winsor and Newton oil colours frome the tube.

A variety of paper materials can be used for crafting flowers and plants, especially when those papers are dipped into the melted wax mixture. Crepe paper and construction paper come in a multitude of colors, making them particularly adaptable. White bond (typing) paper can be painted with oil paints and used. But even facial or bath tissue, paper towel, or onion skin tracing paper can be used, particularly when a leaf or petal is to appear wilted or dying.

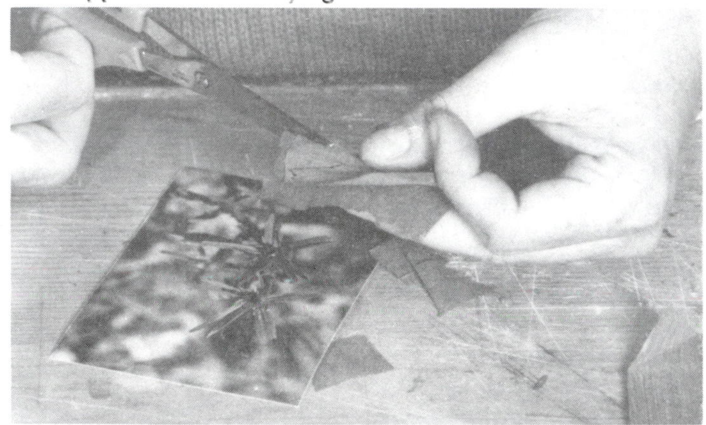

Using reference photos as a guide, Mozelle cuts out the shapes of flower petals from the appropriate color of crepe paper.

She dips each petal in the matching colored waxes and as they cool, she shapes accordingly. Once they are cool, she welds the seams with a wood burner or soldering iron dipped in the correct color wax to seal the points of contact. The flowers and leaves are attached to the stem of the plant by touching the point of the soldering iron to a hot melt glue stick, and transferring the adhesive to the stem.

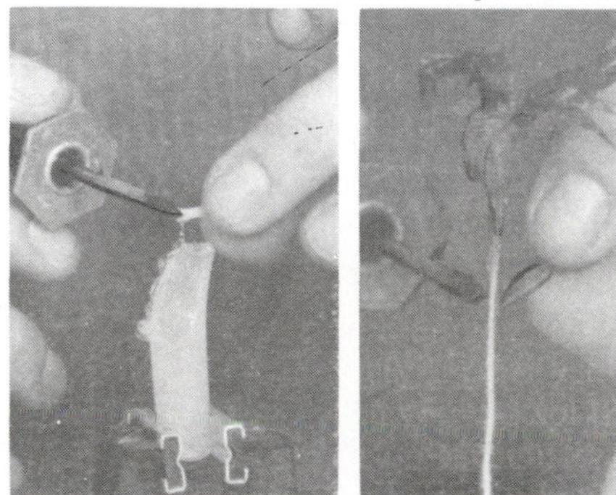

To create the central portions of a flower containing the "filament, stigma, style, stamen and pistil," the use of strings dipped in wax is recommended. Braided nylon used for sewing capes can be ideal for the larger stems, and monofilament fishing line in a variety of weight tests can be used for even more sizes.

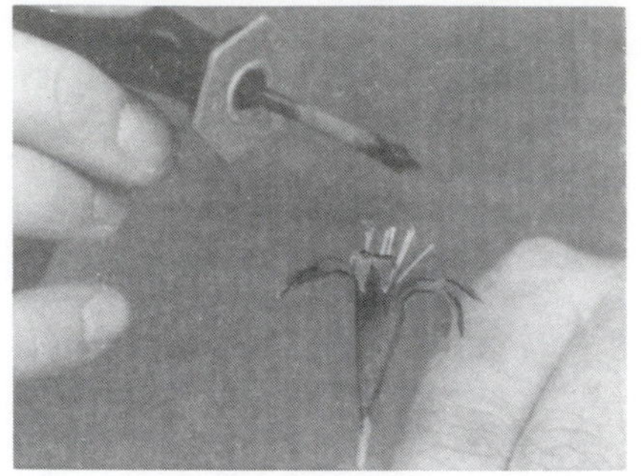

To add the pod or "anther" to the tips of the pistil or stamen, simply add a bead of wax or hot glue to the thread after it has been waxed or glued into the flower's center.

As an alternative to paper and wax flower construction, Mozelle also makes flowers out of Sculpey. Sculpey is a clay-like sculptable product available at most art supply stores. After an item has been formed, it is placed in a kitchen oven to cure in about a half hour. Sculpey can be rolled out quite thin, shaped,

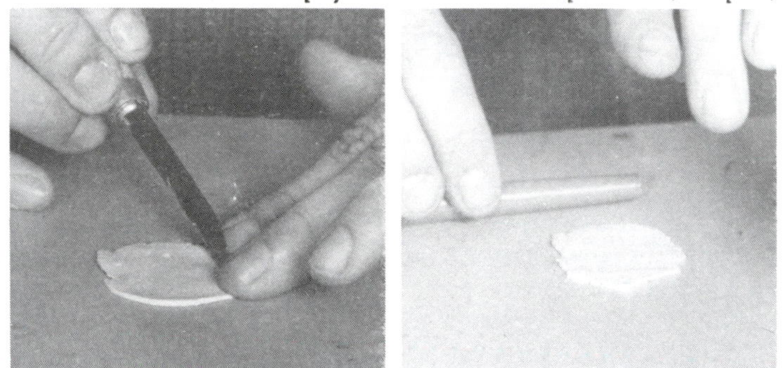

and detailed to look like leaves or flower petals. It can then be placed on a piece of aluminum foil and curled at the edges. After curing in an oven, the Sculpey will become hard. It can then be painted and assembled with other pieces to form a durable flower or plant.

## Using Natural Materials

When using natural materials, such as driftwood, stumps, limbs, shelf or bracket fungus, etc. it is absolutely necessary to treat the piece before making it a part of an indoor display. Most dead or dying plant life plays host to a multitude of insects and their larvae which contribute to the breakdown of the life form back into the eco-system. Unless you want those same insects hatching out into YOUR (or your customer's) eco-system, it's a good idea to fumigate the wood complely.

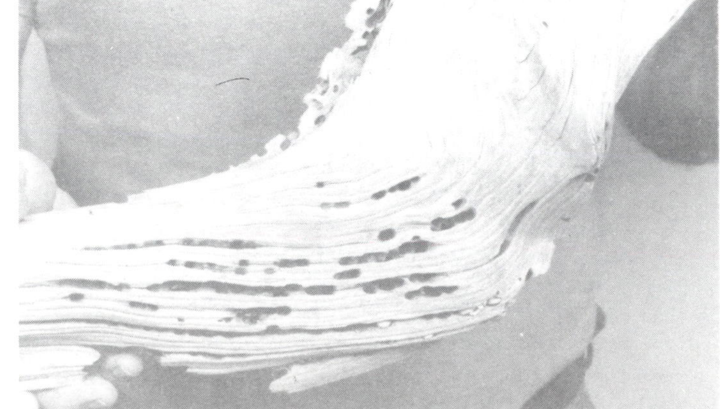

*Natural wood must be fumigated to guard against future insect damage.*

Pest Control companies can provide the best suited products and procedures, as well as advice on the care and handling of the chemicals. In most cases, the pesticides are "TOXIC" and should be handled with caution. In fact, it is advisable to wear protective clothing such as aprons, gloves, and eye guards when using any noxious materials. It may also be well taken advice to wear a repirator to avoid toxic fumes or vapors. To be best prepared, read the "CAUTION" label on the container *before* you use it. Be ready to follow the recommended remedies in the instructions without delay. (It makes simple and logical sense to *know* the treatment for chemicals splashed in your eyes *before* an accident than it does to be trying to figure out what you should be doing with eyes too chemically burned to read instructions.)

Using a plastic bag as large as needed to completely enclose the wood and keep it air-tight, pour or spray the chemical directly onto the piece. Always be careful while handling that you don't perforate or puncture the bag as the affectiveness of its contents depend a lot on keeping their full effect encapsulated and working at full strength, rather than dissiapting into the atmosphere. To better insure this condition, consider using heavy duty-type bags, or doubling or even tripling bags to give you the added strength. Once you have added the pesticide to the bag, *tightly* seal the neck of the bag with wire or nylon tie-wraps. Once it is secured, allow the contents to work undisturbed for about a week or so. This is usually sufficient for the smaller pieces used most by taxidemists.

As an additional safety precaution, include an identification tag listing the chemical inside and add the recommendation that it is for "HANDLING BY AUTHORIZED PERSONNEL ONLY!" If possible, the bagged wood should be kept outside during the fumigation to avoid the possibility of fumes escaping into your room, or insects escaping from the wood.

For larger examples, such as logs, dead trees, etc. it may be necessary and recommendable that they be "tented" and fumigated by a professional exterminator.

Be sure and take advantage of natural materials when building habitats. If a natural item will not deteriorate and does not add too much weight to the exhibit then by all means use it.

Take a cactus for example. The main portion of a cactus will wither and deteriorate if the cactus dies, however, the thorny spines essentially remain the same. In this instance, the main portion of the cactus should be reproduced, however, the actual spines should be used. They will not only look better, but they will save a tremendous amount of time building reproduction spines.

The same holds true for small stones or gravel. Why spend valuable time reproducing them when the artist can use actual stones with very little additional weight?

Many other natural habitat materials can be used. Be sure and bugproof items likely to contain such critters before using them.

# Natural Limbs–Artificial Leaves

In many cases, it may be desirable to use artificial leaves wiht natural limbs. (The procedures for making artificial leaves are described in detail in Chapter 5.) Begin with a fumigated limb and a good variety of leaves to choose from. Mozelle recommends using green florist's tape to wrap and attach the wire stems of the leaf to the limb. Photos or other references of the tree being duplicated should be checked to make certain of the position and number of leaves in each cluster growing from the limb. When the correct number of leaves are attached to the limb with the tape, bend the stem wires to spread the leaves into a pleasing cluster. Remember that most leaves will have their top surface in position to collect the most sunlight.

# Lily Pads

A commercial artist or taxidermist may not be able to justify the expense of time spent in creating from scratch a selection of realistic lily pads. For them, the solution is the Lily Pad Kit available from Wildlife Artist Supply Company.

Three different sizes in a kit of 10 plants come prepainted and ready to attach to the flexible stems and pushed into drilled holes in driftwood or a foam based habitat. The pads are designed to be used either with or without stems (without stems they can be placed between Artificial Water pours for a "floating" effect.

Some artists prefer making their own habitat vegetation. Tom Lenort took the 1984 Best In World Fish honors with a splash scene largemouth bass which included freeze dried lily pad leaves incorporated into the resin surface. Tom merely placed the freeze dried lily pads on the surface of the resin prior to the final pour.

Highly-awarded taxidermist Tom Ridge uses sheet aluminum ("like house siding") for one of his techniques. Cutting out the shape of a pad, Tom leaves a short "tab" at the intersection of the "V" where the stem is attached. This tab is folded over and inserted into the stem later.

To create the creases in the leaf, Tom just folds the thin aluminum over and then flattens it back out by hand. The fold duplicates the details of the leaf. Do this as many times and places as necessary.

Tom makes stems by pouring *Polytranspar*™ PlastiCoat paint (Lily Pad Green HP40) "through" surgical tubing, inserting a wire into the tube, and cutting both to the desired length. He paints the pad with the same color after hot gluing it to the tube where it slipped onto the "tab".

The sheet aluminum is one of the quicker ways of making your own lily pads but is less detailed than most other methods.

# Cactus

There are several methods of creating realistic cactus; freeze-drying, molding, and carving or sculpting being the most common. In almost all cases with the exception of drying which is sometimes less than satisfactory with some forms of cacti, the spines either come off in the mold, or are nonexistent to begin with. To replace them, any of the following methods will work well.

Tom Ridge uses real spines clipped from a cactus in both his plaster molded castings, and his hand-carved foam cactus (which he covers with Sculpall™). These are "pinned" into place and glued with super glue if the cactus is a molding, or just poked into the foam while the Sculpall is still soft if the cactus is carved foam.

Mozelle Funderburk suggests lengths of monofilament fishing line of the approxiamate diameter super glued into holes pinned into the finished cactus *after* it has been painted. Once the spines have dried securely, a sharp scissor is used to cut all the thorns at one time to a uniform length. Cutting at an angle also makes them appear sharper than if cut crosswise and flat tipped.

An alternative to the fishing line spines which is a quicker but less consistent method to use is touching the cactus at the point of a spine attachment with a hot-melt glue gun and then gently pulling away in the direction you want the spine pointing. The glue forms the spines as you lift it away which, like the monofiliment spines, must also be cut to the same lengths with a sharp pair of scissors

Variations between the methods may determine which is most feasible and desirable. The real spines and the line spines are *all* uniform in diameter and shape. The hot glue spines on the other hand are tapered from base to tip and may vary in size depending on how much glue was on the tip of the gun before touching the cactus.

# Molding a Tree

It can be both time consuming and expensive to recreate a complete tree in fiberglass. But when the demand requires, it can be done with a repitition "texture" mold.

Select a tree suited in size and texture to the requirements of the exhibit. Build a Mold Builder (WASCO) or Chicago Latex mold around the entire trunk of the tree and about 2 to 3 feet high.

Over a period of 3 or 4 days, depending on how quickly your mold is drying, apply 8 to 10 more coats using a layer or two of burlap (or similar material) to reinforce the mold.

After the mold is thouroughly cured, use a sharp knife or razor blade to open the mold down one side. Peel the rubber mold away from the tree and rewrap the same area on the tree with heavy duty aluminum foil, pressing the foil into the crevices and details of the bark.

Mix Cab-O-Sil or Aerosil into catalyzed *Polytranspar*™ Lay-up Resin and apply a good even coat to the foil. While the resin is curing but still flexible, cut the casting at the back side and remove it from the tree. Bend the two sides of the cut back together and staple in place.

To use the rubber mold, apply Ceara Mold Release Wax to the latex and then add paste-like consistency catylized resin thickened the same way as previously described to the entire surface of the mold. When it begins to gel, wrap it around the "fiberglass tree trunk" and let it harden completely. Peel the rubber mold away and repeat these steps as often as needed to create as long a tree trunk as required.

Join each section of completed tree with thickened resin (or Ultra Lite) and then prime the completed casting with *Polytranspar* Fiberglass Gray Primer (FP195-G) before painting shadows and details with an airbrush.

When working with long lengths of tree trunks, remember trees have a tendancy to get smaller as they get taller. So with each addition, reduce the diameter slightly.

Exposed joints can be concealed using Ultra Lite or Sculpall to sculpture and texture seams.

Using these methods, entire tree trunks and even their branches can be accurately depicted and detailed.

# Freeze Dry Habitat

***by Terri Chidester***

If a taxidermist has access to a freeze dry machine, it may be advantageous to prepare some habitat materials with this system. Some of the most common habitat item that are freeze dried are the "prop" type specimens. These include small mammals, snakes, amphibians, fish and insects. However, many plants can be well preserved by freeze dying very easily.

The key factors to always remember for success in freeze drying plants are the fiber and moisture content. The more delicate the fiber and the more moisture a specimen contains, the harder it is to maintain the normal structure with the freeze dry process. If shrinkage and brittleness occur, it is an indication that the material is not suited to the freeze dry method. An example of a material that doesn't freeze dry well is cabbage. Although the moisture content is not high, the fibers are extremely delicate, and when dried, it may look perfectly preserved but will crumble with the slightest touch.

With proper use of the specific preservatives designed for freeze dry there should be no danger of insect infestation. The three bath mothproofing system consists of Edolan-U, acidic acid, and denatured alcohol.

Thorough drying of the specimen is another important factor to be aware of. If moisture is still contained in the specimen, the chances are higher that rot and infestation will eventually occur. Also, the risk of brittleness and further shrinkage is high. Experimentation, careful planning, records (such as weight loss each time a specimen is checked for moisture loss) will all determine the success of freeze drying.

As with all specimens to be freeze dried, plants must be completely frozen after the desired position is established. The rigidity maintains the structure and prevents collapsing when the vacuum pressure is resumed inside the chamber. Usually 24 hours is time enough for all habitat materials, and once removed from a freezer the specimen must be immediately set inside the freeze dry chamber. Most materials can be easily positioned on pieces of styrofoam and pinned into place. Simple holders can be constructed to place flowers in (which keep them separated from each other to reduce the chances of damage) while in the chamber.

Flowering and blooming habitat material such as roses and carnations are very easily preserved by freeze drying and the end result is as life-like and much more authentic looking than most silk flowers. Aside from the fiber and moisture content, the color preservation of these specimens is a factor to be controlled. This is done by providing a complete absence of light in the specimen chamber. This is easily accomplished by covering the glass from the outside with dark construction paper.

Some good blooming flowers for freeze drying include most lilies, blackeyed Susan, daisies, baby's breath, and cactus flowers. The time of harvest of a flower is an important factor in determining the final result. Being immature or too mature will result in unsatisfactory fiber stucture and will be too weak which results in shrinkage and/or brittleness.

The following plants are good candidates for freeze drying: brocolli, seeds and nuts, carrots and other root vegetables, berries such as holly and mistletoe, rosehips and similar buds, cauliflower and brocolli, and pomegranates. Mushrooms and fungus freeze dry extremely well (and fast—taking only a couple days to dry completely), but because of their delicate fiber structure, they are very fragile. The preparation time is minimal as they very rarely need painting since they retain most of their natural color. Moss, lichen and similar materials can be air dried with little natural color loss; therefore freeze drying is only a quicker way if a machine is readily available.

CHAPTER 10

# Working With Water

*Scenes which create the illusion of moving water that has been frozen in time are extremely popular with the public. This chapter explains how these effects are created and gives several examples of different techniques used by today's top wildlife artists.*

## Artificial Water

***by Bob Williamson***

Perhaps nothing has created as big of a splash (pardon the play on words) to the taxidermy and carving industry as the recent incorporation of artificial water and ice into a variety of naturalistic renderings. Collectors of wildlife art and sportsmen both are very receptive to the idea and have created a terrific demand for these types of scenes, often paying top dollar for them. We have devoted a large portion of this book to this subject, as we feel that it is an integral part of creating professional quality habitat scenes—and every competent wildlife artist should know how to create them.

There are basically two types of materials that we recommend using to recreate water and ice: epoxies and casting resins. Both have similar properties in that they are two-component systems consisting of a resin and hardener (or catalyst). Basically the two components require mixing with each other either by ratios, percentage of volume, or by percentage of weight. "All" require a very "thorough" mixing with both components being completely blended with each other in order to properly harden (or "cure"). Chemically, what happens is that the two components will "cross link" on a molecular level when they come in contact with each other and change from a liquid to a solid state. This is commonly referred to as "gelling" in the initial stages and once completely cured, it is called "polymerization."

Mixing improper proportions and/or not stirring thoroughly enough can result in a variety of problems such as: having soft spots that will not dry or harden, cracking, yellowing, hazing, warping, and in some cases excessive heat may build to a point where they can actually self-ignite and burn.

Suffice it to say that in order to enjoy any degree of success with this phase of exhibit building, one will have to mix the components "exactly" as dictated by the instructions on the label. This means "specifically" that before starting on any project one should sit down and very carefully read the directions of all materials used and even more importantly read (and observe!) the safety precautions! This holds true with any chemical that is used (see Chapter 1: Safety).

## Measuring

When measuring by weight, the most precise and easiest method to use is a balance (or "beam") scale. These sensitive scales accurately indicate weights in grams and milligrams. To use one, an empty container is first put on the scale and the scale is calibrated back to zero to compensate for the weight of the empty container. Next the scale is set to the desired weight and the container is filled with the liquid (or powder) until the scale indicator reaches the preset weight. There are also several digital electronic scales appearing on the market that are excellent (which work similar to beam scales).

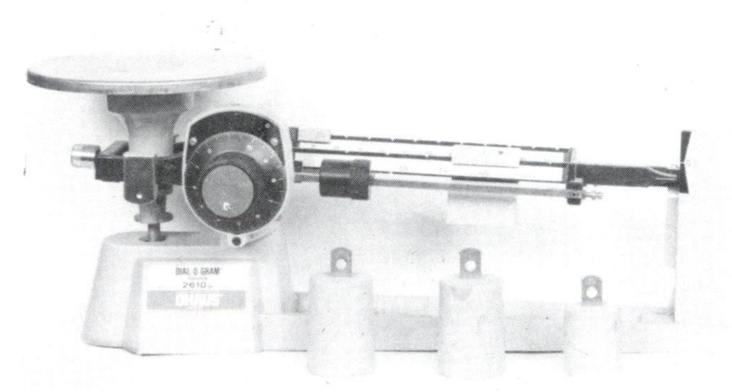

*A balance scale.*

Although these scales are easy to use and very accurate, they are also moderately expensive. For that reason many choose another method of measuring—measuring by volume. Accurate measurements can be made by volume liquid measure.

Measuring devices that are used for liquid measure are much less expensive. Glass syringes, pipettes, graduated beakers, and cups all work well for this method of measuring. Most will measure in milliliters (ml or cc), liters, fluid ounces (fl. oz.), gallons, quarts, pints, etc. A conversion guide to these measurements is provided for your quick reference on page 101.

## Ratios & Percentages Are Easy!

Since epoxies, polyesters, casting resins, urethanes, silicone RTV's, etc., all contain two components (resin and catalyst), one must determine how much catalyst is required to react with the resin. To calculate this, first determine: (1) how much total liquid is needed, and (2) what percentage of catalyst is required to activate and harden the resin. (The directions on the container will tell you the correct ratio and percentage).

Some require a 1:1 ratio. This simply means to use equal amounts of both components. The easiest way to do that is to obtain two identical containers, fill them both with the exact same amount, and then mix the ingredients of the two containers together.

Others will require percentages of catalyst per resin. Think of percentages as a part (or proportion) of a whole. For example, if you desire to know what 50 percent of 80 is, you would be calculating what proportion 50 percent is to 80. Of course, the correct figure is 40. To arrive at the correct answer, simply change 50 percent to its decimal equivalent (.50) and multiply it times 80 (.50 × 80 = 40).

Many materials call for 1 percent catalyst to be mixed with the resin. First, decide how much resin is needed. If, for example, you need 100 ccs of resin, then multiply 1 percent times 100 ccs (or .01 × 100 ccs = 1 cc). 100 ccs of resin would require 1 cc (ml.) of catalyst for a 1 percent mixture.

If 1¼ percent is needed, multiply .0125 × 100ccs = 1.25 ccs of catalyst for a 1¼ percent catalyst mixture for 100 ccs, and so forth. The method of calculating other ratios by volume would be the same.

Take a 7 to 1 ratio for example. This means for every 7 parts of one component, 1 part of the other component is required. The easiest way to accomplish this is to use graduated beakers or mixing containers. Pour component (A) requiring 7 parts to the 7th graduation and fill component (B) to the 8th line to complete the mixture as shown in the illustration below.

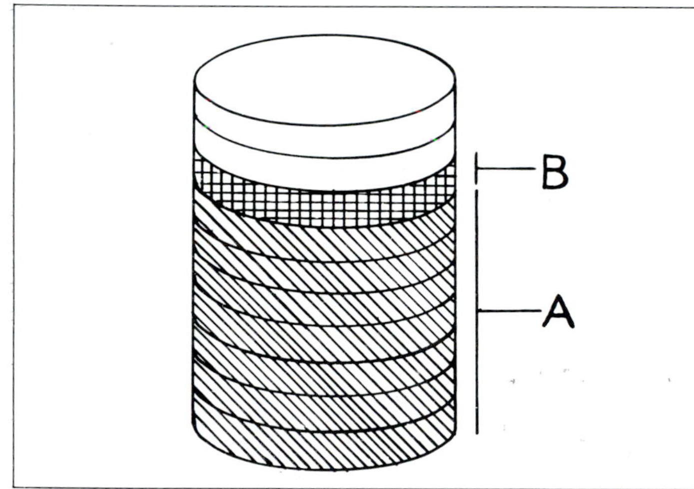


If more or less is required or a container is not graduated, then mark the desired amount on a container and measure the distance. Now divide the distance by 8 and mark the increments on the container as shown in the illustration below.

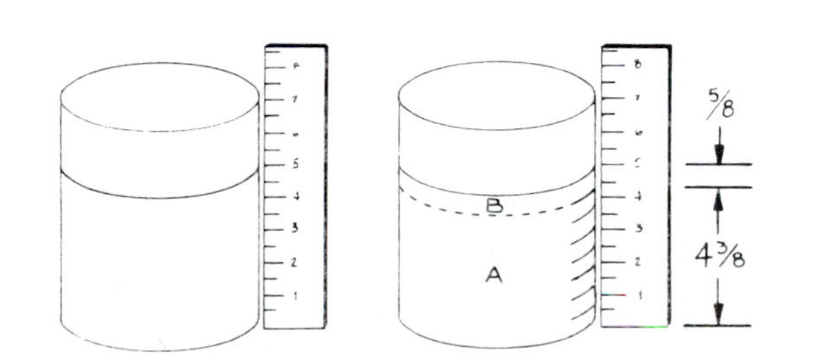


1. Measure the desired amount (= 5")
2. Divide the desired amount by the mixing ratio (5" ÷ 8= .625 or ⅝")
3. Mark lines for component B (1= ⅝")
4. Mark line for component A (7= 4 ⅜")
5. Fill component A to the 7th graduation that you have marked and fill component B to the 8th. This will produce a 7:1 ratio for 5" of liquid. Note: the container must be the same size on the bottom as the top. Tapered containers will vary.

OK NO

A convenient chart for systems with a 7:1 mixing ratio is listed below. *Polytranspar*™ Competition Wet Look Gloss is a 7:1 urethane gloss coat.

## Mixing Ratios of *Polytranspar*™ Competition Wet Look Gloss (FP241)

| Desired Quanitity of Competition Wet Look Gloss (FP241) fluid oz. | Quantity of "A" Component "RESIN" | Quantity of "B" Component "HARDENER" |
|---|---|---|
| 4 fluid oz. | 3½ fluid oz. | ½ fl oz./15 ml |
| 8 fluid oz. | 7 fluid oz. | 1 fluid oz. |
| 12 fluid oz. | 10½ fluid oz. | 1½ fluid oz. |
| 16 fl. oz. **(1 pint)** | 14 fluid oz. | 2 fluid oz. |
| 20 fluid oz. | 17½ fluid oz. | 2½ fluid oz. |
| 24 fluid oz. | 21 fluid oz. | 3 fluid oz. |
| 28 fluid oz. | 24½ fluid oz. | 3½ fluid oz. |
| 32 fl. oz. **(1 quart)** | 28 fluid oz. | 4 fluid oz. |
| 36 fluid oz. | 31½ fluid oz. | 4½ fluid oz. |
| 40 fluid oz. | 35 fluid oz. | 5 fluid oz. |
| 44 fluid oz. | 38½ fluid oz. | 5½ fluid oz. |
| 48 fluid oz. | 42 fluid oz. | 6 fluid oz. |
| 52 fluid oz. | 45½ fluid oz. | 6½ fluid oz. |
| 56 fluid oz. | 49 fluid oz. | 7 fluid oz. |
| 60 fluid oz. | 52½ fluid oz. | 7½ fluid oz. |
| 64 fl. oz. **(½ gal.)** | 56 fluid oz. | 8 fluid oz. |
| 68 fluid oz. | 59½ fluid oz. | 8½ fluid oz. |
| 72 fluid oz. | 63 fluid oz. | 9 fluid oz. |
| 76 fluid oz. | 66½ fluid oz. | 9½ fluid oz. |
| 80 fluid oz. | 70 fluid oz. | 10 fluid oz. |
| 84 fluid oz. | 73½ fluid oz. | 10½ fluid oz. |
| 88 fluid oz. | 77 fluid oz. | 11 fluid oz. |
| 92 fluid oz. | 80½ fluid oz. | 11½ fluid oz. |
| 96 fl. oz. **(¾ gal.)** | 84 fluid oz. | 12 fluid oz. |
| 100 fluid oz. | 87½ fluid oz. | 12½ fluid oz. |
| 104 fluid oz. | 91 fluid oz. | 13 fluid oz. |
| 108 fluid oz. | 94½ fluid oz. | 13½ fluid oz. |
| 112 fluid oz. | 98 fluid oz. | 14 fluid oz. |
| 116 fluid oz. | 101½ fluid oz. | 14½ fluid oz. |
| 120 fluid oz. | 105 fluid oz. | 15 fluid oz. |
| 124 fluid oz. | 108½ fluid oz. | 15½ fluid oz. |
| 128 fl. oz. **(1 gallon)** | 112 fluid oz. | 16 fluid oz. |

Most polyester resin/catalyst systems which use MEKP (methyl ethyl ketone peroxide) as their catalyst, are mixed at ratios from .25% to 2.00% (depending on temperature, amount of mixture, humidity, and length of cure time required). MEKP is normally colorless (clear) for use in Artificial Water situations. When used specifically for Polyester Resin, it is generally dyed a deep purple-red to allow the lay-up operator to see how well the MEKP is mixing with the resin. Although the clear MEKP and the dyed MEKP are the same chemical (and produce the same results of hardening) DO NOT use the dyed MEKP in Artificial Water (unless you desire pink water)!

The following pages contain a table showing the resin/catalyst ratios of 1/2 oz. to 1 gallon of resin. This table can be used with *Polytranspar* Artificial Water, Polyester Resin, Gelcoat, etc.

# Mixing Ratios of Polyester Resin/Catalyst Systems

Note: Catalyst quantities are rounded off to the nearest ⅒ ml. (For measuring purposes, ml is equal to cc.)

| Approximate Quanitity Mixed Resin (fluid oz.) | Quantity of Resin | Quantity of Catalyst @ .25% | Quantity of Catalyst @ .50% | Quanitiy of Catalyst @ 1.0% | Quantity of Catalyst @ 2.0% |
|---|---|---|---|---|---|
| ½ fluid oz. | ½ fluid oz. | 1 drop | 2 drops/0.1 ml | 3 drops/0.2 ml | 6 drops/0.3 ml |
| 1 fluid oz. | 1 fluid oz. | 2 drops/0.1 ml | 3 drops/0.2 ml | 6 drops/0.3 ml | 13 drops/0.6 ml |
| 1½ fluid oz. | 1½ fluid oz. | 2 drops/0.1 ml | 5 drops/0.2 ml | 10 drops/0.5 ml | 19 drops/0.9 ml |
| 2 fluid oz. | 2 fluid oz. | 3 drops/0.2 ml | 7 drops/0.3 ml | 13 drops/0.6 ml | 1.2 ml |
| 2½ fluid oz. | 2½ fluid oz. | 4 drops/0.2 ml | 8 drops/0.4 ml | 16 drops/0.7 ml | 1.5 ml |
| 3 fluid oz. | 3 fluid oz. | 5 drops/0.2 ml | 10 drops/0.4 ml | 20 drops/0.9 ml | 1.8 ml |
| 3½ fluid oz. | 3½ fluid oz. | 6 drops/0.3 ml | 12 drops/0.5 ml | 1.0 ml | 2.1 ml |
| 4 fluid oz. | 4 fluid oz. | 7 drops/0.3 ml | 14 drops/0.6 ml | 1.2 ml | 2.4 ml |
| 4½ fluid oz. | 4½ fluid oz. | 7 drops/0.3 ml | 15 drops/0.7 ml | 1.3 ml | 2.6 ml |
| 5 fluid oz. | 5 fluid oz. | 8 drops/0.4 ml | 17 drops/0.7 ml | 1.5 ml | 3.0 ml |
| 5½ fluid oz. | 5½ fluid oz. | 9 drops/0.4 ml | 19 drops/0.8 ml | 1.6 ml | 3.2 ml |
| 6 fluid oz. | 6 fluid oz. | 10 drops/0.4 ml | 20 drops/0.9 ml | 1.8 ml | 3.5 ml |
| 6½ fluid oz. | 6½ fluid oz. | 10 drops/0.5 ml | 1.0 ml | 2.0 ml | 3.9 ml |
| 7 fluid oz. | 7 fluid oz. | 11 drops/0.5 ml | 1.1 ml | 2.1 ml | 4.1 ml |
| 7½ fluid oz. | 7½ fluid oz. | 12 drops/0.5 ml | 1.1 ml | 2.2 ml | 4.4 ml |
| 8 fl. oz. **(1/2 pint)** | 8 fluid oz. | 13 drops/0.6 ml | 1.2 ml | 2.4 ml | 4.7 ml |
| 8½ fluid oz. | 8½ fluid oz. | 14 drops/0.6 ml | 1.2 ml | 2.5 ml | 5.0 ml |
| 9 fluid oz. | 9 fluid oz. | 15 drops/0.7 ml | 1.3 ml | 2.7 ml | 5.3 ml |
| 9½ fluid oz. | 9½ fluid oz. | 15 drops/0.7 ml | 1.4 ml | 2.8 ml | 5.6 ml |
| 10 fluid oz. | 10 fluid oz. | 16 drops/0.7 ml | 1.5 ml | 3.0 ml | 5.9 ml |
| 10½ fluid oz. | 10½ fluid oz. | 17 drops/0.8 ml | 1.5 ml | 3.1 ml | 6.2 ml |
| 11 fluid oz. | 11 fluid oz. | 18 drops/0.8 ml | 1.6 ml | 3.2 ml | 6.4 ml |
| 11½ fluid oz. | 11½ fluid oz. | 19 drops/0.8 ml | 1.7 ml | 3.4 ml | 6.8 ml |
| 12 fluid oz. | 12 fluid oz. | 20 drops/0.9 ml | 1.8 ml | 3.5 ml | 7.0 ml |
| 12½ fluid oz. | 12½ fluid oz. | 20 drops/0.9 ml | 1.8 ml | 3.7 ml | 7.4 ml |
| 13 fluid oz. | 13 fluid oz. | 21 drops/1.0 ml | 1.9 ml | 3.8 ml | 7.7 ml |
| 13½ fluid oz. | 13½ fluid oz. | 1.0 ml | 2.0 ml | 4.0 ml | 8.0 ml |
| 14 fluid oz. | 14 fluid oz. | 1.0 ml | 2.1 ml | 4.1 ml | 8.3 ml |
| 14½ fluid oz. | 14½ fluid oz. | 1.0 ml | 2.1 ml | 4.3 ml | 8.6 ml |
| 15 fluid oz. | 15 fluid oz. | 1.1 ml | 2.2 ml | 4.4 ml | 8.8 ml |
| 15½ fluid oz. | 15½ fluid oz. | 1.1 ml | 2.3 ml | 4.6 ml | 9.1 ml |
| 16 fluid oz. **(1 pint)** | 16 fluid oz. | 1.2 ml | 2.4 ml | 4.7 ml | 9.4 ml |
| 16½ fluid oz. | 16½ fluid oz. | 1.2 ml | 2.4 ml | 4.9 ml | 9.7 ml |
| 17 fluid oz. | 17 fluid oz. | 1.3 ml | 2.5 ml | 5.0 ml | 10.0 ml |
| 17½ fluid oz. | 17½ fluid oz. | 1.3 ml | 2.6 ml | 5.2 ml | 10.3 ml |
| 18 fluid oz. | 18 fluid oz. | 1.3 ml | 2.7 ml | 5.3 ml | 10.6 ml |
| 18½ fluid oz. | 18½ fluid oz. | 1.4 ml | 2.7 ml | 5.5 ml | 10.9 ml |
| 19 fluid oz. | 19 fluid oz. | 1.4 ml | 2.8 ml | 5.6 ml | 11.2 ml |
| 19½ fluid oz. | 19½ fluid oz. | 1.4 ml | 2.9 ml | 5.8 ml | 11.5 ml |
| 20 fluid oz. | 20 fluid oz. | 1.5 ml | 3.0 ml | 5.9 ml | 11.8 ml |
| 20½ fluid oz. | 20½ fluid oz. | 1.5 ml | 3.0 ml | 6.0 ml | 12.0 ml |
| 21 fluid oz. | 21 fluid oz. | 1.5 ml | 3.1 ml | 6.2 ml | 12.4 ml |
| 21½ fluid oz. | 21½ fluid oz. | 1.6 ml | 3.2 ml | 6.3 ml | 12.7 ml |
| 22 fluid oz. | 22 fluid oz. | 1.6 ml | 3.2 ml | 6.5 ml | 13.0 ml |
| 22½ fluid oz. | 22½ fluid oz. | 1.6 ml | 3.3 ml | 6.6 ml | 13.3 ml |
| 23 fluid oz. | 23 fluid oz. | 1.7 ml | 3.4 ml | 6.8 ml | 13.6 ml |
| 23½ fluid oz. | 23½ fluid oz. | 1.7 ml | 3.5 ml | 7.0 ml | 13.9 ml |
| 24 fluid oz. | 24 fluid oz. | 1.8 ml | 3.5 ml | 7.1 ml | 14.2 ml |

## Mixing Ratios of Polyester Resin/Catalyst Systems (Continued)

| Approximate Quanitity Mixed Resin (fluid oz.) | Quantity of Resin | Quantity of Catalyst @ .25% | Quantity of Catalyst @ .50% | Quanitiy of Catalyst @ 1.0% | Quantity of Catalyst @ 2.0% |
|---|---|---|---|---|---|
| 24½ fluid oz. | 24½ fluid oz. | 1.8 ml | 3.6 ml | 7.2 ml | 14.4 ml |
| 25 fluid oz. | 25 fluid oz. | 1.8 ml | 3.7 ml | 7.4 ml | 14.8 ml |
| 25½ fluid oz. | 25½ fluid oz. | 1.9 ml | 3.8 ml | 7.5 ml | 15.0 ml |
| 26 fluid oz. | 26 fluid oz. | 1.9 ml | 3.8 ml | 7.6 ml | 15.3 ml |
| 26½ fluid oz. | 26½ fluid oz. | 1.9 ml | 3.9 ml | 7.8 ml | 15.6 ml |
| 27 fluid oz. | 27 fluid oz. | 2.0 ml | 4.0 ml | 8.0 ml | 16.0 ml |
| 27½ fluid oz. | 27½ fluid oz. | 2.0 ml | 4.0 ml | 8.1 ml | 16.2 ml |
| 28 fluid oz. | 28 fluid oz. | 2.1 ml | 4.1 ml | 8.3 ml | 16.5 ml |
| 28½ fluid oz. | 28½ fluid oz. | 2.1 ml | 4.2 ml | 8.4 ml | 16.8 ml |
| 29 fluid oz. | 29 fluid oz. | 2.1 ml | 4.3 ml | 8.6 ml | 17.1 ml |
| 29½ fluid oz. | 29½ fluid oz. | 2.2 ml | 4.3 ml | 8.7 ml | 17.4 ml |
| 30 fluid oz. | 30 fluid oz. | 2.2 ml | 4.4 ml | 8.9 ml | 17.7 ml |
| 30½ fluid oz. | 30½ fluid oz. | 2.2 ml | 4.5 ml | 9.0 ml | 18.0 ml |
| 31 fluid oz. | 31 fluid oz. | 2.3 ml | 4.6 ml | 9.1 ml | 18.3 ml |
| 31½ fluid oz. | 31½ fluid oz. | 2.3 ml | 4.6 ml | 9.3 ml | 18.6 ml |
| 32 fl. oz. **(1 quart)** | 32 fluid oz. | 2.4 ml | 4.7 ml | 9.4 ml | 18.9 ml |
| 33 fluid oz. | 33 fluid oz. | 2.4 ml | 4.9 ml | 9.7 ml | 19.5 ml |
| 34 fluid oz. | 34 fluid oz. | 2.5 ml | 5.0 ml | 10.0 ml | 20.0 ml |
| 35 fluid oz. | 35 fluid oz. | 2.6 ml | 5.2 ml | 10.3 ml | 20.6 ml |
| 36 fluid oz. | 36 fluid oz. | 2.7 ml | 5.4 ml | 10.6 ml | 21.2 ml |
| 37 fluid oz. | 37 fluid oz. | 2.8 ml | 5.6 ml | 10.9 ml | 21.8 ml |
| 38 fluid oz. | 38 fluid oz. | 2.9 ml | 5.7 ml | 11.3 ml | 22.4 ml |
| 39 fluid oz. | 39 fluid oz. | 2.9 ml | 5.9 ml | 11.6 ml | 23.0 ml |
| 40 fluid oz. | 40 fluid oz. | 3.0 ml | 6.0 ml | 11.9 ml | 23.6 ml |
| 41 fluid oz. | 41 fluid oz. | 3.1 ml | 6.1 ml | 12.2 ml | 24.2 ml |
| 42 fluid oz. | 42 fluid oz. | 3.2 ml | 6.2 ml | 12.4 ml | 24.8 ml |
| 43 fluid oz. | 43 fluid oz. | 3.2 ml | 6.3 ml | 12.7 ml | 25.4 ml |
| 44 fluid oz. | 44 fluid oz. | 3.3 ml | 6.5 ml | 13.0 ml | 26.0 ml |
| 45 fluid oz. | 45 fluid oz. | 3.4 ml | 6.6 ml | 13.3 ml | 26.6 ml |
| 46 fluid oz. | 46 fluid oz. | 3.5 ml | 6.8 ml | 13.6 ml | 27.1 ml |
| 47 fluid oz. | 47 fluid oz. | 3.5 ml | 7.0 ml | 13.9 ml | 27.7 ml |
| 48 fluid oz. | 48 fluid oz. | 3.6 ml | 7.1 ml | 14.1 ml | 28.3 ml |
| 49 fluid oz. | 49 fluid oz. | 3.7 ml | 7.2 ml | 14.5 ml | 28.9 ml |
| 50 fluid oz. | 50 fluid oz. | 3.8 ml | 7.4 ml | 14.8 ml | 29.5 ml (1 fl. oz.) |
| 51 fluid oz. | 51 fluid oz. | 3.8 ml | 7.5 ml | 15.0 ml | 30.0 ml |
| 52 fluid oz. | 52 fluid oz. | 3.9 ml | 7.7 ml | 15.3 ml | 30.7 ml |
| 53 fluid oz. | 53 fluid oz. | 4.0 ml | 7.8 ml | 15.6 ml | 31.3 ml |
| 54 fluid oz. | 54 fluid oz. | 4.0 ml | 8.0 ml | 16.0 ml | 31.9 ml |
| 55 fluid oz. | 55 fluid oz. | 4.1 ml | 8.1 ml | 16.2 ml | 32.5 ml |
| 56 fluid oz. | 56 fluid oz. | 4.2 ml | 8.3 ml | 16.5 ml | 33.0 ml |
| 57 fluid oz. | 57 fluid oz. | 4.3 ml | 8.4 ml | 16.8 ml | 33.6 ml |
| 58 fluid oz. | 58 fluid oz. | 4.3 ml | 8.6 ml | 17.1 ml | 34.2 ml |
| 59 fluid oz. | 59 fluid oz. | 4.4 ml | 8.7 ml | 17.4 ml | 34.8 ml |
| 60 fluid oz. | 60 fluid oz. | 4.5 ml | 8.8 ml | 17.7 ml | 35.4 ml |
| 61 fluid oz. | 61 fluid oz. | 4.6 ml | 9.0 ml | 18.0 ml | 36.0 ml |
| 62 fluid oz. | 62 fluid oz. | 4.6 ml | 9.1 ml | 18.3 ml | 36.6 ml |
| 63 fluid oz. | 63 fluid oz. | 4.7 ml | 9.3 ml | 18.6 ml | 37.2 ml |
| 64 fl. oz. **(1/2 Gal.)** | 64 fluid oz. | 4.8 ml | 9.4 ml | 18.9 ml | 37.8 ml |
| 96 fl. oz. **(3/4 Gal.)** | 96 fluid oz. | 7.2 ml | 14.2 ml | 28.3 ml | 56.6 ml |
| 128 fl. oz. **(1 Gallon)** | 128 fluid oz. | 9.4 ml | 18.9 ml | 37.8 ml | 75.5 ml |

# Customary to Metric Conversions

(APPROXIMATE)

## Units of Capacity

| CUSTOMARY | METRIC |
|---|---|
| fluid ounce (fl. oz.) = | 29.573 ml |
| pint (pt.) = | .473 liter = 473 ml |
| quart (qt.) = | .946 liter = 946 ml |
| gallon (gal.) = | 3.785 liters = 3,785 ml |

| METRIC | CUSTOMARY |
|---|---|
| milliliter (ml) = | .034 fl. oz. (liquid) |
| 1000 ml = 1 Liter (l) = | 1.057 quart (liquid) |
| 100 liters = 1 hectoliter (hl) = | 26.418 gallons (liquid) |

**CUSTOMARY**

1 tablespoon (tbs.) = 3 teaspoons (tsp.) = 1/2 fluid ounce
1 cup = 8 fl. oz. = 1/2 pint
1 pint = 16 fl. oz. = 1/2 quart
1 quart = 32 fl. oz. = 2 pints
1 gallon = 128 fl. oz. = 4 quarts = 8 pints

**METRIC**

1 mililiter (ml) = 22 drops (of water)
10 mililiters (ml) = 1 centiliter (cl)
10 centiliters = 1 deciliter (dl) = 100 mililiters
10 deciliters = 1 liter (l) = 1,000 mililiters
10 liters = 1 dekaliter (dal)
10 dekaliters = 1 hectoliter (hl) = 100 liters
10 hectoliters = 1 kiloliter (kl) = 1,000 liters

## Units of Weight

| CUSTOMARY | METRIC |
|---|---|
| Grain = | .065 gram |
| Ounce (oz.) = 437.5 grains = | 28.530 grams |
| Pound (lb.) = 16 oz. = 7000 grains = | .454 kilograms |

| METRIC | CUSTOMARY |
|---|---|
| gram (g) = | .035 ounce |
| kilogram (kg) = 1000 grams = | 2.205 lb. |

**METRIC**

10 milligrams (mg) = 1 centigram (cg)
10 centigrams = 1 decigram (dc) = 100 milligrams
10 decigrams = 1 gram (g) = 1,000 milligrams
10 grams = 1 dekagram (dag)
10 dekagrams = 1 hectogram (hg) = 100 grams
10 hectograms = 1 kilogram (kg) = 1,000 grams
1,000 kilograms = 1 metric ton (t)

# Safety

The materials used to create artificial water are potentially dangerous to use. Thoroughly heed all safety precautions and directions on the containers of materials used and observe all of them. Small children should not use these materials without adult supervision. Adequate ventilation is a must! Proper storage and fire precautions must be followed to the letter.

Please read the safety chapter (Chapter 1) in its entirety prior to actually constructing any of the projects listed. Building an artificial habitat with water scenery incorporated into it can be fun or it can turn into a nightmare. Observe the safety precautions and it will be fun!

# Mixing Resin and Catalyst

Once the correct proportions of catalyst and resin have been measured, they must be mixed together. Measuring the correct ratios of catalyst and resin will be worthless if the material is not properly mixed.

Using a stirring tool such as a spatula, tongue depressor or paint mixer attached to a drill, thoroughly stir the materials together. (Note: if using a drill to stir flammable materials use an "air" powdered drill or Foredom tool with motor away from the mixture. Electric drills create sparks and could possibly set off a fire.)

Be sure and stir the sides of the container as well as the bottom (see diagram below).

REMEMBER, if the resin and catalyst are not completely mixed together, the mixture will not harden properly!

# Factors Affecting Curing Time

The mixing together of the two components will cause a chemical reaction which will form a tight molecular bond. Once this process is complete the mixture will have changed from a liquid to a solid.

Several factors can affect how fast that this process occurs. Catalyst/resin ratio, heat, and humidity are three very important factors.

As a rule, the artist will obtain the best results by keeping the temperature of the room and all materials used at a temperature level of 70° to 77° F.

Materials will vary but *Polytranspar*™ Artificial Water generally works best at a 1% ratio of catalyst to resin. Read the directions of any materials and use the suggested ratios. (Note: Epoxies such as Envirotex and Ultra Glo are best used strictly at the recommended 1:1 ratio as they self level and cure at very slow rates anyway.)

Humidity should be low for best results. High humidity can slow down the dry time substantially. This is one reason that artists in the humid south will not obtain the same length of drying time as an artist in the dryer west even though they use the exact same mixing ratio of catalyst and resin.

To add more to the confusion, the shelf life of the catalyst and resin can also be a factor. It can become less potent and will eventually harden with age. Always store this type of product in a cool dry place; never store them directly in the sun. Always tighten the lids securely to avoid moisture problems. Don't buy more than you will be using in a two or three month period of time. Most products of this nature will have a 6 month to 1 year shelf life if properly stored; less if not.

# Modifying the Curing Time

It is often desirable to speed up the curing time of artificial water in order to create special effects. It is also helpful at times to slow down the curing time to allow more working time or reduce the heat generated by the chemical reaction. Artists use a variety of techniques to adjust the speed of the curing time up or down.

Before attempting to adjust the curing time, it will be necessary to first make a test mixture at the recommended catalyst-resin ratio in order to determine how fast it will gel and harden under normal conditions. Start with a very small mixture with a 1% catalyst ratio, and have all the materials at the same room temperature. Write down both the gel time (when it begins or thickens) and the time that it takes to completely harden.

Next try a 1½% ratio of the same sized mixture at the same room temperature and time it. By increasing the catalyst, the mixture will gel and harden quicker and, of course, reducing the amount of catalyst will slow it down. The maximum percentage of catalyst is approximately 2% and the minimum is .025% at a 70° to 77° temperature. A good rule to follow is to "always" test the material first, *before* using it in a display. Do not add excessive amounts of catalyst or it will overheat and self ignite. If you don't add enough, it will not completely harden.

Another method of adjusting the reaction time is with the temperature. Heating the resin and catalyst will speed up the process; cooling it will slow it down. Don't use open flames or excessive heat as this material is definitely flammable. Also more fumes will be emitted if the material is first heated, so use adequate ventilation.

# Adding Color to Resins

Many wildlife artists choose to add color to artificial water pours to create specific effects. The best way to accomplish this is with the transparent dye kit available from WASCO. These dyes produce "see-through" color and are available in a kit that contains 6 colors. They can also be blended with each other to create additional colors. Very small amounts of the dye are all that should be added. These are very intense colors and a very small amount will go a "long way." Adding oil colors, paint shading paste, and the like are not recommended as they are not transparent and will cause the resin to lose clarity.

When coloring water, keep in mind that most often the natural water itself is clear. Often the "hue" that we see is coming from what is underneath the water. For example, an unfilled swimming pool will have a white bottom. Fill it with crystal clear water and it will look blue. The color comes from light reflecting off of the white subsurface through the water.

This is the same principle that applies to paint. Take three sheets of metal. Paint one a base coat that is white, paint another silver and another gold. Allow them to dry and then spray a transparent red over them. There will be three entirely different shades of red. The white base will yield a bright florescent red; the silver will yield a metallic look; and the gold base will yield a "candy-apple" red color. This is caused by the light penetrating through to the base coat and reflecting its color back through the red. The base of an artificial water exhibit works much the same way. Simply coloring the water will not in itself always create the desired effect. The base color under the water will also determine the hue.

Use the base color to your best advantage to obtain the desired coloration. Try placing a piece of artificial water the same size and thickness as used in the exhibit over several different colored surfaces. Additionally, experiment with materials such as tin foil, burlap, cotton, etc., to create varied effects. Often the artificial water can be left completely clear and the proper hue can be created entirely by properly coloring the base. Also try placing the artificial water pour at varying heights to create varied effects (see diagram).

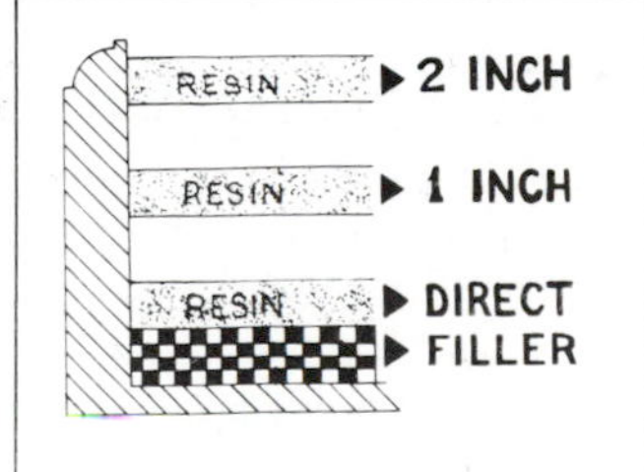


# Dust, Lint, Insects

Artificial water scenes will "attract" dust and lint, and insects often get trapped by the sticky surface. It is especially susceptible to these problems while in the curing stage. Whenever possible, cover the water scene with a protective cover during the initial gel and curing process.

One easy way to prevent dust, lint, or bugs from landing in the sticky resin, is to place an empty cardboard box over it until it cures. (Be sure and thoroughly clean out the box with an air compressor before placing it over the scene.)

Upon completion, water scenes should be permanently enclosed in glass cases, plexiglass cubes, glass domes, etc., to protect them from dust and lint.

If the exhibit needs to be cleaned, blow it off with an air compressor, then wipe it down with a damp lint-free cloth.

# Sticky Surface

*Polytranspar*™ Artificial Water and Lay-up Resin will dry to a hard "non-tacky" finish if properly mixed. Many similar products will not, however, and retain a sticky surface even after the mixture has fully cured.

There is a product called Surface Curing Agent that can be added to the resin to eliminate a sticky surface. It is added to the resin prior to (and in addition to) the catalyst when pouring the final layer of the casting to give a hard "tack-free" surface. Some artists will also rub talc (ordinary talcum powder will work) on the surface to help remove the stickiness. This is fine in cases where the remaining powder residue will not be detrimental to the finished product.

# Surface Coat

Often the final pour of a casting will have imperfections, fingerprints, scratches, etc. These can easily be eliminated by spraying with a coat of Resin Craft Surface Coat Spray or *Polytranspar* Competition Wet Look Gloss FP 241. These imperfections can also be eliminated by painting a thin coat of catalyzed artificial water over the surface area.

Sometimes a casting will turn milky-white looking, especially if used with water or water-based products. (Moisture causes the resin to turn cloudy.) There are a couple of solutions to this problem. One corrective measure is to paint a thin coat of catalyzed resin over the surface. The heat generated from the resin will boil the water out of the resin and clear it back up. Another method is to simply place the object in bright sunlight for a couple of hours. The heat from the sun will usually clear the resin. A heat lamp will suffice if the sun refuses to shine, but be careful to keep the heat lamp far enough away from the casting to prevent it from overheating.

# Embedded Objects

It will be desirable on occasion to embed various objects in artificial water pours. Several factors must be considered prior to beginning the project and special precautions should be taken in order to succeed. Be aware that additives usually retard and slow down the "cure" time of the resin.

One of the most important factors is the "type" of material that is to be embedded. Metals, for example, will often expand when heated by the catalyzed resin and will crack or damage the resin that surrounds it. To avoid this problem, the artist would use a much lower catalyst/resin ratio and further cool the mixture by placing the object in a sealed container and immersing it and floating it in icy cold water while it is curing.

Some embedded objects such as natural plants, insects, mounted crayfish, frogs, or fish, etc., can be easily damaged by the heat generated from the catalyzed resin.

To avoid this problem, first, make sure that the object is both clean and dry and then "precoat" the item with a thin (but thorough) coating of catalyzed artificial water that can be applied either by dipping the item or by painting it on with a brush. Be sure and use a low catalyst/resin mixture and use the cooling technique described for metal embedments on any object that could be damaged by the heat. It will also be desirable to embed objects such as driftwood, porous rocks, seashells, wooden items, and other porous materials. These objects contain trapped air pockets that can produce unwanted air bubbles in the finished artificial water pour if they are not first completely sealed.

Once the subject has been sealed, it will be ready to embed. There are two methods of embedding; using gel promoter with one pour; and layered pours.

The gel promoter and one pour method is most risky, but fast. By adding gel promoter to the resin before catalyzing, the gel time can be reduced to approximately 1½ minutes. The object can be placed in the pour and it will set up before the object can sink to the bottom.

Layered pours are more controlled. Make an initial pour, let it gel, then place the precoated item in position and pour another thin coat over it. Repeat until the desired thickness is reached.

## Troubleshooting Artificial Water Problems

| Problem | Probable Cause | Corrective Measures (*NOTE: More than one method often can be used to correct the "same" problem.*) |
|---|---|---|
| FINGERPRINTS | Handling before resin has completely cured | A. Spray with *Polytranspar* Competition Wet Look Gloss FP241<br>B. Spray with Surface Coat Spray<br>C. Paint thin coat of catalyzed Artificial Water on the area |
| CLOUDINESS | Moisture | A. Allow to sit in sun<br>B. Apply heat lamp<br>C. Apply Artificial Water with heavy catalyst ratio |
| WARPAGE | Excessive Catalyst<br>Excessive room temperature | A. Reheat with heat lamp, hair dryer, or hot water and reshape<br>B. Prevent by pouring resin in 70–77° room temperature or reduce catalyst ratio. |
| YELLOWING | Excessive Catalyst<br>Inferior Resin<br>Contamination | A. Cut back on catalyst ratio<br>B. Use *Polytranspar*™ Artificial Water<br>C. Replace lids securely in place after usage. |

# Reference

Before attempting any scene using Artificial Water, the artist must first have a clear understanding of what they are trying to accomplish. The key to this understanding—and the ultimate realism of the final outcome—is the same factor which determines quality mounts and carvings: REFERENCE.

A few years ago, instructions were published explaining how to produce a splash by pouring Artificial Water over plastic garbage bags. Suddenly, many of the splash scenes at major competitions looked like plastic bags instead of splashes. It is not enough to simply follow the directions of how someone else made a certain effect—the artist must have the knowledge of what a splash *actually looks like!* This can only come from careful collection and study of reference materials.

Gather reference materials for splashes, streams, pools, ripples, turbulence and the like, and file them in a photo album for easy access. Good reference material will save both time and money. An artist needs reference material to recreate accurate habitat exhibits. Obtain a good camera and take reference photos as well as clipping photos from magazines. Always be on the lookout for good reference material from any source.

An excellent source for water reference is a video casette recorder. Record fishing or nature-oriented television programs and replay them while looking for good reference material. The pause button can then be used to "freeze" the image of a jumping fish, a diving duck, or a waterfall. The water effects can then be studied at leisure and carefully analyzed to determine which procedures would best suit their duplication in a habitat scene.

Decide first how the completed scene should look, and then decide on the best methods to accomplish it.

Don't be afraid to experiment with Artificial Water. Most of the methods in this chapter were discovered by innovative artists willing to try new ideas. No doubt there are new and better methods yet to be discovered by someone reading this.

# 1. Creating a Shallow Pool (A Solid Resin Pour)

The simplest method for reproducing Artificial Water is the shallow pool scene. A shallow pool consists of a solid section of resin representing the surface of a body of water. The pour (or pours) are made directly into the base with the total thickness of the resin depicting the total depth of the water.

The resin may be left clear (allowing the base to be seen beneath the water) or the resin may be colored (even opaque). If the resin is left clear, the bottom surface must be prepared and colored to look like the bottom of a shallow pool (see Wendy Christensen-Senk's Nyala base in Chapter 7). If the resin is colored to the point where the base cannot be seen, an effect of deep water may be suggested (as is the case in Tom Sexton's largemouth bass mount).

The basic steps for constructing a shallow pool are as follows:

**Step 1.** Build or buy a suitable base with a contour arrangement that will hold a pool.

**Step 2.** Seal the base by painting with a coat of Ultra Seal or Fiberglass Resin mixed with sand or dirt. The sealer coat must be thorough enough to prevent air bubbles from escaping into the pool.

**Step 3.** If the bottom of the pool is to be painted now is the time to do it.

**Step 4.** Mix a quantity of Envirotex or Ultra Glo according to the directions on the container. Air bubbles will dissapate if you blow on them or use a small torch as directed on the container. Continue making pours until the desired thickness is reached. (Do not exceed 1/2" as it will begin to lose clarity beyond that.)

**Alternate Step 4.** *Polytranspar*™ Artificial Water can be used in the same manner as Envirotex. Simply catalyze it according to directions and make pours of 1/8" to 1/4" until the desired thickness is reached. Again, don't exceeed 1/2" for clarity reasons. If the water is to be colored, dyes must be mixed in the Artifical Water.

# Tom Sexton's Best in World Largemouth Bass Habitat Scene

In 1985, Tom Sexton won Best of Show at the World Taxidermy Championships with his portrayal of a largemouth bass jumping from the water and breaking its line. The Artificial Water technique used was the "shallow pool" method, in which Tom dyed the water so that the bottom of the base coudn't be seen, to give the impression of deep water. Here is Tom's account of the construction of the habitat scene.

***by Tom Sexton***

The mount is designed to be viewed from 360° with the bass pedestal mounted (two-sided mount) and suspended over a base. As with most world champion pieces, it started with a very unique idea: creating a scene depicting a Largemouth Bass "re-entering" the water after it had successfully jumped out of the water and broken a fishing line. (Most competitors have chosen to take the approach of a fish "emerging" from the water.)

By choosing a "re-entry" pose, I eliminated much of the difficult splash construction problem, choosing to reproduce the ripples left from the exit splash on one side of the piece, and having to construct only a very small splash where the Bass's mouth was re-entering the water on the other side.

I chose an oak base obtained from Sinclair's panels and gathered the needle grass from a nearby swamp. Ultra Seal was applied to the needle grass. After it had dried, the grass was airbrush painted with a heavy coat of *Polytranspar* Bright Yellow FP/WA 260, then lightly misted with FP/WA 30 Black from a distance of approximately 12" to 15" away to obtain the green hue.

I inserted wires into three blades of the grass after first cutting slits into it where the wires would exit the grass and enter the fish. Approximately 1½" protruded through the bottom of the grass for attachment to the base.

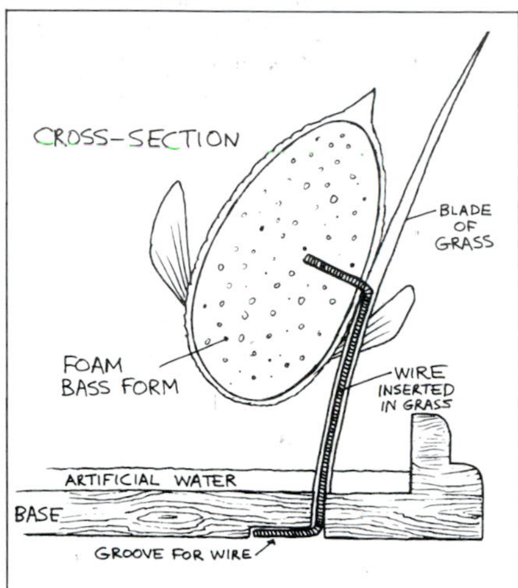


After locating the approximate position that the fish would ultimately be in, I then drilled a hole through the base for the first blade of grass. The bottom of the base was grooved to accept the wire. Next, a hole was drilled into the fish and the first wire was inserted into the fish. At this point, the fish was still very wobbly. The location for the second wire was positioned in such a manner that it would give the most support. A hole was drilled in the fish to accept the wire, and it was inserted into the fish. The final support wire was done in likewise fashion. NOTE: The fish was positioned in such a manner as to allow for an initial 1/8" pour of Artificial Water, i.e., the jaw of the fish was suspended 1/8" from the oak base on the three wires.

Next, the rest of the needle grass was attached to the base. It was first cut to the desired length and then hot-melt glued to the base. The fish was removed from the wires and the first pour of Artificial Water was made. Black, red and yellow dyes were added to the artificial water to make it resemble brackish water. At this point, approximately 1/8" of the resin was poured into the base. (It is advisable to seal all cracks before making this pour. If the base is not completely tight, it will leak.)

After the resin had cured, the support wires for the fish were roughed up with wire cutters to enhance adhesion. The three drilled holes in the fish were filled with Sculpall™ and the wires were inserted. The excess Sculpall that squirted out was cleaned off and was then allowed to harden. This gave the fish a fairly sturdy base.

Next, a paper cup was cut in half lengthwise and filled half full with catalyzed Artificial Water. This was held at a 20° angle by hand until almost gelled. Before it had completely hardened, I used a tongue depressor to "roll" the thin ends of the splash up and back over the resin to shape them. Several of these "splashes" were made using different shapes and designs and then allowed to cure.

I then selected the splashes that looked the best and positioned them beside the jaw where the fish was re-entering the water. Aluminum foil was used to keep the next pour of Artificial Water from going into the gill area. Another 1/8" pour was made securing the splash pieces and jaw to the base. By blowing a hair dryer on the gelling resin as it cured, the ripple effects on the water surface were formed.

# 2. Creating a Deep Pool (Using A Resin "Ledge")

Constructing a base for a deep pool will differ somewhat from a shallow pool. A deep pool consists of a "ledge" or 'shelf" of Artificial Water which represents only the surface of the body of water. The shelf can be constructed entirely out of resin or a resin pour can be made over a thin sheet of plexiglass. Ripples, turbulence effects, or splashes can then be added to the Artificial Water surface for added realism. Deep pool surfaces are usually left transparent (or very subtly colored) to allow the viewer to look *through* the apparent water and see what is underneath.

The water surface can vary from completely smooth (still water) to extreme turbulence or splashes. The basic steps for constructing a simple deep pool are:

**Step 1.** A shelf must be built to suspend the water pour above the bottom of the pool.

**Step 2.** A 1/4" piece of plexiglass is cut to fit the shelf. The bottom of the pool should be painted an appropriate color to create the desired effect, if a natural bottom is not to be used.

**Step 3.** Catalyzed Artificial Water is poured over the surface until approximately 1/4" thick.

## Creating Ripples on the Surface

Only in rare cases will a water surface appear completely smooth. One of the features of a water surface that makes it "look like" water is ripples. As water currents move, as the wind blows upon the surface, or as a bird, fish, or mammal swims through its surface, ripples and waves will be formed. It is therefore necessary to add ripples to these type scenes in order to produce a realistic exhibit.

## Air Pressure Ripples

Many wildlife artists use air pressure to make ripples on Artificial Water surfaces. This method does not produce extremely accurate ripples, nor are the results easy to control. However, it is the quickest and easiest method for creating ripples and can produce pleasing results with a little bit of practice.

**Step 1.** To use this method, catalyzed resin is poured into the scene.

**Step 2.** As it approaches the gelling stage, air pressure is directed upon the surface with a hair dryer, air compressor, or vacuum cleaner blower. This pressure will create small ripples across the surface. Steady pressure is applied until the resin completely sets up.

## Carving Ripples in Plexiglass

Many wildlife artists prefer to carve ripples patterns with a Foredom Tool (or similar device). Greater control over the final outcome may be obtained with this method as opposed to using compressed air.

**Step 1.** Use a 1/4" plexiglass base and draw the ripple pattern with a felt-tipped marker.

**Step 2.** Using a cutter designed for plexiglass and a Foredom Tool, carve the pattern into the plexiglass surface.

**Step 3.** Catalyze the Artificial Water and pour a shallow layer over the surface to smooth out imperfections and restore the crystal-clear appearance.

## Creating Ripples from a Mold

This is the most accurate method for creating individual ripples (and then having the option of arranging them on the water surface).

**Step 1.** Sculpt or carve the image of the individual ripples. Materials commonly used for this are plastilene clay or parafin wax.

**Step 2.** Use Mold Builder or RTV silicone to make a mold of the sculpture.

**Step 3.** Fill the mold with catalyzed resin. When it has fully cured remove the ripple piece from the mold.

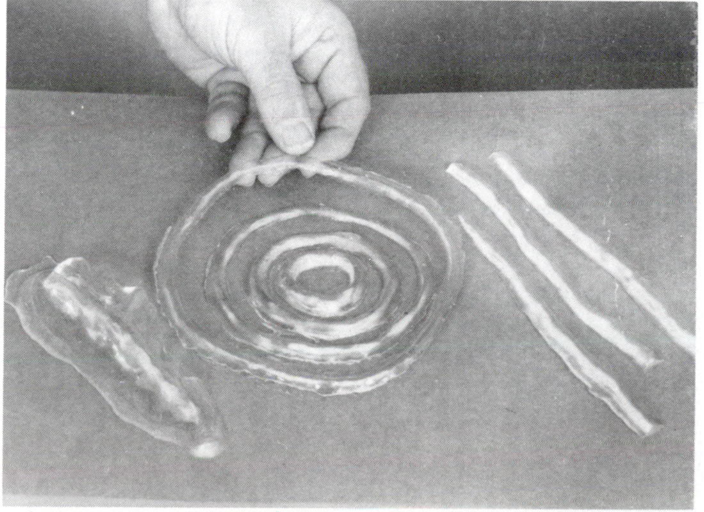

*Individual ripples can be made from a mold or purchased from a supplier.*

**Step 4.** Install the ripple by placing it on a plexiglass base or Artificial Water surface and gluing it with instant bonding glue.

**Step 5.** Pour catalyzed resin completely around the ripple to hold it in place and blend it into the scene.

NOTE: A series of ripples can be installed in this same manner to recreate a flowing stream current.

# Molding Ripples In Clay

*by Ken Edwards*

Master Taxidermist Randy Nelson of St. James, Minnesota creates his water surfaces by sculpting the ripples in clay, making a parrafin wax mold of the clay, and pouring the Artificial Water in the wax mold. The following is Randy's step-by-step proccedure for making a stream surface.

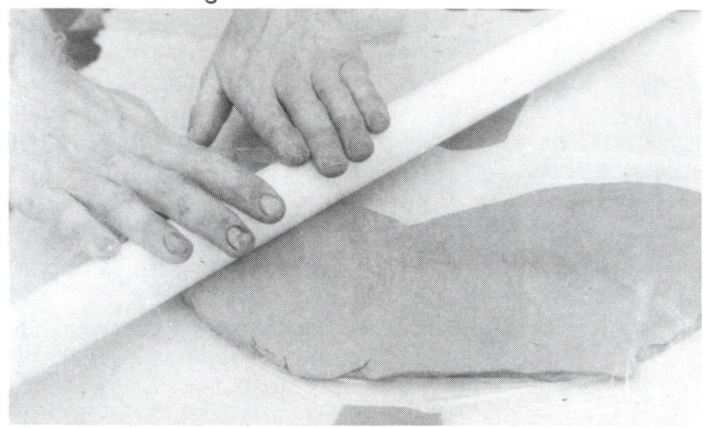

Randy begins by shaping oil-based clay into the approximate contours of the water surface needed. The clay is rolled flat for a smooth surface upon which to work.

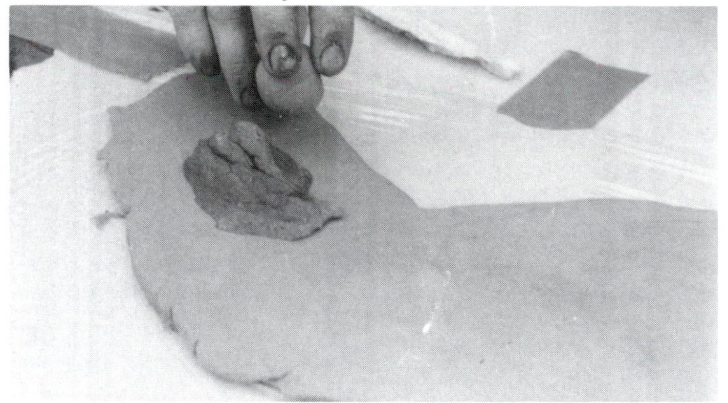

This particular habitat will include an artificial rock breaking the surface of the water. Randy locates the correct position for the rock and models a small piece of clay to approximate its size and shape.

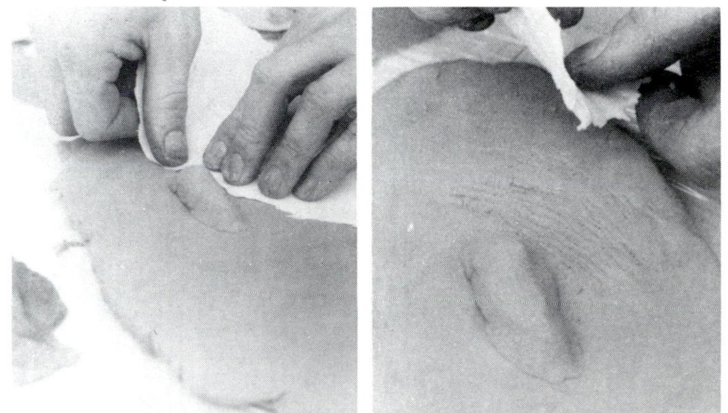

The ripples in the water surface are formed by pressing latex texture molds into the surface of the clay. (For more on Randy's texture molds, see box on page 69.) The molds for the ripples on the upper current side of the rock are made from woodgrain textures. Randy prefers to use natural textures rather than meticulously sculpting the individual ripples. "Why bother with sculpting them," asks Randy, "when God has provided the perfect textures in nature?"

For the water turbulence downstream from the rock, a latex mold of tree bark is used. "There are only so many textures that exist in nature," Randy states. "The differencea are in the materials themselves. The same texture used to create tree bark can be used on an artificial rock or to create water turbulence." The clay surface is further detailed with additional latex molds.

When satisfied with the surface, a clay dam is attached around the edge to contain the wax. Surface Coat spray gloss is applied to the clay to seal it and hide any fingerprints. The finished clay mold is now placed in a freezer until thoroughly cooled. This will prevent the hot parrafin from disfiguring the clay when it is poured.

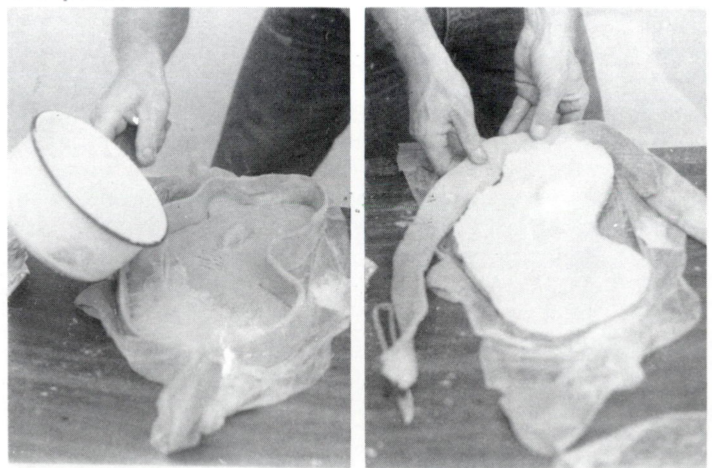

Melted parrafin wax is then poured into the frozen clay mold. When the wax has cooled, the clay is removed.

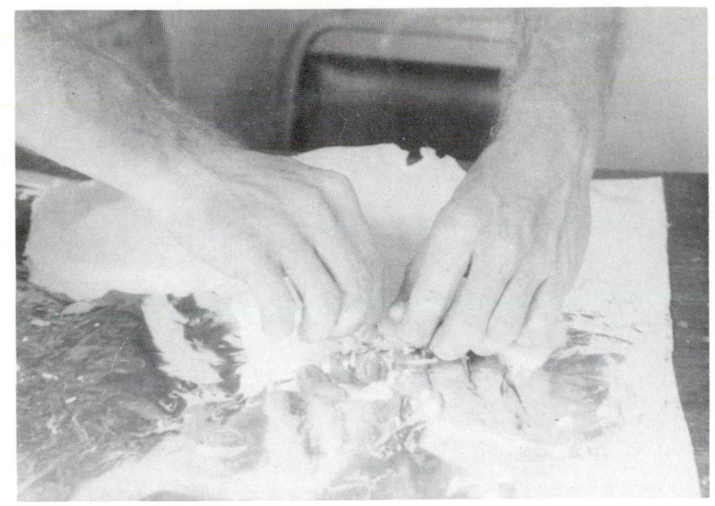

The wax mold is inverted and a dam is constructed of aditional parrafin. Randy uses a hair dryer to slightly melt the edges of the wax to keep the dam from leaking.

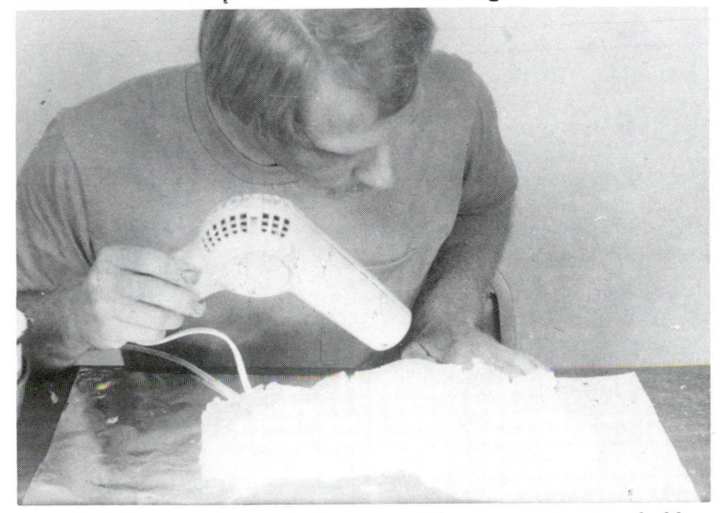

To highlight ripples, turbulence, and flowing water, Randy likes to add color to selectively "paint" his Artificial Water surfaces. *Extremely* small amounts of transparent dies are added to catalyzed Artificial Water, which are then applied in the wax mold and arranged with a tongue depressor.

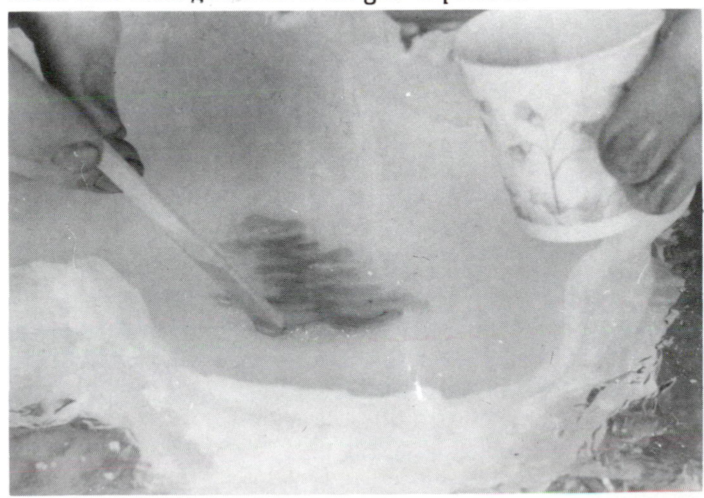

Randy makes several small batches of varying colors for different areas of the water surface. By changing the amount of time between pours, Randy is able to control the amount of blending between the colored resin and clear resin. On this project, Randy used a combination of blue, green, and black dies to color different areas of the water surface.

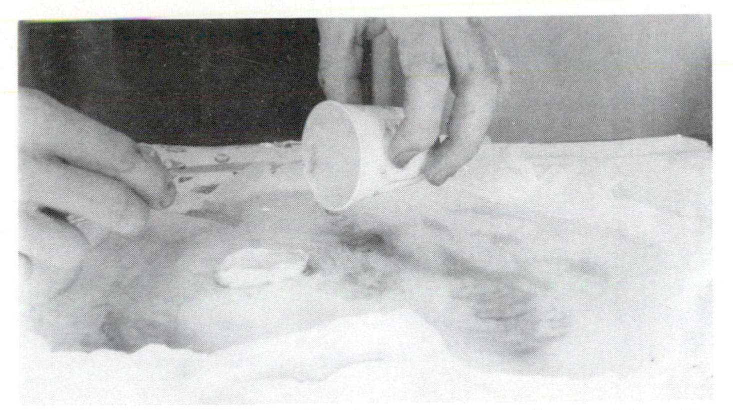

A final pour of clear artificial water was made to achieve the desired thickness.

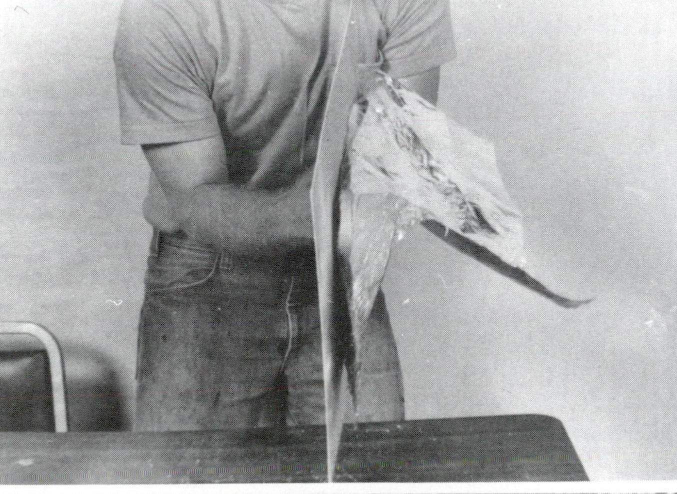

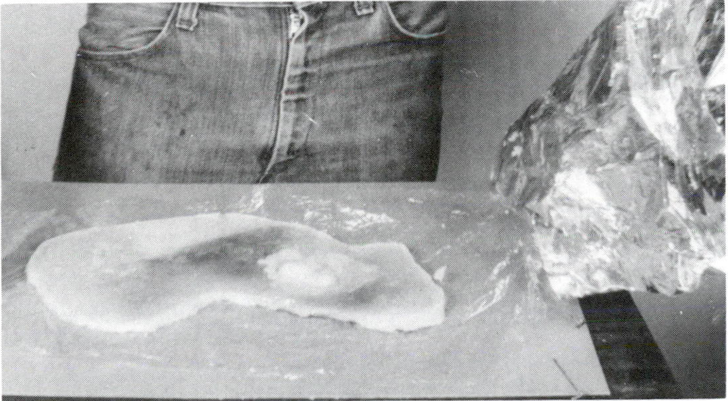

When the resin has begun to gel, Randy removes the dam and carefully flips the resin and wax over onto a piece of plastic wrap.

The wax is separated from the Artificial Water, exposing the surface.

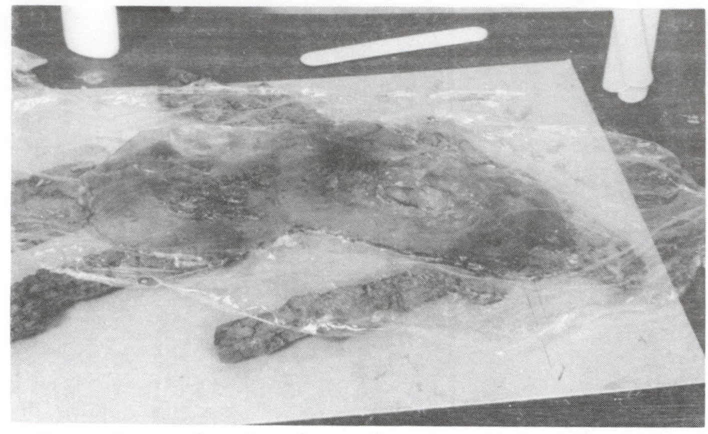

Supports are placed underneath certain areas of the Artificial Water, so that it will cure with an undulating surface (instead of a flat surface).

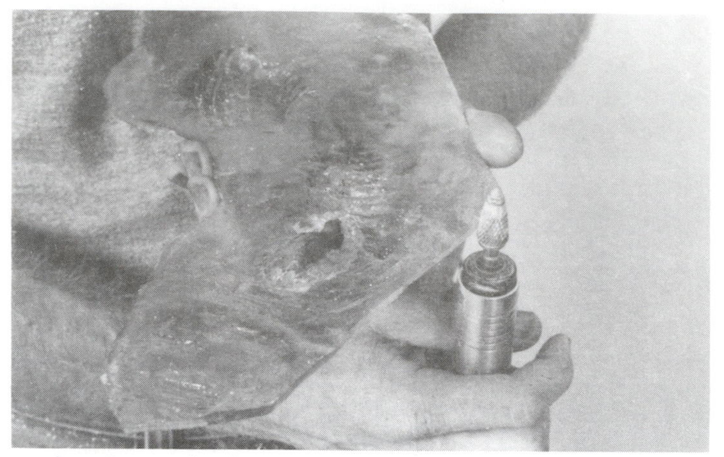

After the Artificial Water has completely cured, it is test-fit within the base. A Foredom Tool is used to fine-tune the sheet to the exact shape.

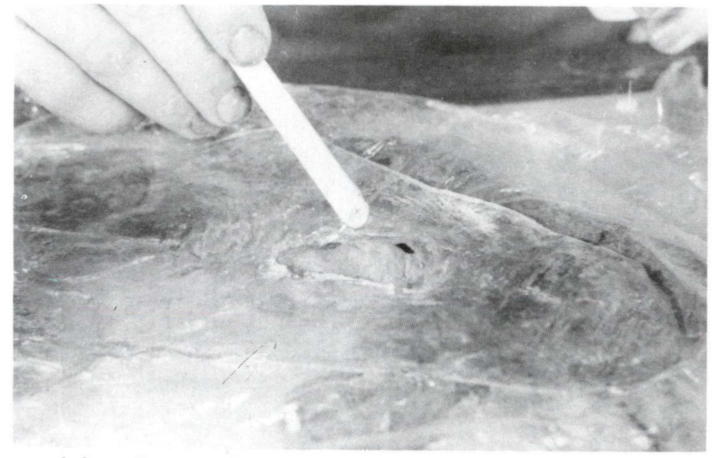

While still in a gelatenous state, Randy further modifies the water edges, checking the rock hole for a proper fit.

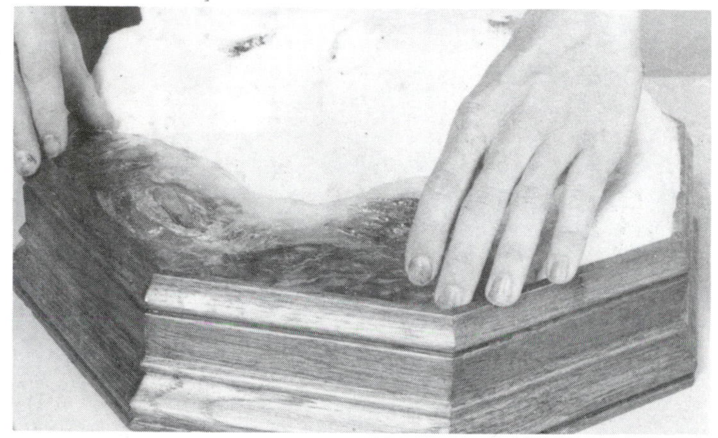

The water is positioned in the base and two-part epoxy adhesive is used to permanently attach it.

# 3. Splash Construction With Artificial Water

One of the most intriguing facets of habitat and exhibit work is building an Artificial Water splash. There are many different ways to create a splash scene. As with other forms of exhibit work, obtaining good reference materials is necessary before beginning the project. Find a good photo of a water splash and keep it handy.

Perhaps the best and most accurate method of recreating a water splash is to sculpt a model of the splash components. Then, use mold builder or RTV Silicone to make molds and cast the finished parts from the mold.

Once all the components have been cast, they are bonded together with instant bonding glue and finally coated with a thin layer of catalyzed resin.

Another technique which is less time consuming and yields good results is what I call the "pour sculpture technique." Basically all that is required is to pour catalyzed resin over a piece of plastic that is draped over a bed of sand that has been formed with varied contours. The resin may be lifted with a tongue depressor and shaped.

Once cured the pieces may be glued together in a pattern similar to the artist's reference material.

It is very easy to reshape splash pieces. They may be heated under a heat lamp or hair dryer and then bent into the desired position. When the piece cools, it will remain in the new position.

Droplets landing in water are best made by making a mold of the desired shape and casting a droplet piece. This may be installed by pouring catalyzed resin around it to hold it in place.

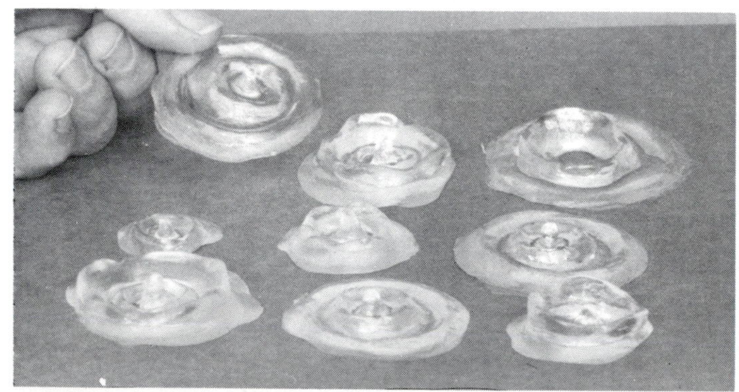

*Individual splash droplets made from a mold of a clay sculpture.*

Creating artificial water splashes is not very difficult. There are several different methods that artists use to create them. Probably the best method is the "molding technique." As with most things the "best" is generally the most expensive and most time consuming and this molding technique of making splashes is no exception.

## The Splash Molding Technique

To begin this method, first find a good reference picture of a splash similar to the one that is going to be recreated.

Next, sculpt or carve a model of the main portion of the

splash. (Individual splash pieces may be created and added later.) Once the splash model has been sculpted, make a mold of it using either Mold Builder or an RTV silicone material that is reinforced with a "mother mold" of plaster or Ultra Lite (see Chapter 5: Mold Making).

Once the mold is completed, catalyze some Artificial Water and make the first pour of the splash. Note: Do not pour very large sections of resin with each pour. Pour a thin section, let it cool, then pour another until the desired thickness has been obtained. Never pour a thickness over 1/4" of Artificial Water at a time, as the heat that is generated from the catalytic reaction can crack and/or discolor the scene.

If the mold has steep sides and the Artificial Water tends to "run" down to the bottom of the mold and will not stay on the sides, keep brushing it back up until it reaches the gelling stage. Another method is to keep the catalyzed resin in the container (or pour it out on a piece of plastic) until it first begins its gelling stage and then apply it to the sides with a paint brush.

Surface-type artificial air bubbles (bubbles on top of the water) should be placed into the mold "prior" to making the first pour. Air bubbles depicted as being underwater are added in the second or third pours by mixing with the resin.

Add additional pours until the water splash has been built to the proper size and thickness. The splash can either be a "solid" mass, or it can be left hollow on the inside. Embedded objects should be embedded with the second pour by simply adding them to the mixture and then making the pour (always precoat embedded objects with resin first).

Once the splash has hardened, remove the mother mold and then the flexible mold from it by peeling it off. The splash can then be placed in the water scene base and a thin layer of Artificial Water can be poured around it to hold it in place. It is often helpful to use instant bonding glue to temporarily hold it in place while the pour is being made. Once the main splash is in place, the individual splash pieces may be attached in the same manner by first gluing them in place and then pouring Artificial Water around them. They can be adjusted by heating with a hair dryer or placing in hot water and applying pressure to bend the piece to the desired position.

## Ralph Lehrman: Hot Water Splash Construction

Ralph Lehrman is an accomplished flat artist as well as an award-winning taxidermist. The production of this spectacular splash scene can be easily incorporated into various habitat scenes.

***by Ralph Lehrman***

For me, creating a resin splash with a fish has bridged the "gap" between taxidermy (the art as we see it) and the public's view of art, particularly sculpture.

I have received many positive comments of appreciation from both sportsmen and non-sportsmen after they have viewed such an art piece and feel, therefore, that I have made favorable contact with a wider area of the public. This in turn has given me a great deal of satisfaction and I always find myself stopping to take a look, again and again after completing one of these works.

I have learned a great deal and have so much more to learn. Yet through a sharing of information, we can all open new areas of skill and improve others. Hopefully, this section will stimulate just such a positive response.

Before beginning, just a few notes on Artificial Water:

1. It is flammable, therefore—*NO SMOKING!*
2. Use it in a well-ventilated area, (paint booth, outdoors, etc.).
3. The hardener or catalyst is Methyl Ethyl Ketone Peroxide, which is a strong oxidizer, meaning it can cause a spontaneous fire if it comes in contact with certain combustible materials. Therefore, it is suggested that towels and rags that are contaminated with resin or hardener be put in a bucket of water to prevent a fire.
4. *ALWAYS* keep your work area clean.
5. *ALWAYS* wear rubber gloves, goggles, and rubber apron—**No Exceptions!!**
6. Always use resin from Wildlife Artist Supply Company (*Polytranspar™* Artificial Water), so you don't need to use promoters, etc., as this resin is ready to use when catalyzed. (Some are not.)
7. Keep the resin and catalyst at 70 to 75 degrees when using or as close to that temperature as possible. Warmer resin will set up much faster, and cold, much slower.
8. Either use the Chemco measuring cups with graduated lines (available from Wildlife Artist Supply Company) or buy some clear plastic disposable drinking cups and measure out 100 cc's of water and pour it into the cup. Mark it and put in another 100 cc's and mark it again. Now mark two or three extra cups by holding them beside the first one. This will provide some "pre-measured" mixing cups that can be disposed of when used.
9. As for measuring catalyst, a glass syringe works great. One can get very accurate measurements of 1.25%, 1.50%, 1.75%, etc., for mixing with the resin. For example, if a 1.5% solution is desired, simply pour the resin in the cup up to the first line (100 ccs), then draw 1.5 ccs into the syringe and put it into the resin. It's as simple as that. Obviously, if a larger batch is desired, use the same proportion and more of both ingredients.

I will describe in this section how to make a splash scene complete with a mounted brown trout. An explanation of how to do the pedestal fish mount can be found in the *Breakthrough Fish Taxidermy Manual* and painting instructions can be found in the *Breakthrough Fish Painting Encyclopedia*.

To begin with, the artist will need to come up with an idea or a photo of a fish jumping out of the water. Using a fishing tape on a VCR that has a single frame (stop action) capability is an excellent way to get a picture of what a fish should be doing in a jump and how the water is shaped as it forms a splash.

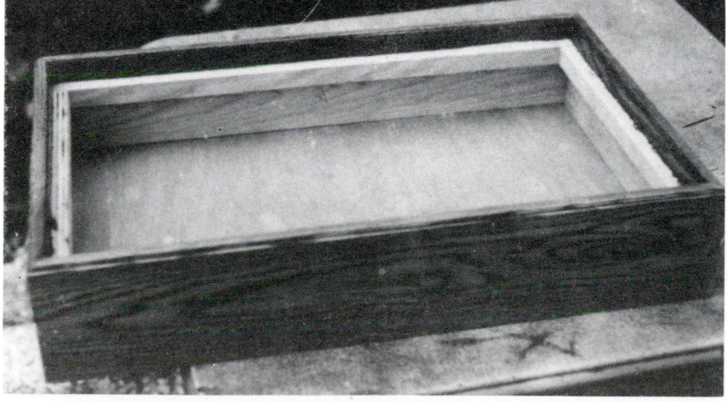

Once an idea or model is decided upon, one must then decide how large of a base will be needed. I chose a 16" × 26" hardwood base. It will be desirable to place a ledge about 1¼" down from the top. That way, if 1/2" plexiglass is used as a surface support, and two 1/8" layers of resin are poured, there will still be enough of an edge to hold the glass case.

When obtaining a plexiglass surface, don't hesitate to use a piece that has been scratched. It will be much cheaper to buy and the scratches will disappear as soon as they are coated with resin. I use 1/2" plexiglass for a base the size of this one.

That way, when the resin cures, it does not warp the surface, which can be a real problem with thin sheets of plexiglass.

Always remember to leave some room around the edge of the plexiglass so that it can be lifted out when needed. This will allow the artist access to the bottom of the case for finishing purposes. It is difficult to determine how much of the stream bottom will be visible through the water surface until the surface is completed and in place.

The next phase will be to start pouring splash components. It will be necessary to have on hand some acetate sheets that are fairly thick. Lay these in a plastic pan with about 2½" to 3" of hot water in it. (Rectangular plastic pans made to hold cat litter work very well.) Make a 100 cc mix of resin, using 2 percent catalyst and stir well. The components should be at about 70 degrees F. Watch the resin carefully. After about 7 or 8 minutes, it will begin to thicken or gel but will still be flowable. When it reaches this stage pour it into the hot (or quite warm) water on top of the plastic sheets. Use a knife, paint stirrer, or tongue depressor to scrape, lift, and otherwise move the resin around. Continue this as it begins to set and has become very gummy and does not flow anymore. At this point, a very irregular piece of soft resin should be in the water, similar to very stiff Jello.

Now remove the sheet of plastic with the resin on it and set it aside to cure overnight in a dust-free area. (Placing a cardboard box over it will eliminate dust problems.) Using this technique the artist can make 5, 10, 15, or more pours and stockpile them. After they are cured there will be a nice selection of splash pieces from which to choose from when creating different effects.

It is also advisable to make some 200 cc and 300 cc pours for support pieces. Be sure to use proportionately more catalyst. However, it may be advisable to reduce the catalyst percentage to a 1.25 or 1.5 ratio when using this much resin because if too much is used, it may generate too much heat in the larger quantity and start discoloring. I've seen it look like it was going to catch fire and actually start smoking, *so be careful!* If this happens, try pouring cool or cold water on it until it settles down. Some of the pieces might become cloudy or milky looking but don't worry, as this can be corrected later.

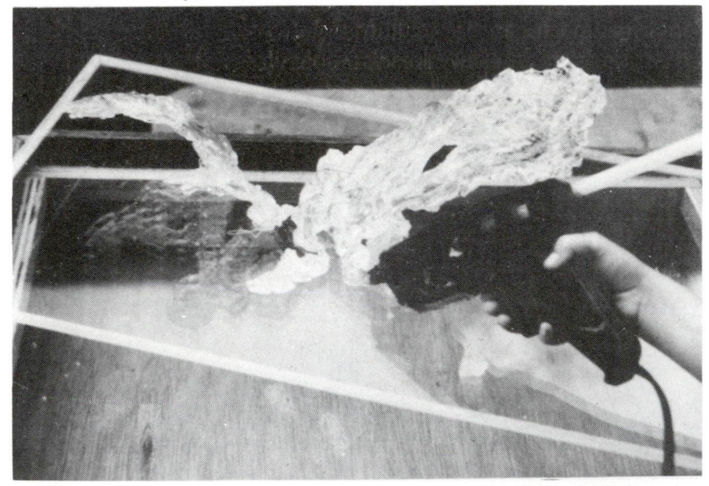

Take the fish mount or carving which has been carefully posed and place it in the position and approximate location that it will go. It may be desirable to temporarily support it with blocks, cups, or whatever so that a support piece of resin can be found that looks as if it will do the job. To install it, first peel it off the plastic and put it in a bucket of 140 degree water (hot tap water) for a few minutes. Pull it out and slowly bend it to fit under the curve of the fish. Using clear gluesticks, such as those from florists, or using clear 5 minute epoxy, glue the piece onto the surface (photo above).

Next, begin the rest of the splash. Peel off appropriate pieces from the plastic sheet.

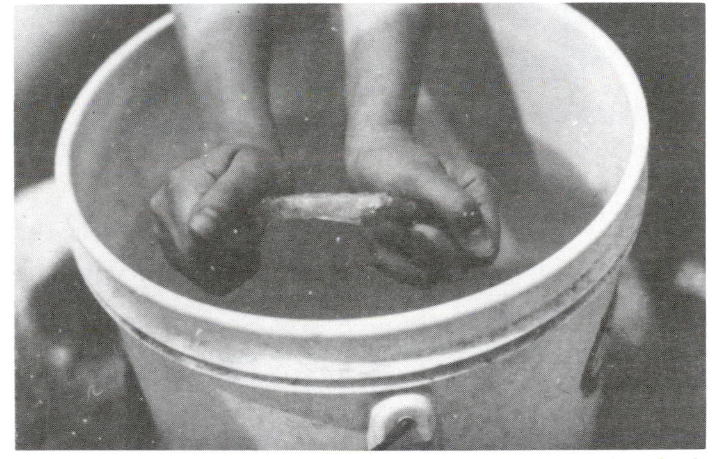

Put the pieces in the water and shape them until they look right. (This step will make the pieces cloudy and fingerprinted, but *don't worry!*)

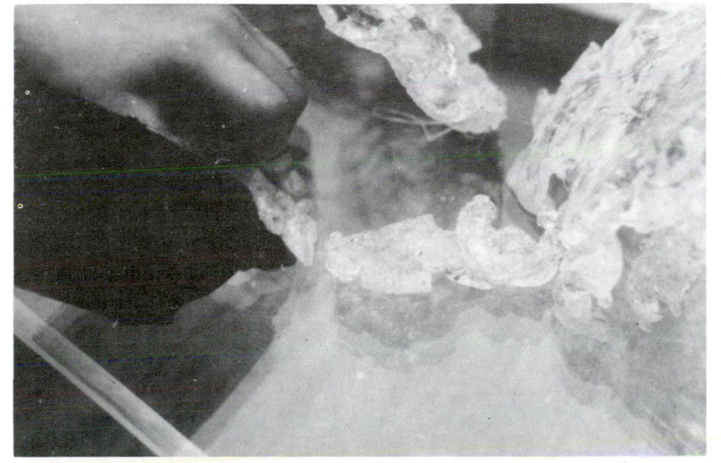

Glue these pieces into position.

The outcome will be a *simple* support with a splash.

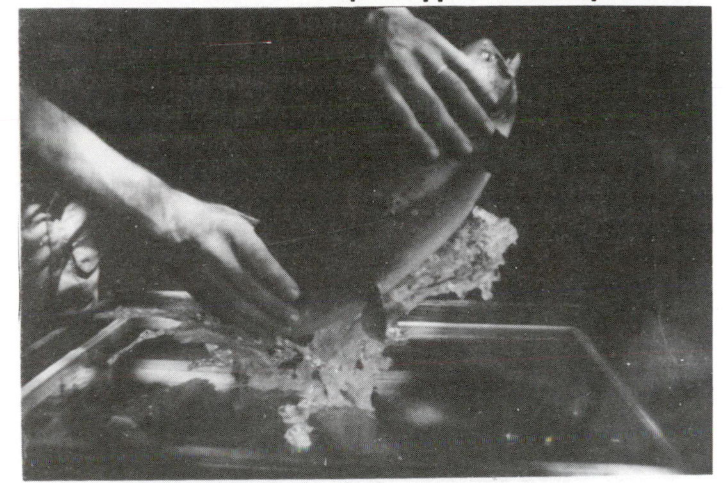

Now, put the fish in position and glue it down. Add more splash pieces until the desired look is obtained.

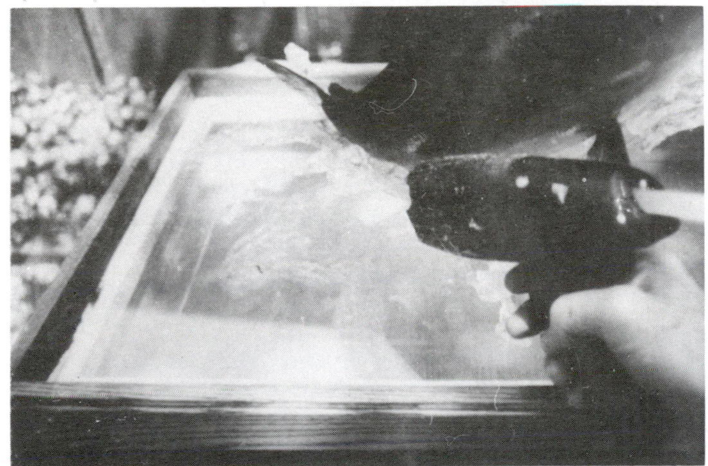

Also, put in smaller pieces between the splash and the fish to fill in the gap. (Be careful though not to overdo it!) When satisfied that the proper pose has been obtained, then make a 1.25 to 1.50 percent catalyst mix and pour over the pieces that are next to the fish allowing some of it to get onto the fish. This will weld it all together. Also, using a disposable brush, coat the rest of the pieces lightly, top and bottom. All the fingerprints, milkiness, etc., will disappear when this is done. Be sure to *not* use a stronger mixture of resin and catalyst—otherwise the excessive heat generated during the curing stage could damage the fish.

If there are any gaps remaining between the pieces, catalyze some resin (1.50 percent mix), wait until it begins to get fairly thick, and then pour it into these areas. One must be careful not to put a large amount of resin in a small area, again because of the curing heat which can cause warping, cracking, and/or discoloration. This step will eliminate the possibility of the scene looking like a "group of pieces."

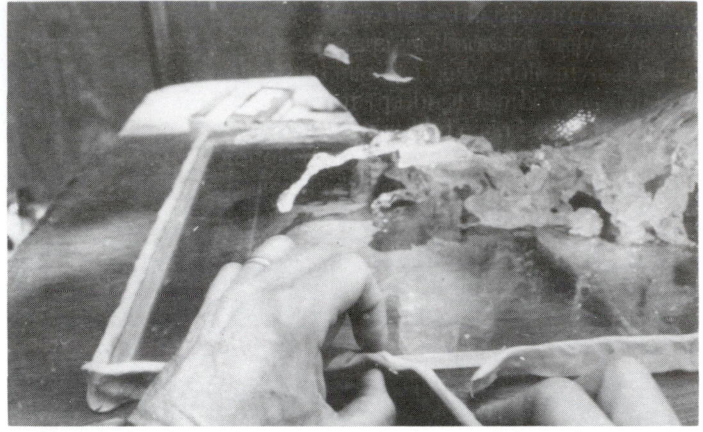

After this is done, the scene can be removed from the wooden case and a dam can be built all around the plastic surface to prevent the resin from running off. I use Plastilene modeling clay for my dam but other possibilities exist. Just be sure to leave an edge about 1/2" high, *with no leaks.*

Now, catalyze the main part of the water. About 300 to 500 ccs of resin will be necessary for a pour the size of mine using a 1.5 percent catalyst solution. At this time there should be a few splashes on hand that were poured on the plastic sheets that look like ripples. To install the ripples, pour the resin out for the main water flow making sure that it reaches all surface areas. Now lay in the ripples. They need to be *very* flat on the bottom so that the edges won't show through the 1/8" or so of surface resin that was just poured. Leave all the "outside" edges submerged.

In about 8 minutes (at 75 degrees temperature), start scraping the surface with a cardboard spatula. Keep doing this all over the surface until the resin does not flow out any more. Once it has set, don't scrape anymore or the surface could be fractured creating unnatural cracks in it. This scraping technique gives the surface a good texture which looks like water (as can be seen in the photos). The viewer can see through it, but it will not look like clear plastic.

If some of the ripples have fingerprints or cloudiness on them, it will be necessary to brush fresh resin on them—but "only" on the ripples, not on the water, or it will flow out and will completely destroy the effect. Experiment with different techniques for making small ripples or water surface textures. Once the ripples are in place, put the scene in a dust-free area overnight to cure.

For a nice effect, put some drops of catalyzed Artificial Water on the fishing line. Choose an appropriate lure and attach fishing line to it. Put the lure in the fish's mouth and bring the line down around a ripple. Continue to one edge of the case and tape it down with masking tape.

Glue the line to the ripple with 5 minute clear epoxy. Mix a very small batch of resin (2 percent solution) and let it set for a few minutes. When the resin starts to thicken, put a small amount on a tongue depressor and scrape it off on the line, a droplet size at a time. The resin may run down the line or fall off but just keep doing this until it no longer runs and the proper effect has been achieved. Then, let it cure, preferably overnight. Now cut the excess line off at the spot where it was anchored to the ripple.

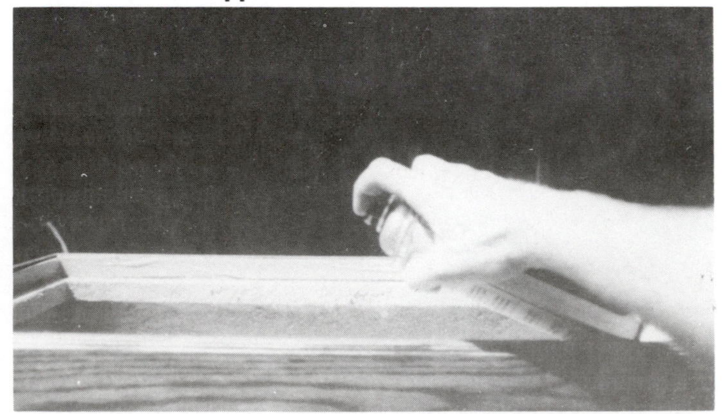

Now remove the plexiglass and spray the inside surface of the base with a spray adhesive.

Then, coat the base with either sawdust, ground-up moss, sand, etc., and then paint with whatever color is desired to show through the water surface. After setting the scene back in the case, put in the molding pieces to cover the edges of the scene.

As can be seen in the above photo, one should always take advantage of what artists call "happy accidents." In the support piece that I chose there just happened to be a slit which I used to put a fin through, creating a very realistic looking effect.

This should complete the splash scene. Other items such as leaves, driftwood, lily pads, etc., can also be put in if so desired, but be careful to not clutter it up. Now obtain and place a glass case over the scene right away. Dust seems to be attracted to, and is almost impossible to remove from even a very hard-cured water scene.

# 4. Molding Artificial Water for a Waterfall Habitat

World Champion Bird Taxidermist Kelly Seibels, of Huntsville, Alabama, is well known for his masterful habitat scenes and outstanding waterfowl taxidermy. This harlequin waterfall scene won Best in World honors in the 1986 World Taxidermy Competitions. Here is how he created this masterpiece.

***by Kelly Seibels***

The first step in building any habitat scene is to visualize and determine exactly what the end result is to be, prior to starting the project. The best way that I have found to do this is to go back to nature and "study" the "real thing" for basics such as rock formation, water flow, etc. Use as much reference of natural habitat as possible for all aspects of your work. Select a pose for the specimen that is compatible with the scene, i.e., don't put a winter plummage bird in a summer scene, or vice versa. By the same token, don't put a sandstone rock in a scene where granite would be more appropriate.

Once the subject's pose has been formulated in your mind, the next step is to design the scene based on the location and pose of the subject which in this case was a harlequin duck. I sketched several different versions until I was finally satisfied with the design and composition of the piece. Three different levels were used: the upper level was used for background and origin;

the middle level was the main impact which was used for the bird and primary focal point; and the lower level was used as the base and foreground. Since I had already established a good mental picture and sketches of the desired rock formation and water flow, it enabled me to calculate dimensions based upon the actual size of the subject. This was done using a previously mounted harlequin of the same size as a model. Remember to always allow the bird enough room and try not to "smother" him with habitat or make the scene too "cluttered" looking.

After the exact shape and size of the waterfall was determined, a hardwood base was designed and constructed in such a manner as to gently follow the rocks from top to bottom. The base was constructed with red oak except for the floor which was made from 1/2" plywood, as it would be covered by the scene. I contracted a cabinet company to actually construct the hardwood portion of the base. The base was slightly elevated to allow for optimum viewing. In any habitat such as this, I always make the base first so that I will have a reference for fit when shaping the styrofoam rocks. The base actually serves as the foundation of the entire piece and the rock structure is carved to fit it.

The artificial rocks that I carved were made from four separate 4" sections of styrofoam that were glued together. The type I used was "open-cell" styrofoam (commonly used for carving fish bodies) available from boat dock and insulation companies as well as several taxidermy supply companies. I now recommend the use of polyurethane foam as it is not nearly as reactive to fiberglass resin, heat, paint lacquer thinner, etc., as the styrofoam is. (This was, indeed, a learning experience.)

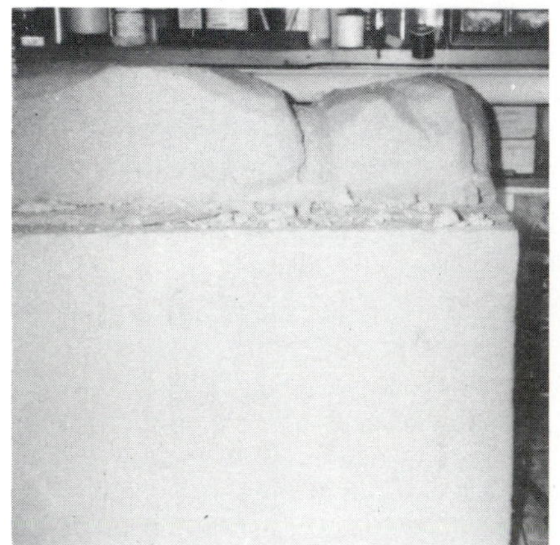

The trick to creating natural looking rocks is to make the rocks appear individual and random; however, at the same time they must lend the appearance of controlling and restricting the water flow as in a natural waterfall. Another "practical" consideration for this scene was that separate rocks would look great but resin would run everywhere when the Artificial Water pour was made if they were used, so a great deal of thought had to be put into the design of the rocks and water flow.

To aid in assuring a good fit of the foam rocks into the wood base, I traced poster board patterns from the base and pinned them to the sides of the foam while the carving of the foam took place. As the carving progressed, it was frequently placed in the base just to be sure that it was fitting properly.

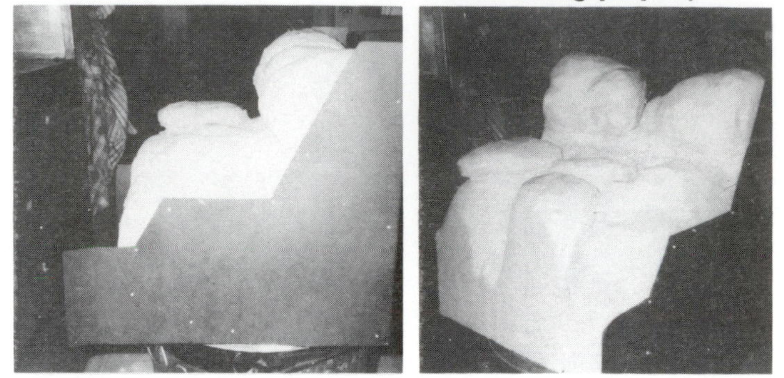

The lower section of the falls was designed to be a pool with a stream bed beneath it. A cut was made horizontally into the foam just below the mid-level. This cut was made to accommodate an 1/8" sheet of acrylic plexiglass which slid into the cut slot and extended out of the foam to the front of the base and rested on the ledge of the base. Knowing exactly where to make the cut falls back to the original design and sketch of the base in which the ledge was included.

The carved rocks (with acrylic plexiglass temporarily in position) were placed in the base for a final fitting, and then the acrylic was removed so a pattern of the rocks could be drawn on the floor of the base. This pattern showed me what area would be covered by the foam rocks and exactly what area was to be visible through the acrylic plexiglass. This area would be the stream bed, which required a relatively simple procedure to create, especially when considering that it would be barely visible later, due to the Artificial Water effects that would be used on the water surface. To create the stream bed, Elmer's glue was spread on the plywood base with a 1" paint brush over the stream bed area. Various sized natural rocks were randomly placed in the

glue, and then fine sand and gravel were sprinkled over the whole area. Once dry, the base was emptied of excess sand and gravel. A good coat of Grumbacher's Tuffilm "matte" fixative was sprayed over the bed for added security and to lend a soft matte "underwater" looking finish.

Now that the base was completely ready, the carved rocks of the waterfall were ready to be textured, sealed, sculpted, and painted. Several heavy coats of Polytranspodge FP191 were brushed onto the foam which helped to seal and close up the porous surface.

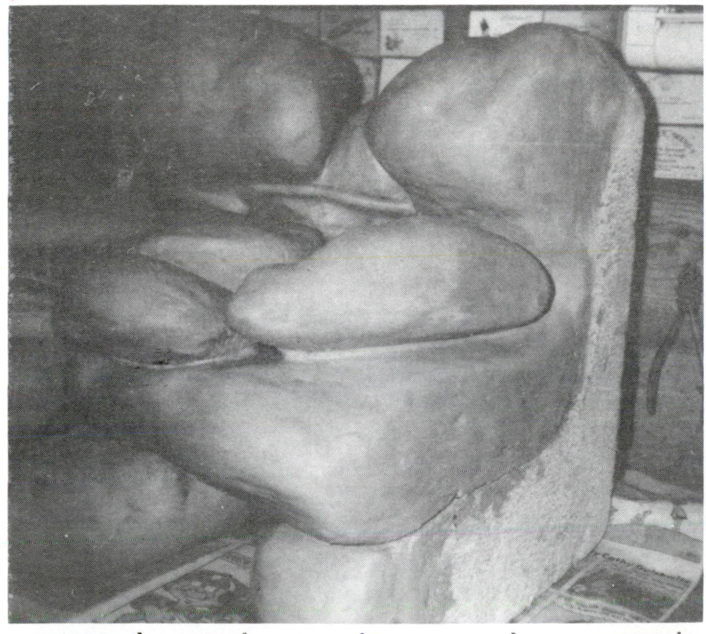

Potter's clay was then spread in an even thin coat over the entire surface that was to be exposed. When the clay dried, it formed cracks all over the surface adding a nice, realistic effect to the rocks. Most of these cracks were filled when additional sealer coats and paint were applied, but enough remained to be noticed and to lend authenticity to their appearance. After the clay had completely dried, it was sealed by another moderate coat of Polytranspodge FP191 before proceeding to the next step. As an added bonus, the clay also provided an excellent natural base color to paint over, (much better than blue foam).

One must keep in mind that a complex project like this has several smaller projects working within and related to each other, and it will be necessary to work on each of them simultaneously because of their dependence on each other. Preplanning the project will aid in this end. It is extremely helpful to list the various components of the scene on a piece of paper and mentally construct each phase from beginning to end.

The carved rocks were next used as a base for sculpting a model of the main water flow. I used Plastilene sculpting clay for this, as it is an excellent material to work with and comes in four grades. I used the No.1 which is the softest and easiest to work, and when slightly heated it will become even softer.

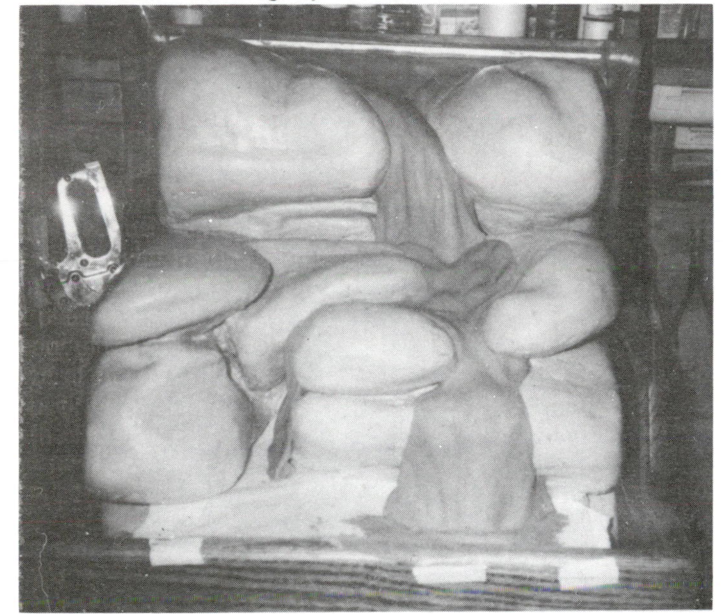

By making a clay model, the artist has the advantage of being able to sculpt into the scene any effect desired, i.e., ripples, turbulence, rings, waves, etc. All of these effects can be modeled right into the sculpture, molded, and finally poured with casting resin (Artificial Water). If the effect desired is to recreate water running over the duck's foot, as in my scene, then simply sculpt that in as well using a spare foot from a shop bird that is the same size.

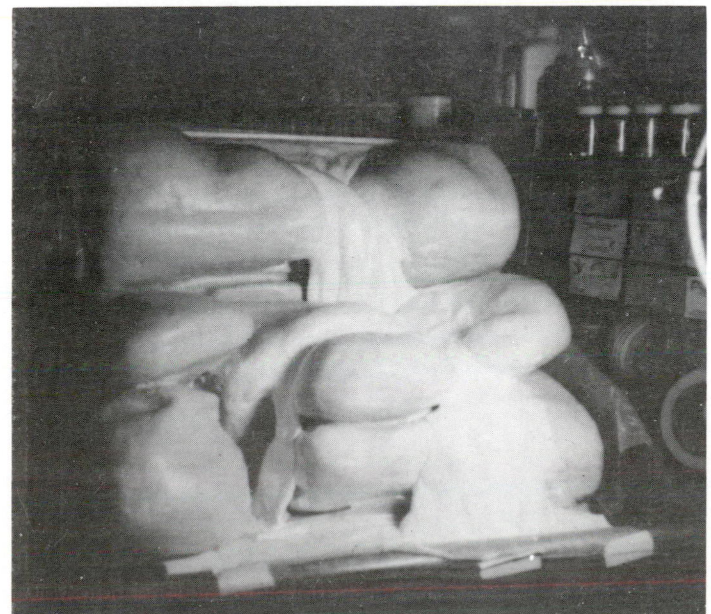

After the sculpting of the water surface was completed, a thick rubber mold was made of it by brushing on about a dozen coats of Mold Builder liquid latex rubber (brushing on one thin coat at a time, allowing it to dry and then another, etc.).

I would recommend a silicone rubber material as being best because of its strength and versatility, however, the difference in the high priced silicone is significant when compared to latex. If the mold is a one-time mold and not intended to be used in several future scenes, then latex would probably prove best. Always brush enough rubber over the edges of the sculpture to form a good flange. Once the rubber mold completely dried and cured, I cut it with an X-acto knife into sections. I made cuts for the upper pool, the first waterfall, the middle section, and the two lower waterfalls. The flimsy rubber molds need reinforcement jackets to support them during casting. To make jackets or "mother molds" for the latex surface molds, I used Bondo mixed with excelsior (wood wool) for strength. This proved to be a quick and easy method.

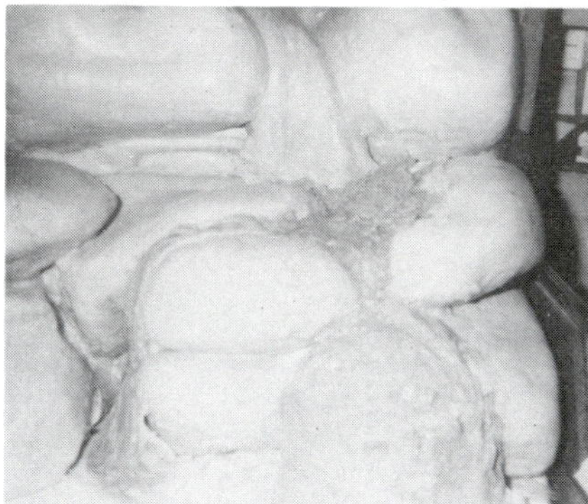

After each jacket was made and then removed, each section of rubber was peeled away from the clay and carved rocks. (Note: If you're wondering about the lower pool, a different molding technique was used and will be covered later.)

Once the water "surface" molds were removed, the bottom or "underwater" molds had to be made. The separate sections of clay for each level were very carefully removed one section at a time. Some fell apart slightly but were easily put back together. The sections were flipped over and the surface molds were put back in place. This exposed the underside of the sculpture which was rough and irregular. This had to be smoothed out enough so as not to lock-in the next mold and create problems releasing later. These bottom sections were molded using the same techniques as the surface molds, i.e, using latex detail molds with Bondo jackets for support.

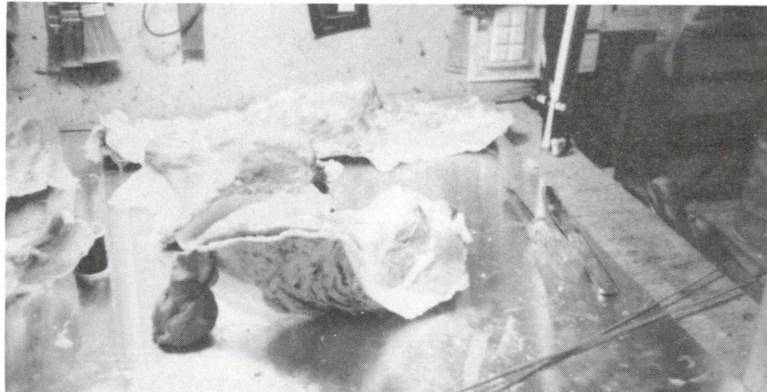

After the molds had completely dried and cured, the clay was carefully removed from the molds. The molds were then rinsed out with soap and water, allowed to dry, and a cardboard box was set up using wire supports to hold the molds upright for pouring the Artificial Water.

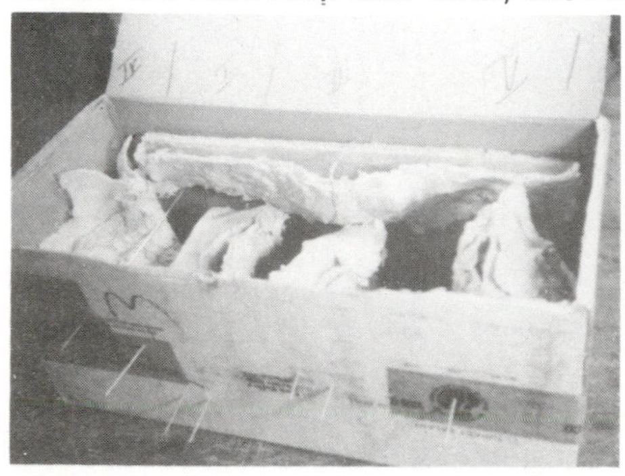

Some of the molds had a large entrance for the resin, others only a small spout, depending on the nature of the section.

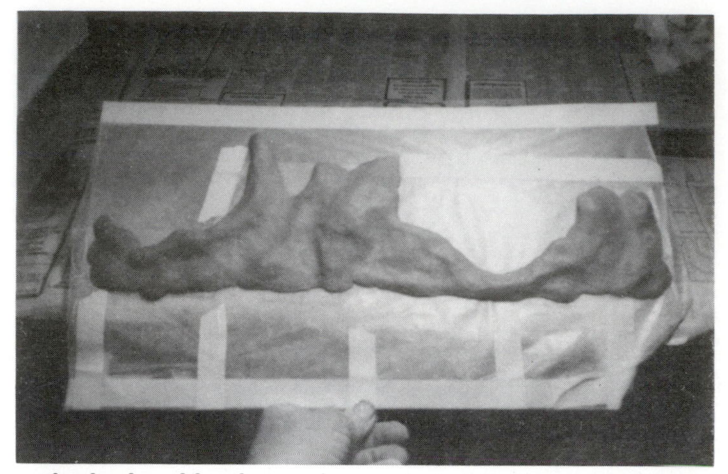

The final mold to be made was that of the lower pool water surface. This section was modeled in Plastilene clay (on wax paper). The sculpture was done directly on the acrylic sheet, and then removed prior to making the mold.

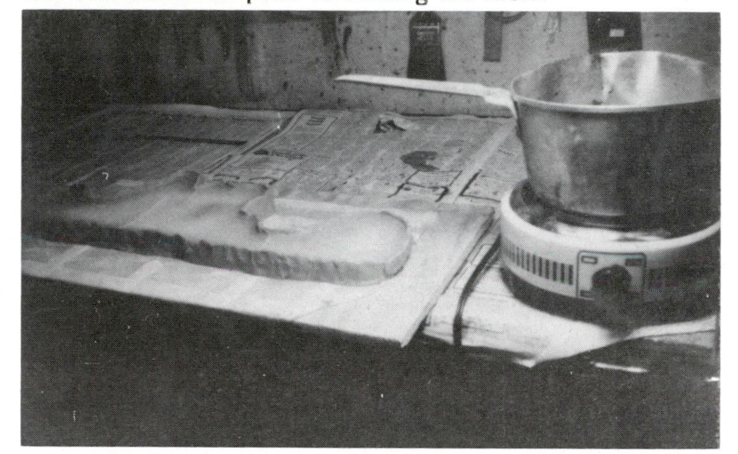

To make the mold, it was placed on a flat surface and a clay wall or dam was built around it. This time, a parrafin wax mold was made because the underside would remain flat and rest on the acrylic and the surface detail was all that was actually needed. I used parrafin wax from WASCO which was heated until melted and then poured into the mold until it was full. The mold had been first lightly covered with Vaseline for a release agent.

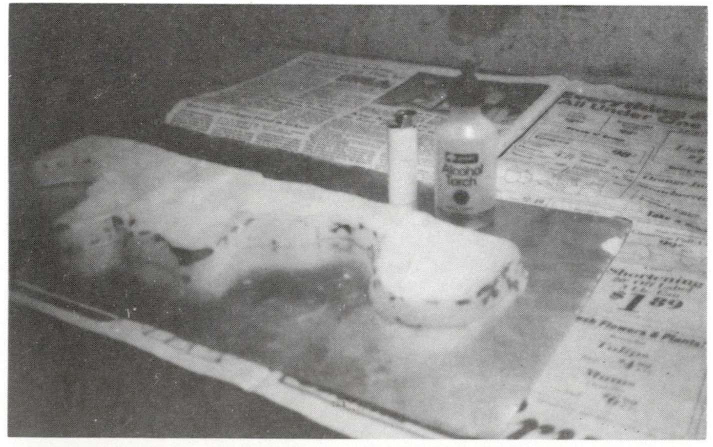


When the wax cooled, it was flipped over, and the clay was removed. This resulted in some scarring in the mold but was easily smoothed out by heat from a small alcohol torch. I used *Polytranspar*™ Lacquer Thinner (after the torch was extinguished) to help remove the clay (it had no ill effects on the wax). This was the easiest mold to make and it was set aside along with the others for the pouring.

With all the molding out of the way, I turned my attention back to the rocks which were now ready for the final sealer coat. This was done with Olympic polyurethane liquid plastic No. 1607, which is available at some of the larger hardware outlets (Handy City, Moore Handley, etc.). This material provided a tough finish, and acted as a final back-up in case any resin managed to work its way through the coats of paint and acrylic.

The paint job was a combination of airbrushed *Polytranspar*™ Water/Acrylics and a variety of application techniques; with a sponge, a toothbrush, and a paint brush.

The first "airbrush" coat was a mixture of White (WA10) and Black Umber (WA29) and was used for shadowing the depressed areas and circles between the rocks. This was a very important step in lending a realistic look to the rocks. Shadowing and highlighting are very important in every painting operation, as a natual rock has both. Highlights occur where light directly hits, and of course, shadows occur in areas not exposed to light. Also, the color texture of rocks varies considerably; being light, medium, and dark in various areas. The next "airbrush" coat was a light overspray with Yellow Ochre (WA141). I applied this a little heavier in sections where the water would flow. This was followed by a moderate spray of Hooker's Green (WA63) which has a good mossy look to it, and was applied primarily in areas of heavy water flow.

With the airbrushing out of the way, splattering and sponging the *Polytranspar* Water/Acrylic paints was next. A variety of colors were used: White (WA10), Gill Red (WA160) mixed with Sienna (WA200) and Sailfish Blue (WA300), Burnt Umber (WA71), Black Umber (WA29), Hooker's Green (WA63) and Shimmering Blue (WA440).

These paints were thinned with water a good deal, as it's better to apply too little paint with light washes than to apply too much and end up with too opaque of a coat. Splattering was accomplished by dipping an old toothbrush into the paint and either flicking the brush at the rocks or holding the brush in front of the airbrush and spraying air on it, which also sent the paint flying off the bristles onto the rocks. If paint was applied too heavily in certain areas, a damp sponge lightly pressed and stippled into the area helped to lift the excess paint and remove any harshness. (These techniques will require some experimentation.) The results are well worth it, leaving an excellent speckled effect with good inlay color blending and shadowing. I did a great deal of stippling with a small stiff-bristled paint brush using Black Umber (WA29) in cracks, crevices, and depressed areas for shadowing. For highlighting, the "scrub" side of a sponge was raked across the rocks from all angles. This scratched and removed enough paint from the high points and edges for the desired effect. The result was very pleasing, lending an old, worn-out, rough look to it, which was quite natural and gave the appearance that the rocks had been there for awhile.

The rocks were again sealed with a light coat of Polytranspodge and then lightly stippled with a coat of Ultra-Glo Epoxy Resin (a clear acrylic material that is used on cypress clocks, counter tops, etc., to give them an extremely high gloss). By stippling this acrylic as it gels, one can form tiny air bubbles which will remain in the finish, lending an authentic look to the scene. This coating of Ultra Glo also added further protection to the carved styrofoam rocks from the casting resin.

The harlequin duck for the scene had been mounted prior to the painting of the rocks and was placed on a temporary base during the painting, carving of the rocks, and water work construction. The carved rocks were next set in place in the hardwood base with the acrylic plexiglass sheet intact. The edge between the carved rocks and the hardwood base was then sealed with strips of clear hot glue to keep resin from seeping into the floor of the base. Instead of applying hot glue directly which would melt the foam and damage the paint job, strips of hot glue were cooled in water. This was accomplished by squirting the hot glue across a cooking pan lined with tin foil and filled with water. As the glue is laid across the water, it cools, hardens, and maintains its shape. (This is how some people make catfish whiskers.) I next take these glue strips which are still somewhat soft and press them into the edge between the rocks and the wood base for a gasket. Next, the wood base was covered with masking tape to keep it free from the messy resin pour.

Finally I was ready for the artificial water pours. The molds were poured one at a time with *Polytranspar*™ Artificial Water that had been very slightly tinted with black dye. (It doesn't take much, so be careful.

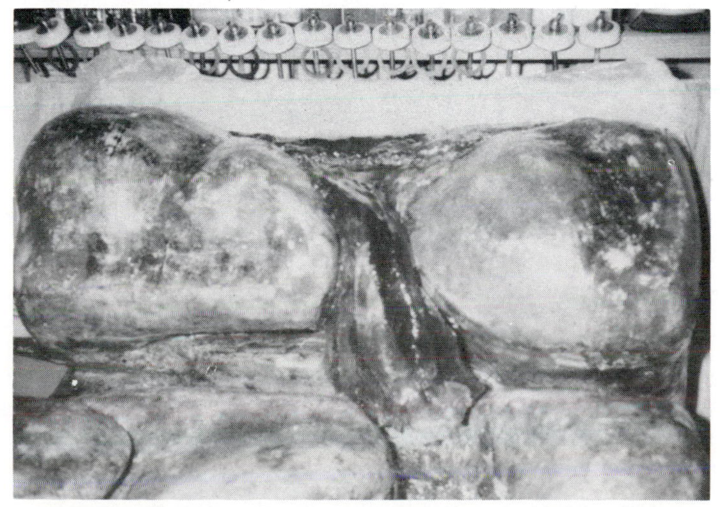

Several different colored dyes are also available, and I've

also tried using oil paints to do this, with success.) After the first section that was poured had cured, it was removed and placed into the scene, adding a small amount of catalyzed Artificial Water to the rocks to hold it in place.

Some of the sections were not a perfect fit which caused a great deal of anxiety for me until I discovered that the resin can easily be ground (with a Foredom tool or whatever) and polished on a degreasing wheel. When fresh Artificial Water is brushed over these areas, the scratched area becomes invisible and will look crystal clear.

This method of modifying the shape of cured resin without it being noticeable will aid in repairing and achieving various water and ice effects. I recommend wearing goggles and doing this far away from your habitat base. It's best to wear latex gloves too. The resin can also be modified and bent into various different shape by heating it with a hair dryer or placing it in hot water. Each section mold was now poured and allowed to cure, then set in place in the same manner.

Once the sections were all in place, the next step was to blend each section into the next and fill in around the edges of the rocks. This was done by mixing small amounts of tinted Artificial Water with catalyst and dropping small clumps of resin into the cracks and voids. Finally, after all of these were filled, several thin coats of resin were brushed over the entire water flow to smooth any unwanted lumps, and to give it a continuous water flow effect., Catalyzed Artificial Water was stippled and splattered randomly elsewhere on the rock and the appearance of air bubbles was created and added to the surface. These bubbles were made from the clear plastic coating which enwraps Tohickon glass eyes. Each artificial bubble was cut out with an X-acto knife and placed in the final coat of resin as it began to gel. Additional resin was slightly brushed over the plastic air bubbles to blend them into the water. Drops were formed by allowing the resin to begin to gel and then small amounts were stippled to the rocks. I accomplished this by dipping a wire into the catalyzed Artificial Water until it would hardly fall from the wire. I then lightly touched the area where I wanted a drop to appear and it left a perfect droplet. This same technique was used to apply the drops to the bird also. (I recommend that an old mount be used first for experimentation until the technique is perfected.) Remember to always wait until the artificial water drops bead up exactly to the desired form before placing them on a bird.

The final step in the creation of this piece was to place the bird in position. I attached the bird to the rocks by applying clear 5 minute epoxy under the feet and then all voids around each foot were filled with clear catalyzed Artificial Water.

Artificial Water was also stippled on the feet and bill for a wet look. The final coat was a thin layer of the Ultra-Glo epoxy resin for a soft, durable, non-tacky finish.

There was much trial and error involved in the creation of this piece and a good deal of patchwork before reaching desirable and acceptable results. This information has been offered to you as "one" way of doing it (not the only way). These methods are intended for the reader's use, to experiment with, and modify, and hopefully improve upon according to individual needs. If nothing else, it is my hope that a great deal of time and frustration will be spared the readers which always seems to go hand-in-hand when reaching and searching for new and better methods.

# 5. Construction of Underwater Habitat Bases

Dan Blair of Peterson, Iowa, accomplishes for the first time an underwater habitat scene using a solid core method and then demonstrates an easier process, the hollow core method. Despite the obstacles, Dan has opened the door for many new and innovative ideas for water diorama bases.

***by Dan Blair***

## A Solid Core Underwater Habitat Using Artificial Water

The challenge of creating this, never before attempted, art piece (by me at least) was great, but the rewards seemed far greater; great enough to present the possibilities to *Breakthrough* Publications Editor/Publisher Bob Williamson. I proposed building an Artificial Water scene complete with fish habitat and a duck swimming on the surface. This wasn't unusual. However, I planned to pour my scene completely solid with Artificial Water.

"Do you know what you're doing?" challenged Bob, and my reaction would have been to question his apparent lack of confidence in my ability. However, before my ego took charge of my mouth, Bob followed up his question with an answer of his own. "I have never heard of anyone trying this before. You could be starting a whole new trend in habitat displays, if it works."

My own eagerness and excitement from my prior experimental plans, coupled with Bob's encouragement to attempt the unattempted, gave me the surge of incentive and inspiration to carry out the effort. And so I did! Little did I know what was in store for me!

My first concern was to decide upon the best material and technique to use creating the Artificial Water pour. The mold would need to be smooth-surfaced and clean; preferably made from a non-static material so as *not* to attract dust or the hair and lint particles so common to a taxidermy studio. Secondly, the shape of the mold from a standpoint of esthetics was also an important consideration. I ruled out constructing a mold made from sheets of plywood or plexiglass as they would have been square with sharp corners making handling somewhat hazardous. From an artistic standpoint, rounded corners would be more appealing to the eye.

Upon discovering an ideal subject for my mold, a Rubbermaid wastebasket, I was both excited and surprised. Making the mold was going to be *considerably* less time consuming than I had anticipated; and the total cost of the mold in terms of time *and* money was 10 times less as well. This combination alone could easily justify the use of the wastebasket for the mold. Additionally, I could justify it with the following:

**1.** The wastebasket has an excellent shape, being rectangular, but with gradually diminishing dimensions, plus the appealing quality of beautifully rounded corners.

**2.** It was readily available in a variety of sizes from several stores.

**3.** It was reuseable, with only a small amount of maintenance required.

**4.** By using the wastebasket there were virtually no waste materials from its construction.

**5.** It was easily removable from the finished casting *without* the added expense of time or money for messy mold releases, etc.

Of course, with all that going for it, there had to be at least *one* flaw and I recognized one immediately. Simply put; "static cling!" But for that the answer was simple. All that was necessary was to wipe it with a "tac" cloth to remove unwanted dust, etc., before making the resin pouring. (A much greater problem would arise later but wasn't suspected during the initial examination of my potential mold.)

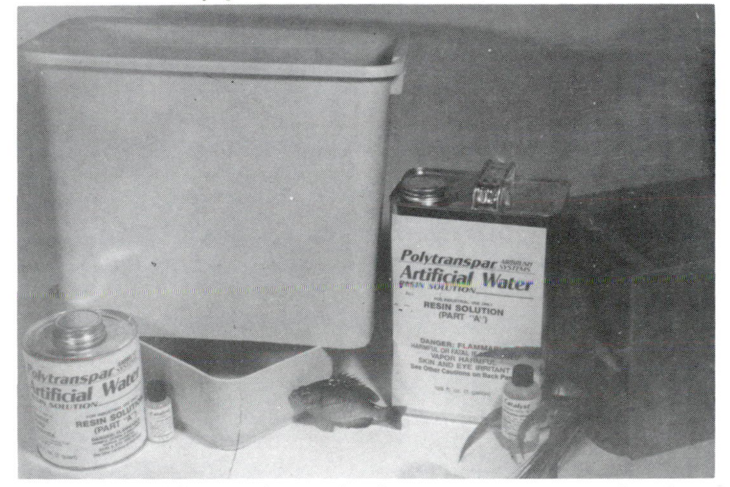


I used a Rubbermaid wastebasket 7½ × 13 × 15 inches from which I cut off 2 inches of the bottom. (This cut-off section can be used as a handy storage tray, a container for making half-body plaster castings of panfish, or as I did, use it for mixing batches of Artificial Water.) The cost of this item is only slightly more than $5.00 and the one I bought on sale was less than $3.00. Even at full retail prices, it was a very inexpensive mold.

To begin actual construction of my habitat exhibit, I needed a foundation to build upon. Since I intended to "light" the column from underneath its platform, I had to consider a transparent base. The answer was simply to use a sheet of plexiglass 1/4" thick for a base. This was cut slightly larger than the base size of the wastebasket so that it would rest on top of it rather than around it.

Next, using Plastilene clay as my sealer, I laid a line of clay the diameter of my little finger all the way around the outer edge of the wastebasket. By pressing the lip of the wastebasket and clay down firmly onto the plexiglass, it completely sealed the two together, reducing the chance of any resin leaking "out" or air leaking "in" during the pouring process.

Once I'd mated the base and mold together, I was ready to determine the locations of my habitat materials. I had previously sketched out an approximate design of what I wanted to create, so it was easy to begin hot-gluing the driftwood snag and aritifical rock to their ideal locations on the plexiglass base. Once I'd done so, I could pinpoint exactly where to locate the six hidden "windows" that would be used to project light into the underwater diorama.

From the beginning it is important to remember to *not* allow anything from the habitat scene to touch the inside of the mold for a *solid core* casting. Contact to the side of the mold could cause a dimple, or may even captivate an air bubble, thus creating a flaw difficult to perfectly conceal. It is easier to prevent this potential problem from occuring, than correcting and repairing the actual problem after it happens.

Next, 1" square windows had to be cut from the same 1/4" plexiglass used for the foundation base. I used a saber saw with a blade designed for cutting plexiglass to cut the squares.

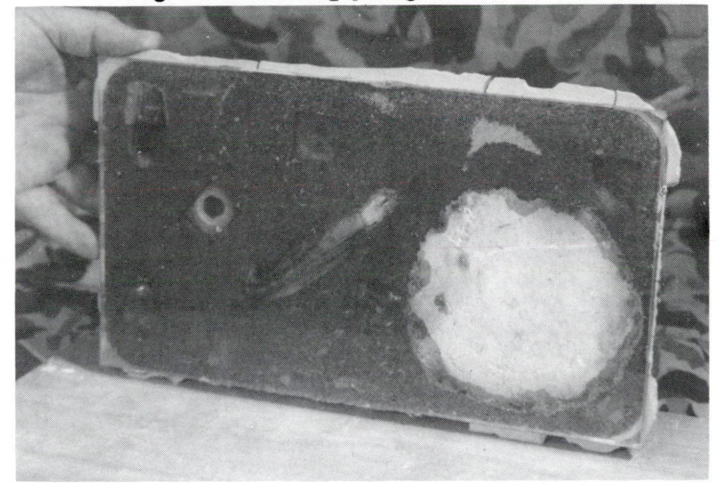

The windows were arranged about the bottom of the exhibit to evenly distribute the lighting. Super glue on the corners of the 1" squares held them securely in place. Once the glued window squares had "set," a 1/4" layer of fine sand was poured into the bottom of the exhibit. Then, using a long-handled artist's bristle brush, excess sand was swept away from the upper surface of the windows. Also using the brush, the surface of the sand was leveled, taking particular care around the rock and wooden snag. (Any imperfections or foreign particles visible in the sand were removed at this time.)

*Polytranspar™* Artificial Water was then thoroughly mixed (about 1/2 quart) with the catalyst and slowly poured directly onto the windows. This was done because the sand could "drift" under the pressure of the flowing resin if the resin was poured directly on it. The resin flowed off of the windows evenly and soaked into the sand, firmly securing it to the plexiglass base.

I then allowed this base coat of resin to "cure" and harden overnight. The next day, I carefully removed my mold from around the base. The Plastilene clay was also removed and put aside to be used later. With the mold removed from the base, it was an easy matter to arrange the vegetation and life forms about the interior of the exhibit. The wastebasket mold could be easily refitted to the base after the arrangement of the habitat.

Additions to a habitat are as limitless as the artist's imagination. In this case, I used three forms of snail shells, two types of clams, and the broken claw from a crayfish. I also used both natural and synthetic aquatic plant life; some of the synthetic being plastic plants designed and created for use in aquariums. I also used long-wired stems and leaves from silk flowers. Natural greenery such as

sheet moss was used about the bottom of the habitat, needle grass was interspersed across the bottom, and long pine needles painted green were added to give the impression of new shoots of grass amid the old. A small maple leaf was added to use as one of the screens to conceal a window from view. Next I added three mounted sunfish. I also included one old corroded Mepps type spinner, a miniature dry fly, and a rusty hook and swivel. I used an empty shotgun shell on the bottom of the exhibit. How I incorporated it into my light system will be explained shortly. (It was my intention to create a connection between the shotgun shell and the bufflehead drake that was swimming across the top of the scene to stir the imagination of viewers of the exhibit.)

When attaching plants and shells, etc., to the base, I did so in such a manner as to screen or hide the light windows from view. I accomplished this by using sheet moss and building a cone (like a volcano) around two of the six windows. One

of the windows was shielded by long-leaved stems of plastic aquarium plants, yet another was covered with an old discarded plastic fish hook box, (shown in the photo above), and still another was hidden by a half-opened clam shell. It is the sixth window which attracts the most attention as therein (or thereon) sits the empty, red shotgun shell. But, there's more here than meets the eye, because rather than letting the light shine "around" the shell, the brass end of this one was drilled and opened to let the light shine "through" the shell. The effect produced is truly unique and must be seen to be appreciated.

After all windows were screened (not meant to be a play on words) with sufficient cover, so as to not be noticeable from the exterior of the resin column, additional vegetation and other life forms were added to give the layout balance and continuity. It's almost like doing a painting in three dimensions with the added feature of being able to view the canvas from four sides instead of only one.

To make it easier to handle the base and view it from all sides quickly without constantly picking it up or continually handling it, a simple kitchen-type "Lazy Susan" swivel arrangement was utilized as a good solution to the problem.

Once the landscape was completed, I was ready to begin the additional pours of Artificial Water. To begin, I replaced the clay sealer ring and wastebasket mold onto the base in the same position and manner as when I first started. Always remember to make certain that a good clay seal is added and does not contact the base. The safest way to do this is by putting the ring of clay around the base itself, rather than on the wastebasket. At this point, it may also be a good idea to secure the edges of the mold to the previously poured base while the final pours are being made. I did this by literally nailing the wastebasket at the base all around the edges of the mold and bending the nails in toward the rim, to hold it both over and down in place.

Another precaution I took at this time, was to build reinforcements to fortify the sides of the wastebasket form. This would prevent the somewhat flexible vinyl mold from swelling outward from the weight of the resin. This was easily accomplished by cutting *two* pieces of masonite (plywood would work equally well) slightly larger than the two widest sides of the wastebasket and attaching them together on both ends with screws to the 1" × 4" support pieces that were cut to fit at the proper angles.

Once I was certain that everything was secured, properly aligned, and reinforced, I began the pours of *Polytranspar*™ Artificial Water. Be sure to follow the mixing instructions on the container exactly, paying particularly close attention to the mixing procedure itself. It is absolutely necessary with *all* two-part components to mix them thoroughly to ensure an even and total cure. The few minutes or more spent on *extra* mixing could save a great deal more than just the cost of the resin itself. We had to invest a tremendous amount of time, thought, energy, *and* money to build this project correctly, and we certainly didn't want to write it off as a loss for the simple mistake of ignoring the warning, "Haste makes waste." (Believe me, I was reminded of this axiom the "hard" way with this "very" project.)

I started making my resin pours at the rate of one inch or so per mixing. From the top down it looked absolutely beautiful and perfect until, after the third pouring I noticed the catalyzing resin got hot enough to smoke slightly and actually began to melt some of the plastic plants. Another problem also developed when the heat generated from the catalyzed resin began to warp the vinyl wastebasket mold.

Additionally while watching the curing of the third pour, I overheard a soft buzzing sound coming from inside the mold which sounded much like the beating wings of an insect which I envisioned landing in my resin. What I actually discovered was much worse. The sound was being created by the contractions of the over-heated resin which, within itself, was creating fan shaped vacuum pockets as it split away from itself. It was literally tearing itself apart. (At that point I was doing much the same to my hair.) I stopped at that level and after the resin was cured up hard, removed it from the mold.

Inside I had "cooked" three perfectly good mounted sunfish and encapsulated a great deal of habitat materials, fishing tackle, and time. I was disheartened but, I couldn't write the effort off as a complete loss because I was learning, and still believed that this experience could be turned into a profit. (I must admit though, at this point I was beginning to wonder if this was the reason that no one had attemped this type exhibit before.)

All of this brings me back to my earlier reminder that "haste makes waste" to which I want to add two more statements that I have wholeheartedly agreed with for some time: "Excellence takes time" and "There's no shortcut to success."

If you attempt the solid core resin base, plan on taking the time to do it right. Further research into casting methods disclosed that for this type of project we would need to pour resin at the rate of *no more* than 1/2" of resin, in a 24 hour period, otherwise too much heat would be generated by the catalyzed Artificial Water resin.

Obviously the resin should be covered during the curing stages to keep dust and other unwanted particles from marring the surface of the soft resin, particularly overnight. (How well I remember another project in which resin was applied to a redwood burlwood table top. The sticky surface captured a big black fly which became a permanent addition to the glass-clear finish.) I used a large cardboard box as a dust cover to protect the resin from lint and dust, etc. while it cured. It was big enough to fit over the entire exhibit without touching it.

As I worked my way up the interior of the mold, I added a few acrylic beads, which come in three or four different sizes, to recreate air bubbles. These were added at several random points and levels in the resin. This was also a good time to include a water beetle and a few freeze-dried minnows.

The fish were positioned to give them the appearance of swimming by suspending them in the resin itself. I accomplished this by tying a string lightly to one of the dorsal fin spines and tying the other end to a piece of dowel laid across the top of my mold. It was easy determining which spine held the fish at the best angle, and just as easily the depth of the fish could be adjusted by moving the string up or down.

Be certain to mark the exact level at which the swimming duck will be located. Remember that the duck's feet will hang down and must be inserted at the proper level. Know the limitations beforehand and be ready.

To avoid a saturated look to the breast and belly feathers and to keep them from "soaking" in the resin prior to its curing, I tried sealing the submerged area with Fin Backing Cream from Wildlife Artist Supply Company with satisfactory results. Use a long bristle brush of about medium stiffness, and apply the cream evenly and smoothly only to those areas which will actually contact wet resin. One complete coat is sufficient and should be thorough but thin. (I doubt that this method would be beneficial if used for downy baby waterfowl.)

Once the resin was within 1/2" to 1" from the top, it was time to consider which method would be best for making waves and ripples on the surface. You may choose one of other means described elsewhere in this manual, but I proceeded as follows.

With only 1/2" or less left to pour, I secured a rotary file in my drill motor and sculptured in the ripple rings around the perimeter of the duck. This was a simple task to perform. All that was necessary was to wipe it from time to time with a clean rag dampened with lacquer thinner. This enabled me to see what the finished waves would look like. Once I had satisfactorily resurfaced my resin water, I brushed on a thin coat of resin to bring back the water-clear appearance complete with ripples. (Be most careful with this final coat that no hair or dust particles be allowed to make contact because it becomes very noticeable.) This concluded the solid core method.

# The Hollow Core Method

One drawback of the solid core casting method of the size just constructed, is the weight. A gallon of resin weighs about 9 lbs. and it takes two or more for one casting. Add the weight of the other ingredients, plus the weight of a wooden box to stand it on and in which to house the lighting system and you've got a half pound duck mounted on a 30 lb. base. Excessive weight, the expense of the resin, and length of time to make the pour may inspire you to consider the alternate, but similar display method, the *hollow* core resin casting.

The hollow core method required using plexiglass, so I built a box just slightly smaller than the wastebasket mold that would fit inside, but which would still allow me to pour about a 1/2" coat of resin "around" it. This gave me those same smooth lines and rounded corners on the outside perimeter of the exhibit as desired. It is important to be sure to have all joints of plexiglass tightly glued (I used super glue) to keep air bubbles from appearing and being seen at the corners. This also prevented resin from leaking into the inside of the box during the pouring stages.

This time, the plexiglass base holding the actual habitat materials had to be smaller than the first one because it had to fit *inside* the plexiglass box. As a matter of fact, I built this one to be removeable so that I could change the decor from time to time, either slightly or completely. (That's a feature definitely not available to the solid core casting.)

I attached the glass window and habitat materials in much

the same manner as I did with the solid core method. One difference though was that I didn't use resin to coat the sand which would have permanently attached my base to the inside of the glass box. Instead, I put habitat vegetation everywhere I wanted it and then painted Fin Backing Cream on the areas I wanted covered with sand or gravel. The nice feature of the Cream as opposed to glue is that it doesn't run off the edges and the sand or gravel can be tamped down into it with a finger tip which secures it better than gluing.

Once my habitat interior was constructed, I placed the *empty* plexiglass box on a layer of newspaper (several sheets thick) inside the base of the wastebasket mold. I then placed a clay seal firmly along and against the bottom edge to keep resin from leaking through it. A seal of clay on the lip of the mold also prevented resin from lifting the mold and preventing resin from leaking out. Likewise it is a good idea to secure the mold with nails, etc., to keep it from moving once the plastic box is centered inside the mold.

Now was the time again to slip on the reinforcement form that I had previously made for the solid casting. (Note: If you begin to notice the wastebasket walls warping inward and away from the reinforcement, use double-backed tape to keep the mold in shape by taping it to the masonite reinforcement walls.)

Once the mold is wiped clean with the tac rag, clay sealed and secured with the sealed plexiglass box centered inside it, we were now ready to pour the resin shell around it. By keeping the glass box large enough to keep the amount of resin used to a minimum, we could make the complete casting in just one or two pours. The 1/2" outer shell of resin won't create the degree of heat that a deeper or thicker pour will. To further ensure against such a possibility, I kept a fan blowing cool air against the mold during both the pouring and curing processes.

To make pouring easier, I found that recycled plastic milk jugs were ideal. On my bandsaw I cut off the top two or so inches straight across. What normally would be the handle became a flexible spout which can be tipped down and directed into any corner or crack for "mess-free" pouring. Just be certain that the container is free of water or dried milk before pouring premixed resin into it.

As I poured the resin around the hollow core, I found that less air is apt to be captured and trapped if I poured into one location and let the resin self level, rather than moving the spout back and forth around the inner perimeters of the mold. Regarding trapped air, remember again the mixing instructions. Be certain that the resin is not mixed so rapidly that air is blended into it as those bubbles build upon themselves and are not likely to reach the top of a 12" casting before the resin begins to gel. (I time my mixing from 5 to 7 minutes of slow but thorough stirring with a rubber cooking spatula.)

Since the duck in a hollow core base won't be cast directly *in* the resin, different steps must be taken. My method was to include a plexiglass top with an appropriate size and shape cut out of it with edges tapered down and inward to accomodate the mounted bird. Using the same rotary file I used on the resin, I shaped ripple rings and waves into the plexiglass before I started the pours. Then, as I poured the resin up to the top, I had only to slightly cover the outer edges flush with the plexiglass lid. Once cured, it was easy to remove the excess resin and blend the two materials together with the file.

Next, I painted a light coat of resin (with a clean brush) over the plexiglass which restored the water shine and covered scratches and chips almost entirely.

Since there is little, if any, resin on the glass top of the casting, it may be desirable to blend the top and sides together with a slight spray of *Polytranspar*™ transparent paints such as Light Bass Green FP/WA50 or Sailfish Blue FP/WA300 applied at extreme angles across the surface very lightly. Tops of ripples or waves can be highlighted with FP/WA440 or 441, the Shimmering Blue and Green paint. I also used the Pearl Essence colors and applied them to the top or inside to slightly fog the transparency.

After allowing the completed mold to cure overnight, I pulled the sides loose from the resin and after removing the nails that were holding down the mold, slipped it off the casting.

It was at this point that I found *one more* inherent problem with the mold. The mold for the solid core method was a cream colored vinyl and produced a clear casting. The mold for the hollow core method was a chocolate brown in color and left a similar tint in the resin. The solution was obvious—Don't use a brown wastebasket.

*The bufflehead mount in this scene is by Tim Hecht of Fairmont, Minnesota*

Once out of the mold, we noticed a slight hazy effect on the outer surface of the resin which is from the texture of the mold not being glass smooth. To correct this condition, I tried two remedies. One method was to pour or paint more resin on the outside of the column. Pouring allowed more runs to occur due to the lack of control over the volume of the pour. Painting a thin coating on the otherhand, was a bit more controllable but still showed some streaking from variations in the thickness of the resin. (It is recommended to apply the resin from top to bottom rather than with a side to side stroke.)

My personal preference turned out to be two or three heavy coats of *Polytranspar*™ Competition Wet Look Gloss FP-241 applied with a DeVilbiss Spray Gun. Once the finish coats were thoroughly dried, the hollow core was set over the habitat base and the duck was placed into the cut out of the hollow core.

The "column of water" needed a base so we built a wooden box of walnut which stood about 4" deep. Into its lid, we routed out a section into which the bottom of the column was placed.

One may elect to remove all but enough wood to hold it securely in place, or with a hole saw remove just the locations of the windows.

Use either cool florescent lighting, or very low wattage incandescent bulbs, with wire sockets for the lights. Place the bulbs in locations that will best amplify the light through the windows. Another suggestion worth consideration is the use of miniature Christmas tree lights of blue, green, yellow, and white in combinations or singly. The type which randomly turn on and off could simulate sun beams bouncing off the bottom as reflections from the waves above.

In any case, we created a "one of a kind" example of indoor art which will enhance the decor of any outdoor enthusiast with a creation both beautiful to see and functional to use. In fact, it can be personalized several times over with contents from your customer's own collection; daughter's first duck and the shell casing from the shot, shells Mom picked up at the

beach, one of Grandpa's old fishing lures, the rusty pocket knife junior found at camp, and so on. The possibilities are completely unlimited.

Stop and think about it. Consider buiding one of your own designs into a solid or hollow core pedestal base using these guidelines to create a truly one-of-a-kind work of art. Who knows, you just might start a whole new trend in taxidermy competition. (If it works!)

# 6. Special Effects: Artificial Water & Lighting

World Champion Taxidermist Bob Elzner of Apache Junction, Arizona describes in detail the techniques that he used to complete the rendering "Genesis: And On The Fifth Day." This beautiful piece has won numerous awards on both the state and national levels. Bob is particularly pleased with the rendering as he recently has become a Christian and he created it in reverence of God. The first part of the title of this piece is "*Genesis*," which means "in the beginning.' The second part of the title, "*And On The Fifth Day*," identifies the day that God created fowl.

***by Bob Elzner***

"Genesis" was created as my artistic expression and interpretation of how the creation of the fowl of the earth by our Lord, the Great Creator might have taken place. Two baby ducklings were chosen to depict the original male and female of the species that were directed by God to "be fruitful and multiply." Water, light, and the earth were also portrayed in this rendering as all were created prior to the fifth day, by our "all powerful" Lord.

The materials needed to create this scene are:

- 12" × 12" × 5" high black walnut case
- Light and socket
- Cord
- Plexiglass (1/4" and 1/8")
- Spray adhesive
- Foil
- Wood glue
- Black paint
- Ultrafine Gel
- *Polytranspar™* Artificial Water
- Plastic
- Name plate

To begin the project two hands had to be made. I used Ultrafine molding material to build the molds. Ultrafine is an Alginate substitute material used by dentists to make a quick set mold.

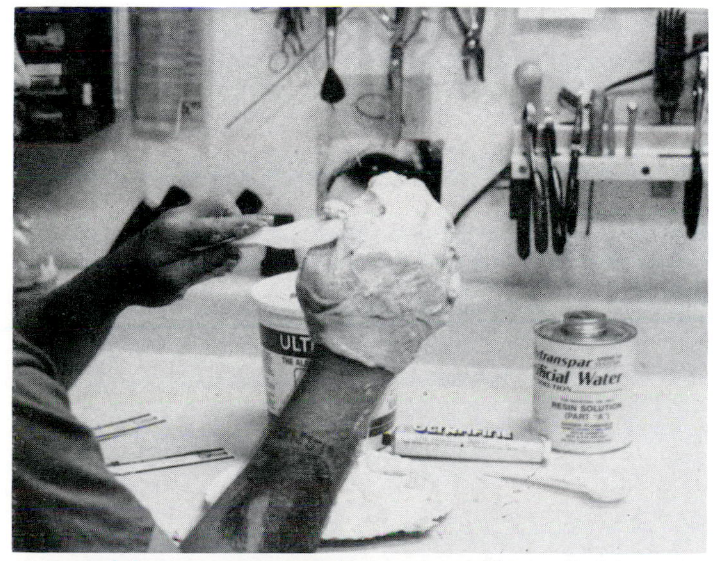

I mixed the two parts together and quickly applied it to one of my hands. It only takes about 30 seconds to start setting up, so I had to move quickly. After waiting about two minutes I gently removed my hand from the Ultrafine mold.

After both the molds of my hands were made I propped them up in a sand filled bucket. Then I mixed up Polytranspar™ Artificial Water and Catalyst to pour into the mold. I poured the hands one at a time and let each hand cure for 24 hours before removing them from the molds.

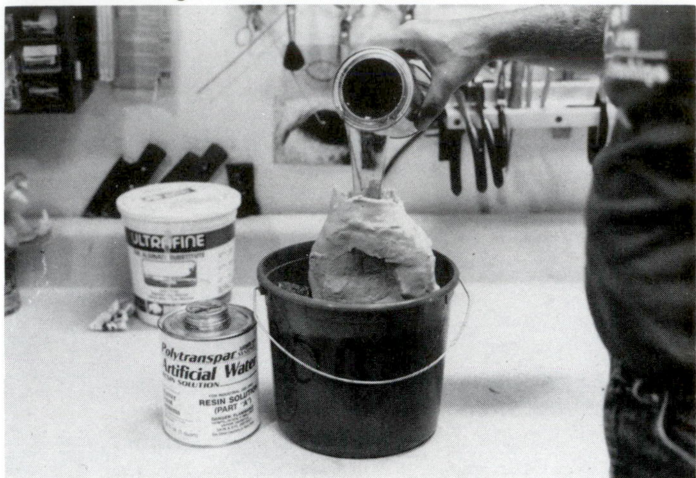


NOTE: Cool water can be added to the sand bed to lessen the amount of heat generated from this much catalyzed Artificial Water. Also a fairly weak percentage of catalyst was used for the same reason.

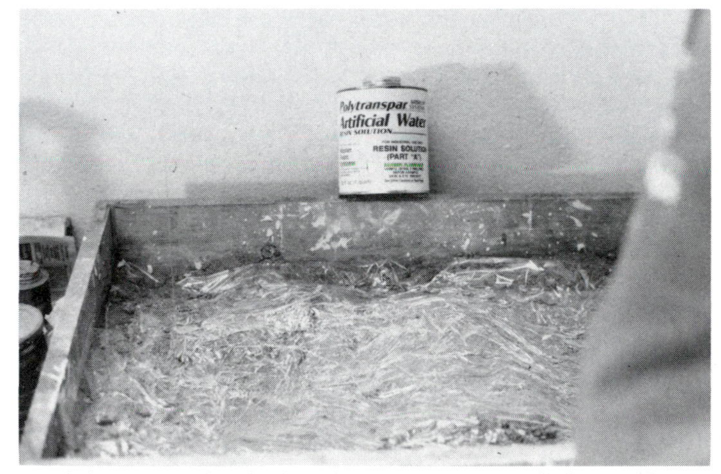


While waiting for the hands to cure I proceeded to make up the splash pieces that were to surround the hands. I used a sand box to make them in. First I shaped the sand in the desired position and then laid a piece of plastic over it. Then I mixed *Polytranspar*™ Artificial Water and poured it all over the plastic. I let it cure for 12 hours before removing the splash piece from the plastic. (This was to be broken into smaller pieces at a later time.

Next I proceeded to drill a 1/4" hole in the back part of the walnut box, to accommodate the light cord.

I then cut four wooden blocks 3½" tall out of 1×2's and secured them to the inside of this box with wood glue. These were used to support the layer of 1/4" plexiglass which supports the water scene.

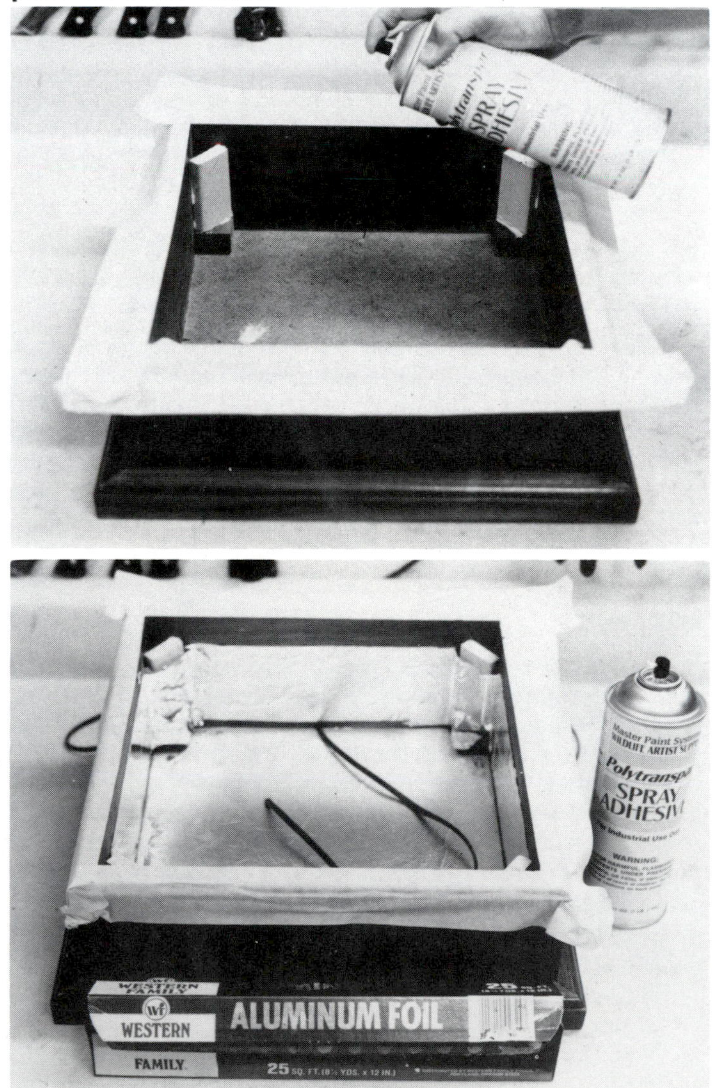


I wanted to line the entire inside of the box with aluminum foil in order to reflect light up and through the splash and hands. I applied spray adhesive to the entire inside of the box, and pressed the aluminum foil to it, as the two above photos demonstrate.

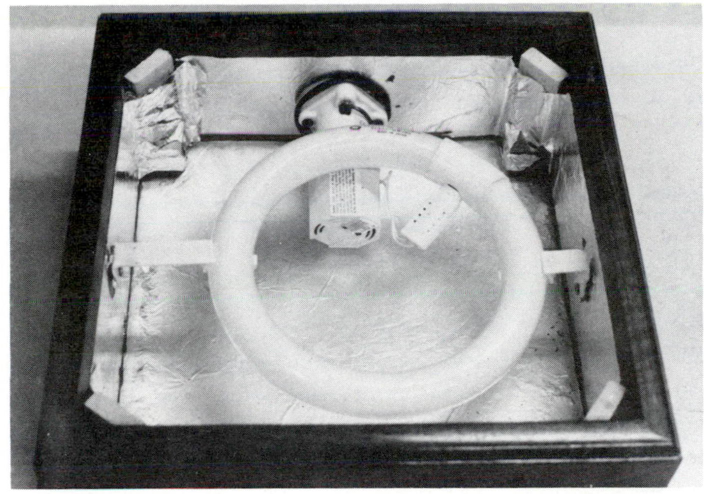

Once the box was lined with aluminum foil, I proceeded to install the light. I used a 150 watt round 8" fluorescent bulb, with a light socket. The light socket was secured to the box by 1/2" wood screws. Then I attached the plastic brackets to secure the light. The brackets were slightly adjusted by heating up the plastic and bending them around the light for a better fit. These brackets were attached to the box with 1/2" screws. Finally the cord was installed and the light was tested.

After completing the box, a piece of 1/4" plexiglass was cut to fit exactly to the inside of the box. The plexiglass rested on the four blocks that were glued in each corner of the box.

Now I removed the Artificial Water hands from the molds. They came out of the mold perfectly with all the detail intact. However, they gave me quite a scare, for as soon as I removed the hands I saw they had turned a milky white color. I accidentally discovered how to solve the problem when I left them outside in the "baking hot" Arizona sun for a few hours after removing them from the mold and it cleared up the milky look.

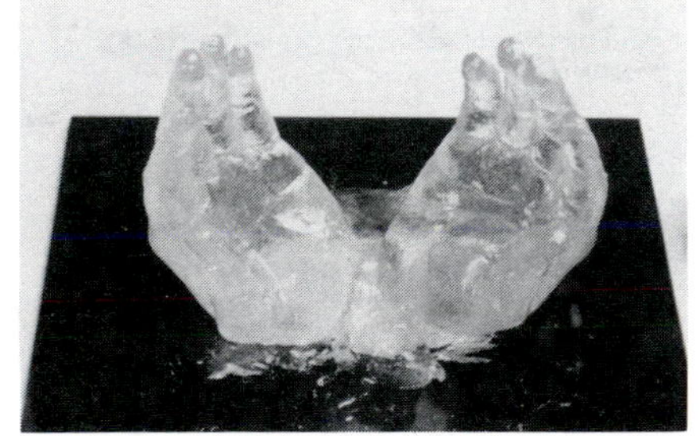

Next I positioned the hands on the plexiglass, as shown in the photo to the left.

I mixed a hot batch (extra catalyst was used to speed up the setting time) of Artificial Water and poured it on the plexiglass and on the bottom of the wrist of the hands. Once the resin set, the hands were anchored securely to the plexiglass base. (See the photo below).

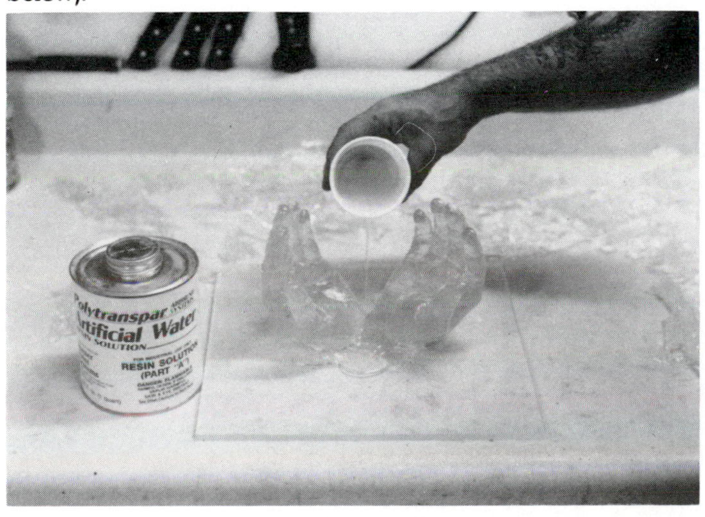


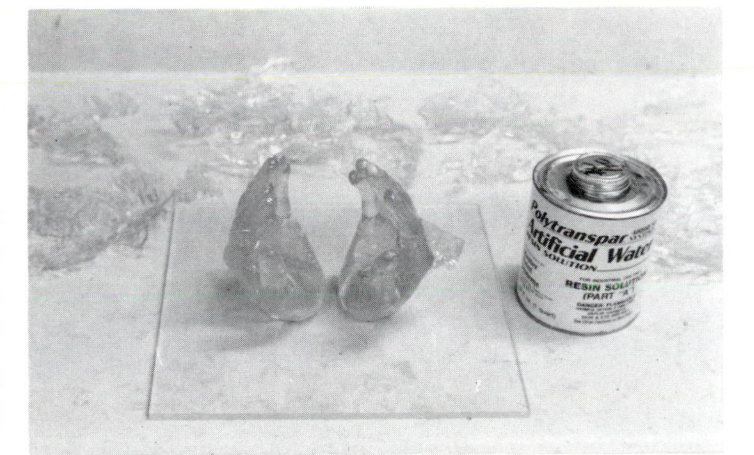


After the hands were secured to the plexiglass, I turned the piece upside down and masked off the hands on the backside of the plexiglass with masking tape. Then I painted the back of the plexiglass with *Polytranspar*™ Black airbrush paint (WA30). Once the paint dried, I used spray adhesive and glued a layer of aluminum foil to the backside or the plexiglass to further block out all light. Finally I cut out the aluminum foil and removed the tape from where I had the hands masked off. This left the area of plexiglass that the hands fit over clear so that the light could pass through them.

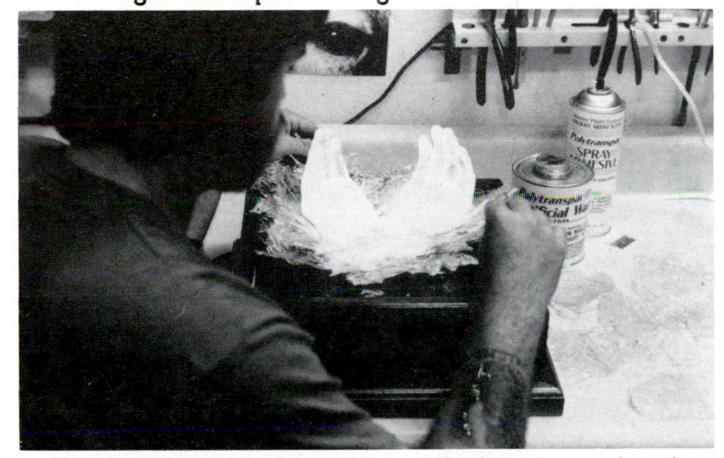


The plexiglass with the hands attached to it was placed on top of the four wooden blocks in the walnut box. I then took the large splash piece that was poured on the plastic earlier and broke it up into several different sizes. Placing them around the hands piece by piece, I designed the splash scene that I desired.

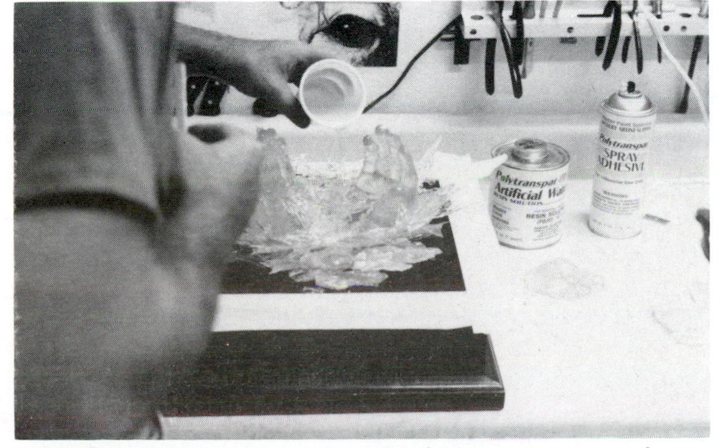


Another "hot" mixture of Artificial Water was poured over the entire splash scene to secure it to the hands and the plexiglass.

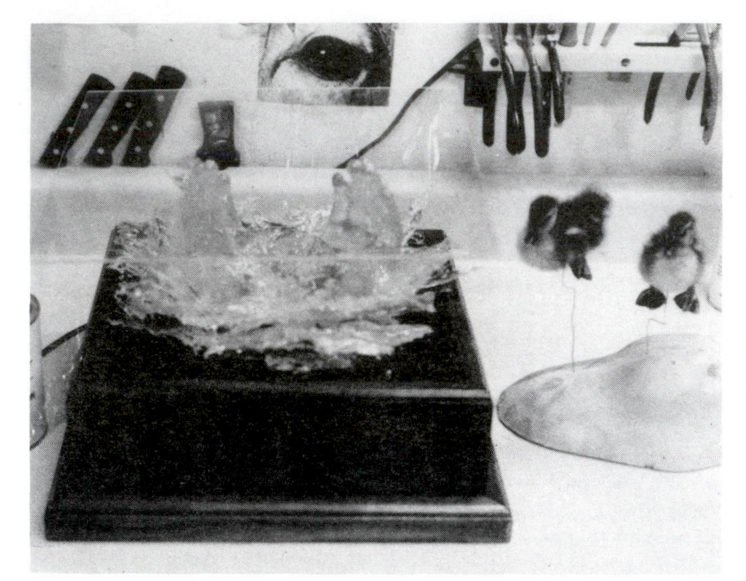

When the mounted ducklings were finished, they were to be "floated" on a water surface made of 1/8" plexiglass held aloft by the two hands.

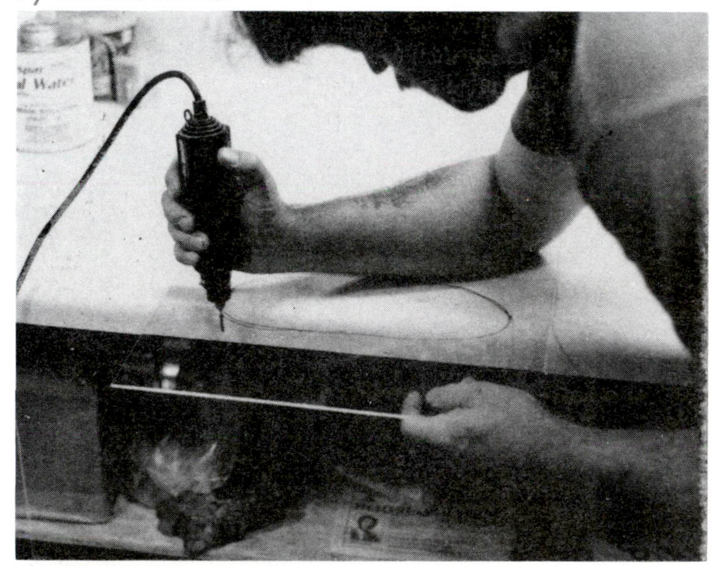

This top layer was designed and marked to fit over the hands. I then cut it out using a Dremel tool.

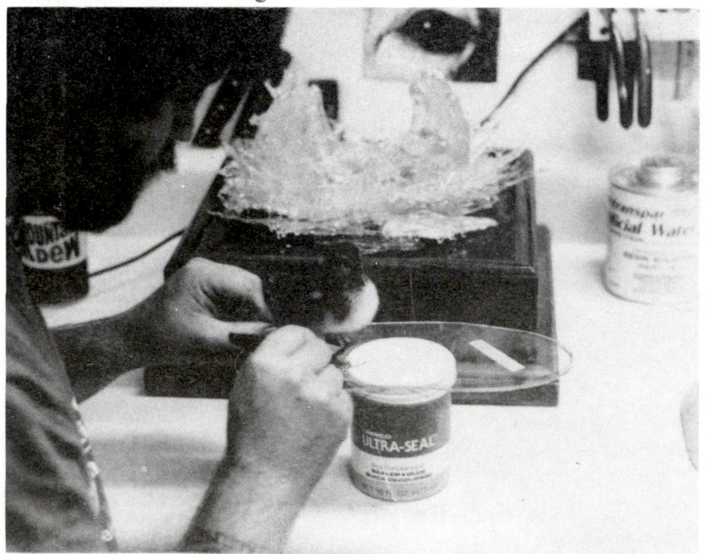


The positions of the two ducklings were traced onto the plexiglass (above photo) and the holes were Dremeled out to accomodate them (following photo).

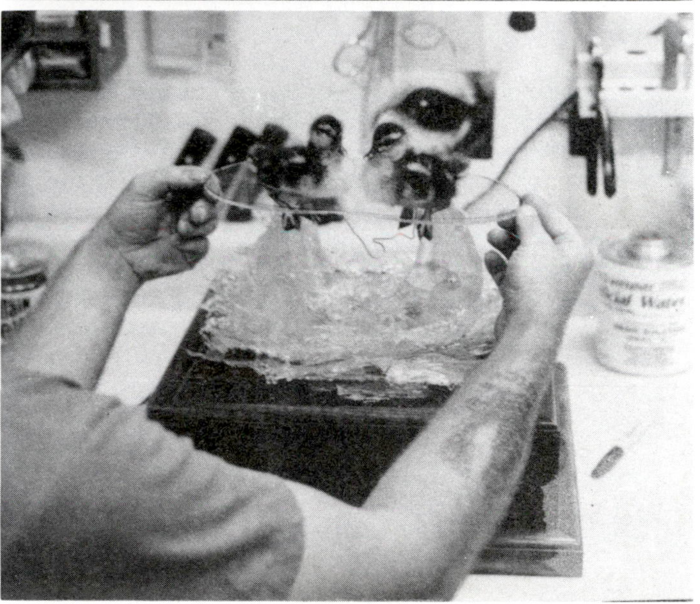

The ducklings were positioned within the openings in the plexiglass.

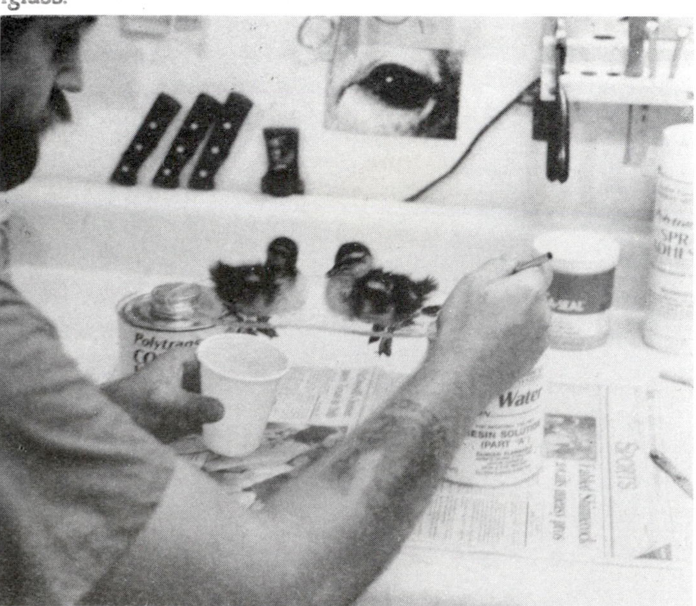

Artificial Water was mixed up and poured onto the plexiglass surface to seal the ducklings in position.

Then I poured the resin and allowed it to spill over the edge when it started to gel. This gave it the appearance of dripping water. Wires were temporarily put into place to help support and balance the top layer until it could be permanently sealed to the bottom layer.

Again, gelled Artificial Water was poured over the edge to connect the top layer to the bottom layer of water. After the Artificial Water had cured the wires were removed.

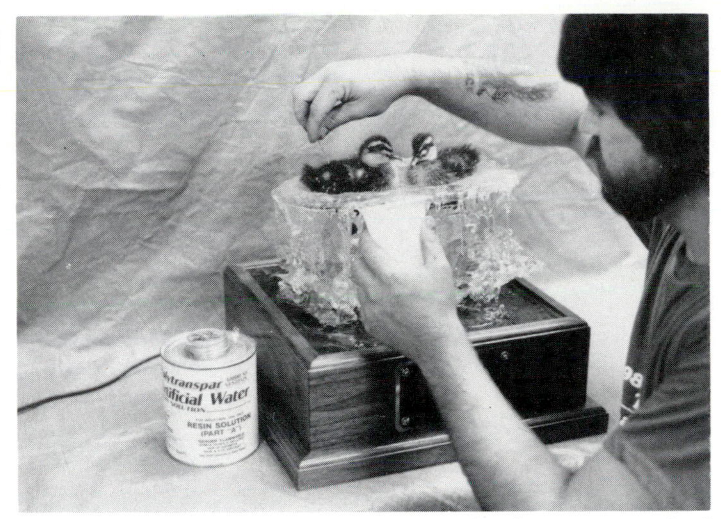


As a finishing touch, Artificial Water droplets were applied to the ducklings. The droplets were applied by using a piece of wire to pick up a drop and apply it to each of the ducklings' back. Applying the droplets just before the Artificial Water begins to gel works the best.

All that remained to do was attach the name plate to the front of the walnut box.

That completed a fun project that tells the story of what happened on that powerful fifth day so long ago.

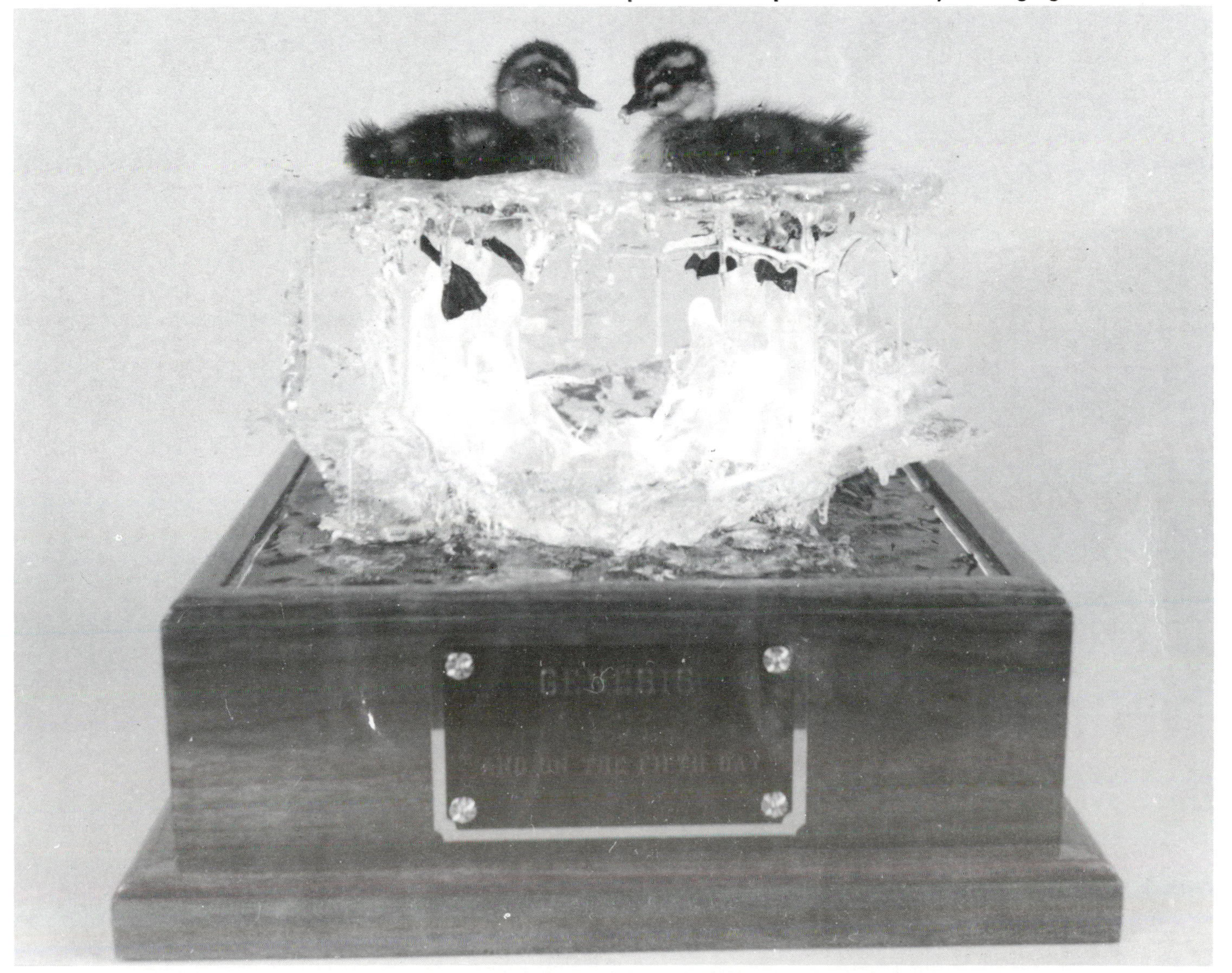

# 7. Actual Running Water in a Habitat Display

Bob Elzner has won a myriad of awards in the short time that he has been competing. He has a very creative attitude and is constantly developing new concepts. This portion of the chapter concerns just such a concept. Bob created this habitat scene complete with actual running water. The rendering was good enough to earn a blue ribbon at the 1986 NTA Convention. One of the most interesting facets of this scene was the effect of the light reflections of the moving water onto the largemouth bass below it. It added a nice realistic touch to the display. Here is Elzner's step-by-step technique for creating a habitat complete with running water.

***by Bob Elzner***

First, there are two important safety rules to always follow: **Number One**—use the proper respirator, and **Number Two**—work in a well-ventilated area when working with Artificial Water. These resins can be deadly if improperly used. Remember these two important rules *must* be followed, or your competition days (and life) will be shortened. Live to enjoy this fine art by observing all safety rules and precautions.

Materials needed to create this running water scene:

- Respirator
- Lumber (plywood 1/2" and 1" × 2" (S1S2E)
- Screws for framing
- Glue for framing
- Newspaper
- Wire
- Staple Gun and staples
- Stucco (available at most any masonry supply company) and sand
- *Polytranspar™* Paint: black, brown and green
- Plexiglass, 1/4"
- *Polytranspar™* Artificial Water and Gel Promoter
- Plastic
- 2 Copper tubes, 1/2" × 6" long
- A length of plastic hose and band clamps
- Submersible pump (the "Little Giant" pump is the brand I used)
- Auto body filler
- Habitat materials

First a water table must be constructed using plexiglass. The water table has three sides approximately 2" high and a bottom.

These pieces are glued together with super glue. The front of the table is left open at this point.

The rough framing consists of a 5/8" plywood base and 1" × 2" supports to house the water table. The table must be large enough to accomodate two streams. A vertical piece was also installed in the center of the exhibit to support the largemouth bass mount underneath the water table.

After the framing was completed, ordinary "chicken" wire was stapled to the frame. The wire will later be covered with stucco, but first must be positioned and shaped to resemble rocks and water falls.

After the wire was shaped and stapled in place, newspaper was used as a filler or backing for the stucco. It is very important to use as much newspaper as possible in all areas, because the more paper that is used, the less stucco will be needed. Stucco is very dense and is also heavy, so the less used the better.

The stucco was mixed with water to a thick consistency and use small amounts at a time. Apply the stucco directly over the wire starting at the bottom.

The front side of the water table must be finished before it can be installed into the habitat. *Polytranspar™* Artificial Water was used to make the front or "show" side of the water table. First use a sheet of plastic and attach it to the underside of the plexiglass with tape, leaving plenty extending out to form the front side of the water table. Lay the plexiglass in a sand box at an angle to form the vertical front side of the water table. Shape the sand under the sheet of plastic to form an approximate 2" high edge front side for the water table. This will make a ripple looking front that looks quite natural when the Artificial Water is poured.

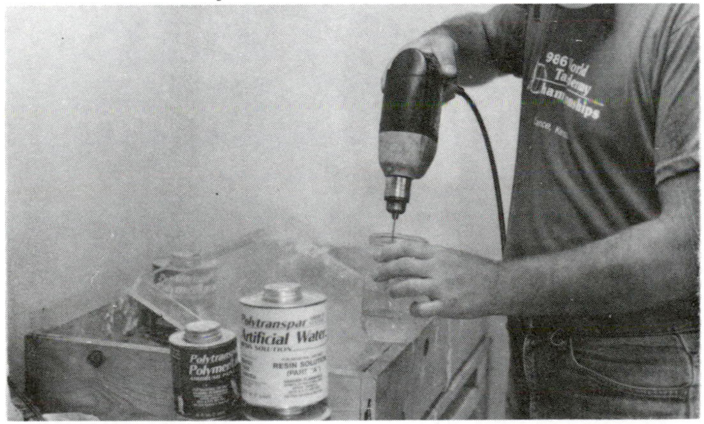

Next thoroughly mix up a batch of resin (use a respirator).

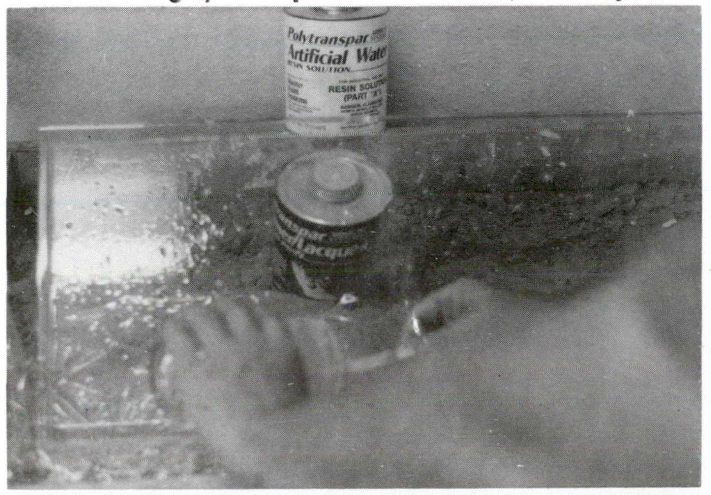


Using a gel promoter to make the resin set faster, Artificial Water is poured over the top of the plastic slowly, to allow the resin to spread over the entire area. Do not pour the resin immediately. Wait until the Artificial Water almost starts to gel, then pour the entire mixture over the area that you wish to cover. Work fast. The Artificial Water should be thick enough to support the weight of the actual running water. If you did not cover the entire area on the first pour, don't worry, just repeat the same step over and over until the desired thickness is achieved. Once it has been covered, Let it stand for 8 to 20 hours before removing the plastic backing.

The next phase is to install a drain into the plexiglass water table. I used a piece of 1/2" × 6" copper tubing as a drain. First I drilled a 1/2" hole in the right back corner of the plexiglass. A piece of copper tubing was cut 6" long and flanged on one end of the tubing so that it could not be pulled through the plexiglass when installed. I then sanded the tubing to ensure a good bond. Next, I mixed up a batch of auto body filler and applied it to the tubing after it had been inserted through the plexiglass. (The flanged end was seated directly on the top of the plexiglass.) Be careful to keep it neat looking, and avoid getting the filler all over the plexiglass.

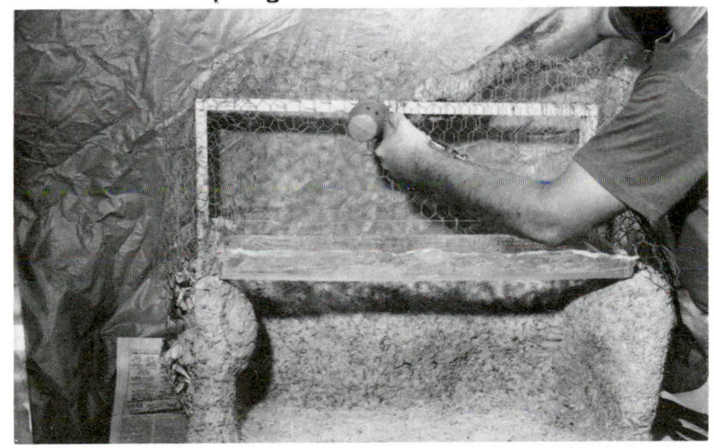

The water table can now be installed on the ledge of the artificial stucco rock. Mix up auto body filler and apply it to the four corners of the plexiglass. This will lock the plexiglass water table securely in place.

Now it was time to finish applying the stucco to the wire and shape it to form natural looking rocks and a waterfall.

The next step was to attach the water hoses. In this scene there were two small waterfalls using actual water. To accomplish this, two plastic hoses were tied together with the pump hose with one 1/2" tee fitting. At the other end of the hose where the water exits, I connected 1/2" elbows.

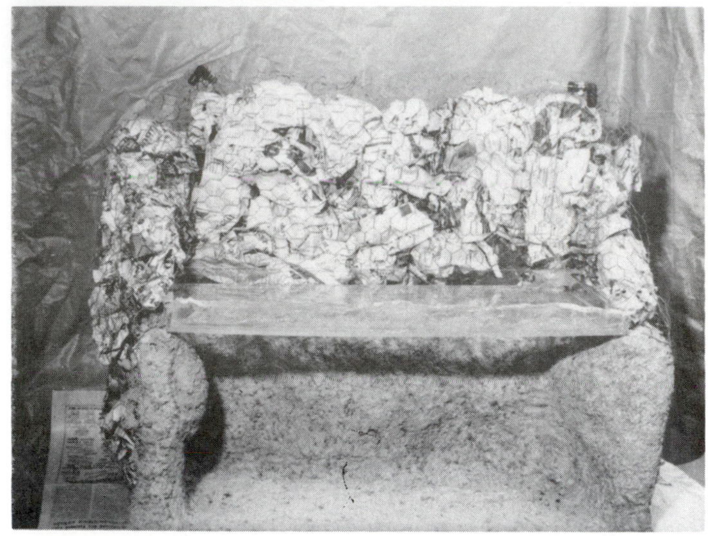

Again, I backed the wire with plenty of newspaper. (Remember to use as much newspaper as possible to reduce the weight of the exhibit.)

A mixture of stucco was applied just enough to cover the wire. At this point, the piece was beginning to look like a giant rock.

The stucco was formed to make the contours of the two small waterfalls. The stucco application was now completed and the base was set aside to dry.

*Polytranspar™* Paint was used to create the colors in the rock. I used three basic colors; Black WA30, Chocolate Brown WA70, and Medium Green WA61. The first color applied was black. I used a pump sprayer to apply the color. Next, brown was applied in the same manner and then green. Be careful not to "over-paint." If too much is applied then remove some of it with a damp sponge.

The next step was to completely seal the stucco rock with *Polytranspar™* Artificial Water. This gives the rock a "wet" look. Do not apply it too heavy (except for the waterfall), or it will look icy. Go back over the waterfall area and put on a heavy coat of artificial water, all the way down the length of the fall. This will give the rock a nice smooth surface.

Next the entire area of the plexiglass water table must be thoroughly cleaned with *Polytranspar* Lacquer Thinner. Once it has been cleaned, mix up more Artificial Water and pour this on top of the plexiglass. This will seal all leaks in the corners of the plexiglass and also seal the portion of the stucco that should overlap on top of the water surface. This will ensure a good seal. Let it dry for 24 hours minimum.

The next step is to create a cavity to house the plumbing. Remove as much newpaper as possible from the back of the rock by simply pulling it out. Clear the area inside the rock and the water reservoir should be ready to install. I used an empty one gallon can for a reservoir, (be sure to use a can that has been "lined" for water based products as unlined cans will rust).

I installed one hose draining into the reservoir from the drainpipe of the rock. I then connected one hose to the base of the reservoir and to the pump. To secure the 1/2" × 6" copper tube in the reservoir I used the auto body filler again as I had with bonding the drain. Finally I connected the hose from the pump to the waterfall. Screw type bands were used to connect the hoses.

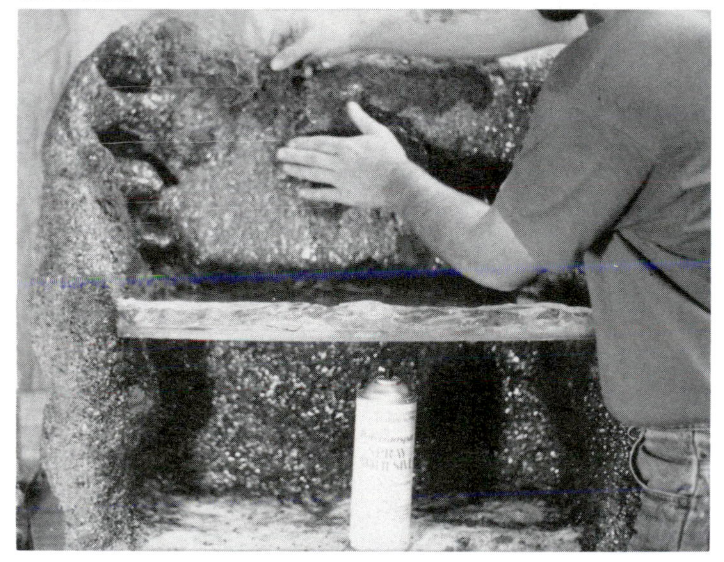

Next I applied the moss and other habitat materials. I used Wildlife Artist Supply Spray Adhesive to glue the moss to the rock. I also used the spray adhesive to secure sand to the bottom of the rock.

A largemouth bass mount was then attached to the rock. A 1/4" hole was drilled through the back of the rock. The fish was then attached with a 1/4" bolt. Finally the piece was ready for the test. I filled the reservoir and the water table with water and turned it on. Congratulations were in order—no leaks! This finished piece was now ready for competition.

# 8. Creating An Over-Hanging Bank Scene

***by Jim Hall***

There will be many instances in which a wildlife artist may wish to show some sort of undercut bank scene, either to display a fish below the water level or some small mammal at, or slightly above the water level. In many cases it won't even be necessary to try and show the water. If you were to use a combination of green and dried grass at the top of the bank scene, and a water-logged root and stream gravel at the bottom of the display, the *illusion* of water will be created in the viewer's eye. A ledge of ice, icicles, or snow could be added to this base to increase the viewer's interest or to change the season of the year.

In the previous photo of the finished bank scene notice how the overhanging root comes out away from the bank, permitting a fish to be positiond *behind* it. This further emphasizes to the viewer's eye that the fish is beneath an overhang, but does not detract from the impact of the mount or carving.

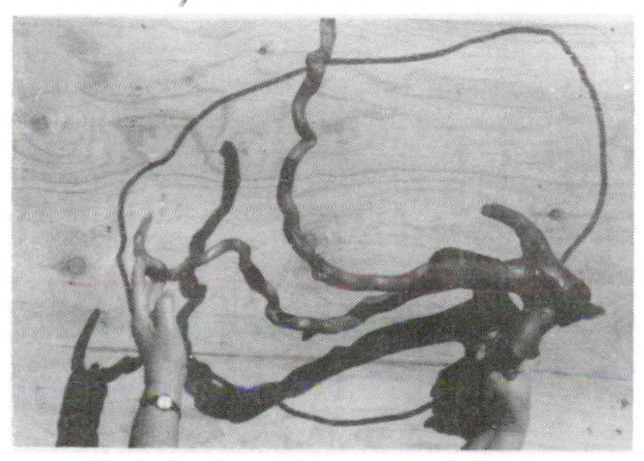

We began creating this habitat piece by selecting a nice, "rooty" looking piece of driftwood from our handy pile.

The photo on the left demonstrates how we outlined this irregular shaped base on a piece of 1/2 inch rough plywood. By holding the driftwood root in various positions, we mentally composed the over-all piece.

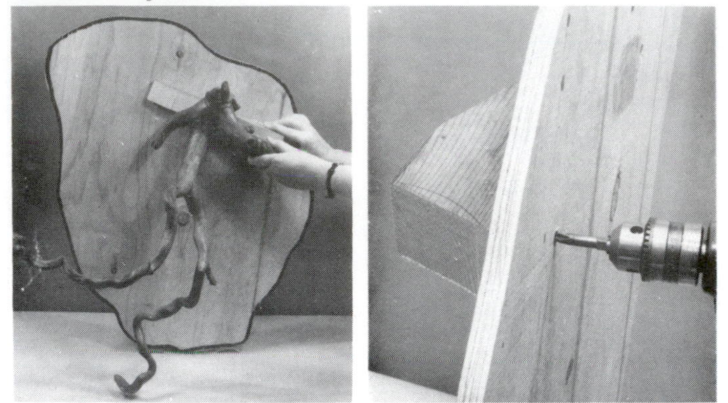

When we were satisfied with the composition, a spacer block was cut to help support the root in the correct position, and fastened to the 1/2 inch plywood with two screws.

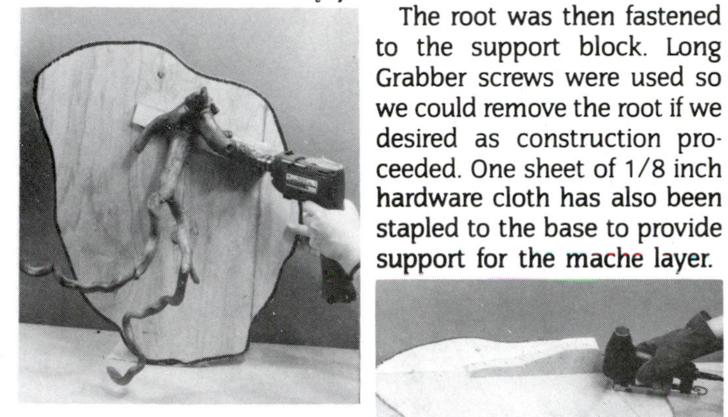

The root was then fastened to the support block. Long Grabber screws were used so we could remove the root if we desired as construction proceeded. One sheet of 1/8 inch hardware cloth has also been stapled to the base to provide support for the mache layer.

As can be seen to the right the hardware cloth is stapled in place at many locations. The irregular surface was formed with our hands to eliminate any "flat" look. (Leather gloves saved our hands from scratches.) We continued applying hardware cloth until the entire base

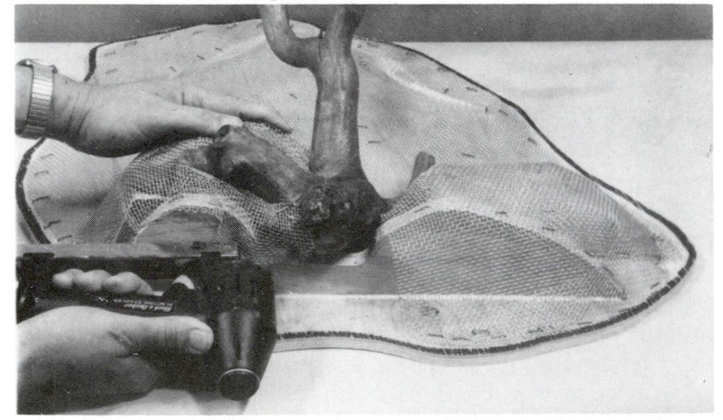

was covered. A wooden mallet was handy to form the hardware cloth easier.

Before the mache layer was applied, we primed the edges of the plywood with white glue or Ultra Seal (photo to the right). This permitted a much better bond between the mache and the wood.

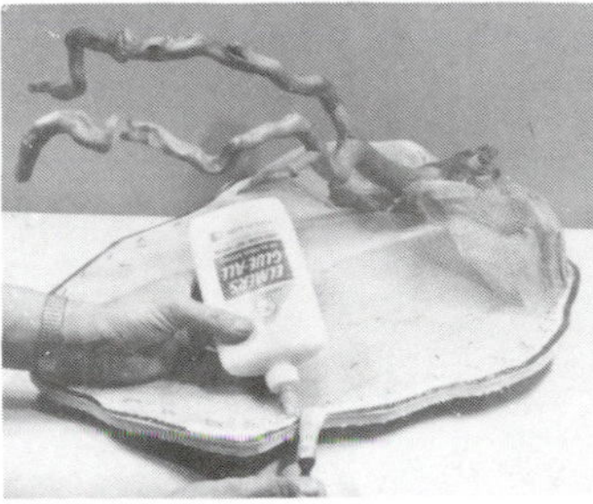

A layer of Jim Hall Mache was then spatulated over the

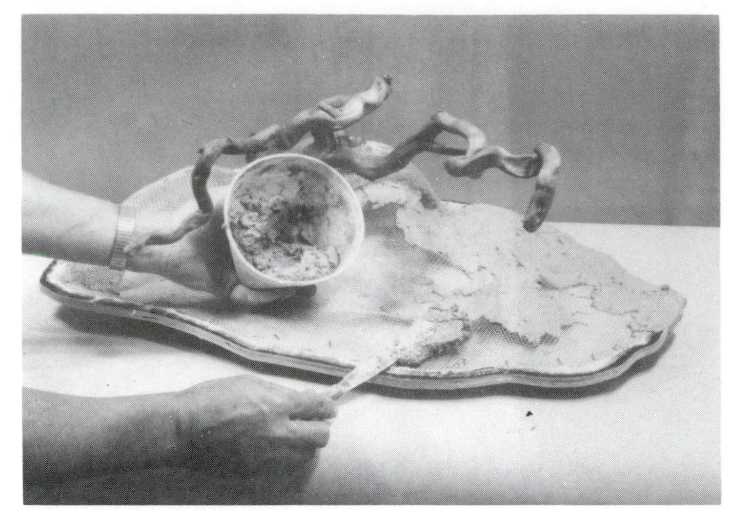

wire surface. It was "worked" into the wire mesh to insure it would stay in place.

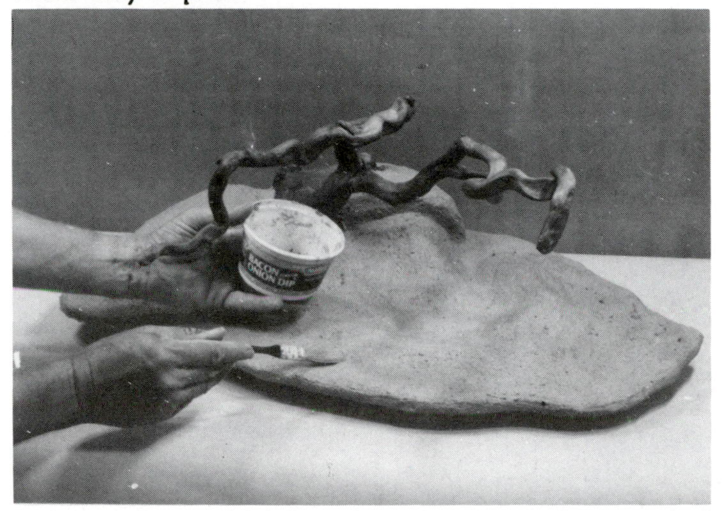

The mache can be left very rough or it can be smoothed somewhat with a small brush dipped in water.

For this bank scene, we decided to cover the mache layer with a darker colored material, and used a regular potting soil mix purchased from a local garden store. This soil was then mixed into a glue made from 75% water and 25% white glue. This darker colored mix was then applied to the base in the same manner as the mache.

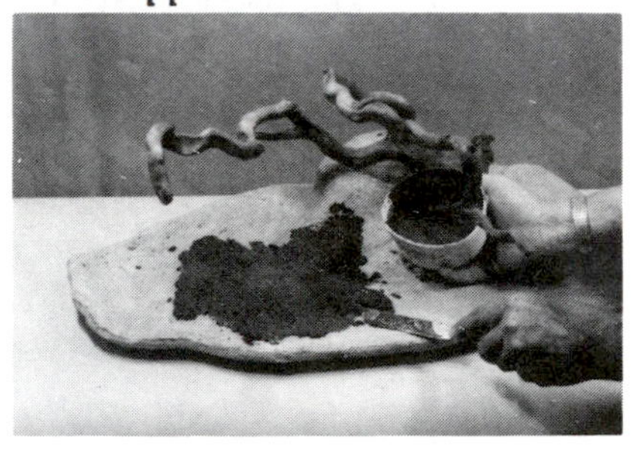

The photo at the right shows the finished ground work on the bank scene. At this point, the scene could be converted to a winter scene very easily by adding snow, ice, and icicles. The white and brown colors would provide a very dramatic color contrast.

To create this bank scene in fall colors, we simply dug up a piece of sod about the size we

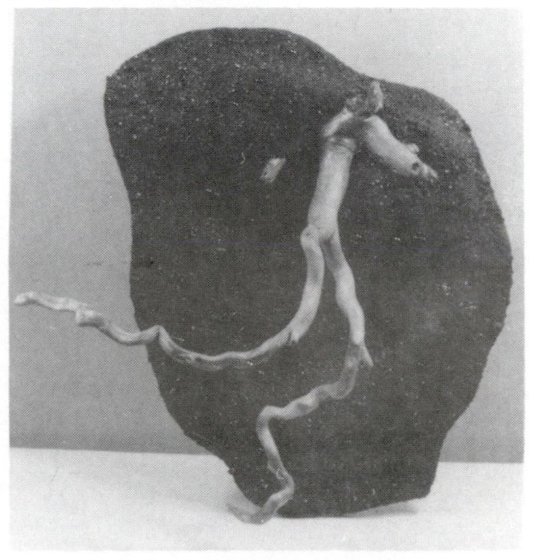

needed to cover thc top part of the piece. By doing so, we were able to create the grass, leaves, and debris effect automatically, eliminating the need to build the scene piece by piece.

The main problem caused by using this technique is that when you dig up a piece of sod, you also dig up mud, worms, and other critters that would be detrimental to your display. We solved that problem by rolling the sod up on edge and washing out all of the dirt and unwanted critters with a garden hose. We were able to wash out all of the mud and dirt, leaving only the root system.

To further guarantee that the sod mat was preserved, we immersed it into a strong Bacteriacide (Wildlife Artist Supply Company) and left it overnight.

The mat was then placed back on the wire rack to drain out as much liquid as possible. By placing the treated sod mat on a layer of newspapers, we were able to draw out still more moisture.

When we had removed as much moisture from the sod mat as possible, it was gently lifted and placed into postion at the top of our bank scene.

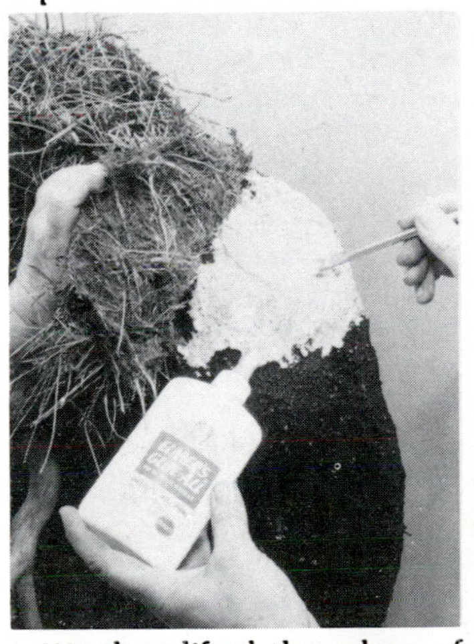

We then lifted the edges of the matt and coated the base with a heavy application of white glue, (above photo on left). We lifted small sections at a time until the entire piece was glued completely. Two inch long bank pins were then randomly placed through the sod to further help hold it in place (above photo on right).

A different fish is shown installed in the undercut bank scene in the photo to the right, to illustrate the versatility of the scene.

# 9. Miscellaneous Water Effects: The "Wet Look"

*by Wayland Adams*

## Making it Look Wet

I wanted my elk mount to stir the viewers' imagination to picture a bull elk in an early September snow storm, with the bull's body heat melting the snow almost as fast as it was hitting.

This melted snow effect was achieved by using diluted *Polytranspar*™ Fin Backing Cream (FC101). The Fin Backing Cream was diluted to the consistency of thin milk with water and applied with a hand sprayer. I first strained the mixture to remove any chunks that did not completely dissolve. (These will cause trouble later as they will clog up the hand sprayer if they are not removed.) If there is trouble spraying the diluted Fin Backing Cream, simply dilute to a "thinner" consistency with more water.

The application was made by spraying from the top of the mount using a fairly large quantity and letting it drip off. (Be sure to use plenty of newspapers under the mount to catch the excess.) While the mount was still wet, I roughed up some of the hair and streaked the mane hair in order to leave it looking as natural as possible. After drying overnight, I applied *Polytranspar* Wet Look Gloss (FP240) with an airbrush to certain areas to give a little more shine or "wet look" to the matted hair.

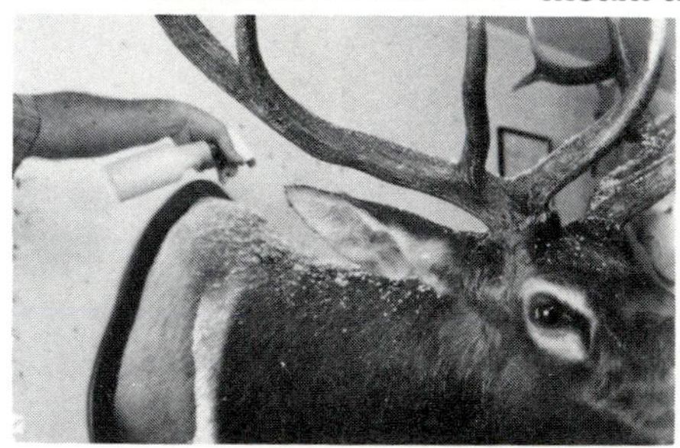

Words of Advice: Always "know your products" and "experiment before actual use." If the diluted Fin Backing Cream had not dried clear, I would have been a very upset taxidermist. But, I first sprayed it on some scraps of elk skins that I had in the shop so that I "knew" exactly what it would do prior to applying it to the actual mount.

## Applying the Snow

After applying the "wet look" of the "melted" snow, it was time to apply the "unmelted" snow flakes to the elk's back and antlers. I didn't want a large amount of snow so I made a sifter out of FCS76 galvanized screen wire and sifted the Wildlife Artist Supply Artificial Snow and used only the large flakes, which were applied by hand, so that I could get a better visual effect. When sifting the snow, I found some flakes that really don't look like snow flakes (don't tell Bob Williamson). I just discarded those and used the natural looking ones.

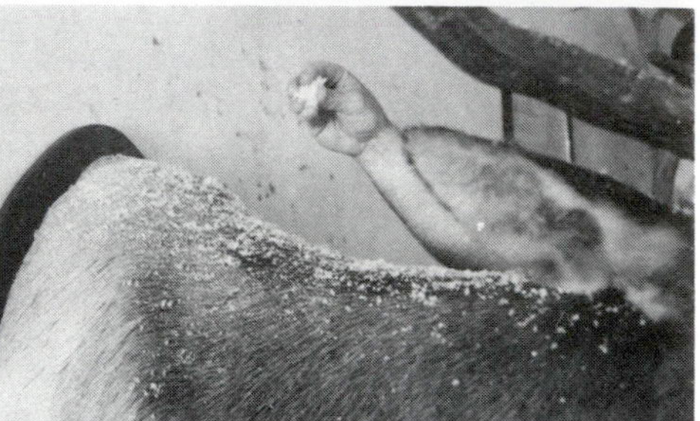

The adhesive I used for the snow was the same as used for the wet look, diluted Fin Backing Cream. This adhesive worked very well on this particular mount in that I used it on the hair and on the antlers. It gave good adhesion for snow and it dried clear with very little shine.

## Adding Artificial Water Droplets

For a realistic finishing touch I recreated drops of water dripping off of the mane, face, etc. I used *Polytranspar*™ Artificial Water applied with a transfer pipet to achieve this effect.

I thoroughly mixed the Artificial Water and Catalyst in small quantities using small clean containers. I then squeezed the bulb of the pipet to draw the catalyzed mixture into it. With the catalyzed resin in the pipet, the application was easy. I simply put a drop of water anywhere I wanted by gently squeezing the bulb of the pipet. To apply the minute droplets to single hairs I simply squeezed a small drop to the end of the transfer pipet and gently touched the hair where I wanted the droplet to appear.

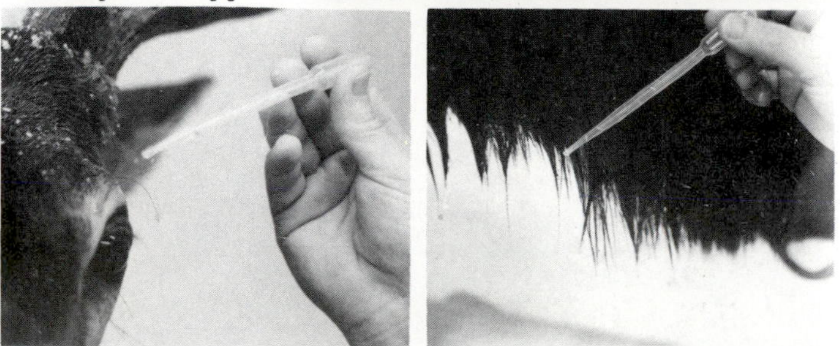

A tip to consider when applying Artificial Water: When the reaction of the catalyst and resin has progressed to the point that it starts to "string," it's best to discard the pipet and mix another batch.

# 10. Frozen Water:Creating Ice and Snow Effects

***by Bob Williamson***

Creating a winter scene is both easy and profitable. Snow and ice scenes are especially beautiful with wide appeal to both wildlife artists and art consumers.

While it is true that virtually anyone can create one of these winter scenes very easily, truly "spectacular" exhibits can be created with little (if any) additional labor by following a few guidelines and careful planning.

It goes without saying that snow and ice exhibits, as with any aspect of wildlife art, require obtaining and using good reference materials. This is an "absolute" necessity. No one can expect to accurately recreate something unless they are totally familiar with it.

## Planning The Exhibit

Several things should be considered when planning the exhibit. Snow varies considerably in the way that it looks. It has entirely different textures due to several factors such as the sun, wind, temperature, and length of time that it has been on the ground.

Snow will appear as fluffy billowy drifts, flakes, fine powder, icy slush, hard-packed granular, or a combination of some or all of the above. Different materials must be used to recreate these variations and by varying the textures, a more realistic effect can be created.

We have already mentioned that the "elements" (sun, wind, temperature, etc.) create the varying textures of snow. Take this into consideration when designing the exhibit. Visualize where the sun's rays would directly hit, which areas would be shaded, and so forth.

One important consideration in the planning stage will be to create a definite wind direction. This is accomplished by adding more snow to one side of the exhibit than the other. Also when the artificial snow is being applied, it should be "blown" on to the exhibit to create an effect similar to the wind blowing. Additionally, the effect of the sun and temperature must be taken into consideration. Areas exposed to direct sunlight will appear partially melted while shaded areas will not.

Another important consideration is the length of time that the snow has been there. Freshly fallen snow will appear fluffy white, while snow that has been there for awhile will appear to be "hard-packed", or fine powder, and often is not as clean and white as fresh snow. Remember that applying "different" textures to the exhibit is not any harder or time consuming than covering the entire exhibit with one texture. The difference in the appearance of the exhibit, however, will be dramatic with the varying textures looking far superior than using just one.

## Creating The Snow

With that in mind, we can proceed to building the winter scene complete with varied textures of snow. Fluffy snow will appear in areas that are uppermost (in regards to the base); fine powder and hard-packed snow will be located toward the lower points and shaded areas. Icy melted snow will be located where components of the exhibit would normally be exposed to direct sunlight, such as icy snow hanging from a limb or an icy glazing covering the uppermost part of a rock.

## Materials For Creating Snow

Recommended materials used for creating the diffferent textures of snow are as follows:

*This winter scene by Master Taxidermist Randy Nelson uses many techniques.*

**1.** Fluffy snow, snowdrifts of freshly fallen snow: Use Wildlife Artist Supply Company Artitifical Snow Kit.

**2.** Fine powdered snow: Use Wildlife Artist Supply Company Snow Flocking.

**3.** Hard-packed snow: Use a cotton base sprayed with Surface Coat acrylic spray, then sprinkle lightly with Snow Flocking.

**4.** Icy Slush: Use *Polytranspar*™ Artificial Water and cotton. It can be lightly coated with Snow Flocking. Saran wrap used with Artificial Water also works well.

**5.** Granular: Use Wildife Artist Supply Company Artificial Snow kit sprinkled with glass beads. Spray lightly with Surface Coat spray.

**6.** Icy wisps: These are often seen pushed up from the ground, around frozen streams, etc. Use strands of angel hair to recreate this effect.

There are a variety of materials that are used to recreate the "sparkle" that is seen in snow. As a rule, these materials are added last, usually by lightly sifting and sprinkling them over the surface. Materials used for recreating sparkle in snow are as follows:

**DIAMOND DUST**—This material is furnished in the Wildlife Artist Supply Company Artificial Snow Kit and is premixed in Wildlife Artist Supply Company Snow Flocking. It can also be purchased separately and added to "cotton" snow, ice, and slush scenes.

**IRIDESCENT GLITTER**—This material recreates the refraction of light common to snow and ice. It produces a nice array of "rainbow-like" colors when applied properly. Note: This material must be used *"sparingly"* for best results. If too much is applied, result. Too much will not look realistic.

**WILDLIFE ARTIST SUPPLY COMPANY GLASS BEADS**—These are used to add sparkle and recreate "crusty" snow with sparkle. Note: Crushed glass can also be used but is very dangerous to handle.

**PEARLESCENT LUSTER PIGMENTS**—These can be added to create varied effects. The fine powder consistency of Pearl Luster Pigments is easy to use by simply sifting or blowing tiny amounts and lightly distributing them over the exhibit. They can also be used with Artificial Water to add a slightly opaque pearl color to ice scenes.

# Other Material

**AQUARIUM FILTER MATERIAL**—This may be substituted for cotton and is very similar.

**SALT**—Ordinary table salt is often used by artists to add a sparkle and crystalline effect to the snow. Salt will turn yellow with age and also is affected by humidity. Seal it thoroughly with Surface Coat spray if this material is used.

**SUGAR**—Sugar is also used to recreate the crystalline effect of snow. Unless totally sealed, it will attract insects.

**STYROFOAM**—Sanding open-cell white styrofoam will produce a good snow material. Use sanding screen, as ordinary sandpaper tends to make it dirty looking, as particles from the sandpaper will mix with the foam flakes.

# Beginning The Project

Plan the base for a winter scene carefully. Don't use summer or spring vegetation in a snow scene. Use rocks, dead limbs, or perhaps an old fence post. If vegetation is to be used, choose an evergreen. They look nice with a snow scene and will be completely natural.

When composing the color scheme for the exhibit try to keep the colors on the cool side. One example would be to use blue-gray colored rocks rather than warmer colored rocks such as brown. The same holds true for limbs and fence posts. Always use weathered blue-gray colors if possible. The viewer should "feel" the cold when viewing a snow habitat. Build natural contours into the scene using methods described in the base making chapter. Permanently secure all rocks, limbs, fence posts, and the like prior to applying the snow.

Always paint any area that is to be "fully" covered with snow with *Polytranspar*™ Super Hide White WA/FP10 for a base coat. Next, use Satin White Pearl WA/FP401 on top of the white. This step will make the base less noticeable if any area is not completely covered with snow. Painting the base white will also help to reflect light.

# Applying The Snow

After the base has been completed and all habitat materials permanently attached, it will be ready for the snow application. The artist should have already decided where the various textures of snow are to be placed during the initial design of the exhibit.

Begin by applying the "hard-packed" texture of snow to the base. It will usually be located in shaded areas, areas directly under melted snow, close to icy water, etc. It is applied by placing a piece of cotton in postion and then "thoroughly" saturating it with Surface Coat acrylic spray. (Some artists prefer Fin Cream or Ultra Seal for this step. Any of the three products will work.) Use a tongue depressor to shape and compress the cotton into a natural setting. Several applications may be necessary before it is saturated enough to resemble hard-packed snow with an icy glaze. Before the final application, apply a very light application of Snow Flocking. (Luster pearl pigment may also be lightly added at this time.) Then, spray another fairly heavy coat to seal the flock and/or pigments and form the icy glaze.

Once the hard-packed snow has been applied, it is time to apply icy slush and icicles. Randy Nelson of St. James, Minnesota has developed the following techniques for making icicles that work extremely well. Note: In addition to cotton, saran wrap can also be used to make small icicles. Use it as the cotton is used in Randy's method.

# Small Icicles

To create a small icicle, simply add catalyst to *Polytranspar* Artificial Water per the directions on the container, and thoroughly stir it in a paper cup, making sure that all parts of the resin and catalyst are thoroughly mixed. (We might as well say in the beginning that this step is "vitally" important. If the Artificial Water is not mixed thoroughly or the wrong amount of catalyst is used, it simply will not dry properly. "Do not" add too much catalyst as it can cause the crystal clear resin to yellow and/or build up heat to the point of smoking and actually catching fire. *Follow the directions!*

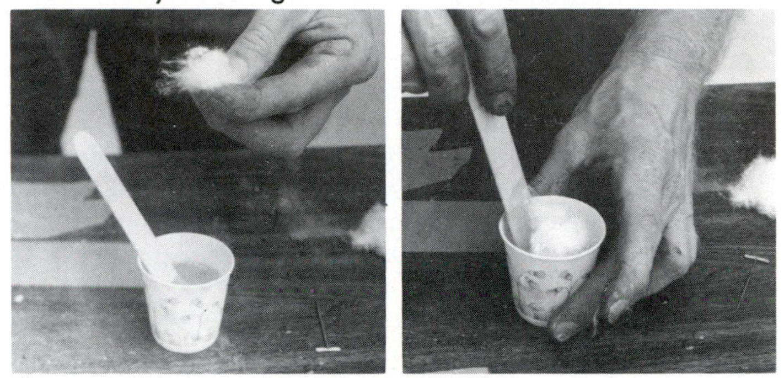

After catalyzing the resin, take a piece of ordinary cotton (available from any drug store) and saturate it with the Artificial Water by stirring it into the resin thoroughly.

Then, using a T-pin, lift it from the cup of resin and let gravity pull it down.

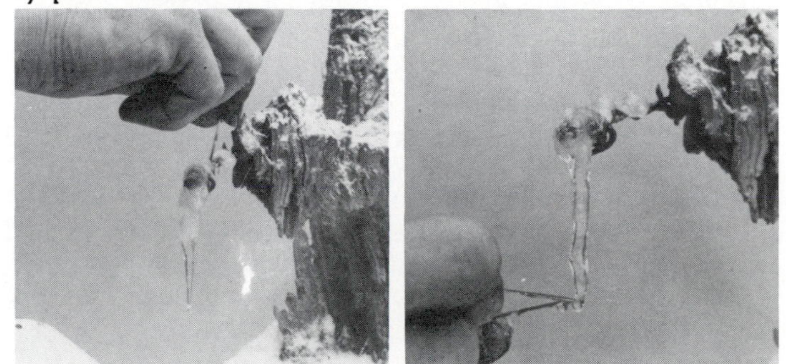

We next pinned it into the stump that we were using for this scene and then used another T-pin to help shape the icicle by pulling it to the desired length.

The saturated cotton not only helps to hold the icicle together but also gives it a wispy, "icy" looking color that adds realism. The stump that we were using had a piece of barbed wire going through it and several small icicles were hung on the wire. These were applied by simply draping the cotton over the wire and then using a T- pin to pull the icicle to the proper length.

On this particular rendering several different sized, small icicles were created using this same technique.

# Large Icicles

Larger icicles require a little more work but again they are not difficult. The main thing is to obtain the proper shape. Using our reference, we selected a picture of an icicle that had been formed in sections for our subject. Icicles such as this are common as the sun partially melts the icicle and then it freezes again forming unique "ball-like" sections.

Two methods were used to reproduce this effect: **(1)** We sculpted an imitation out of clay and made a mold of our clay

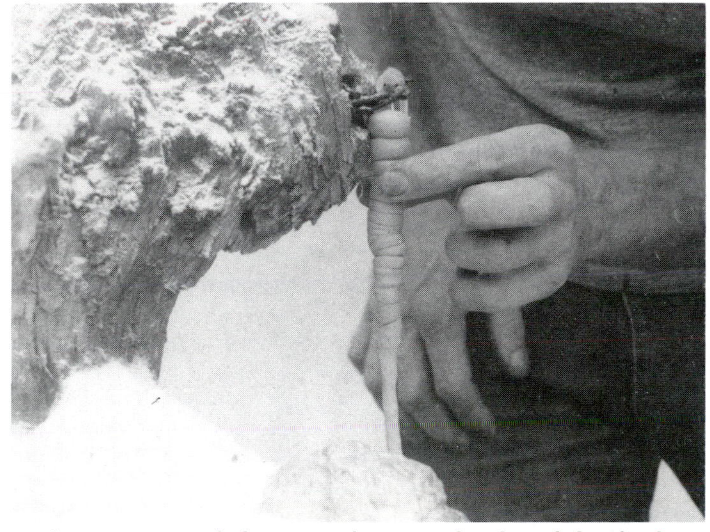

sculpture (pictured above), and **(2)** We also found that both an ordinary carrot and a parsnip obtained at our local grocery store fit the bill with little sculpting necessary. The parsnip was almost an exact duplicate for the icicle that we were recreating. A few slight strokes with a scalpel to shape it up and remove the rough edges and we were ready to mold.

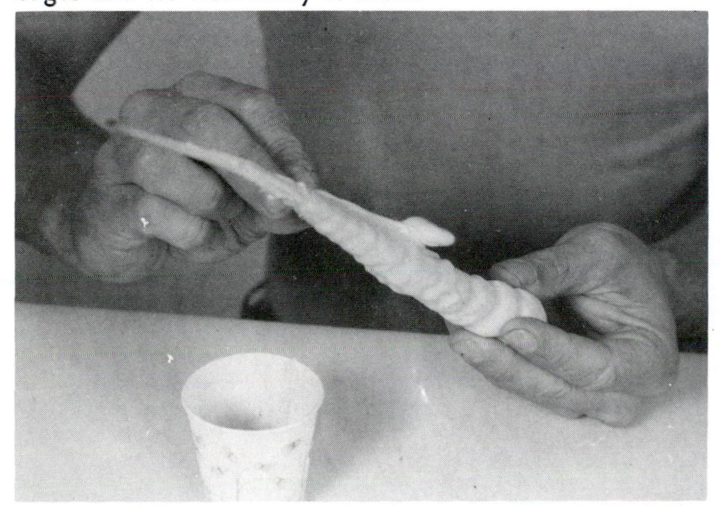

After applying a coat of Vaseline for a separator, we brushed a coat of Mold Builder on the parsnip (a good grade of RTV silicone can also be used). This was built up with several more coats to a sufficient thickness. Once the Mold Builder had cured and was built up to an adequate thickness (6 to 7 coats), the parsnip was slipped out of the rubber mold.

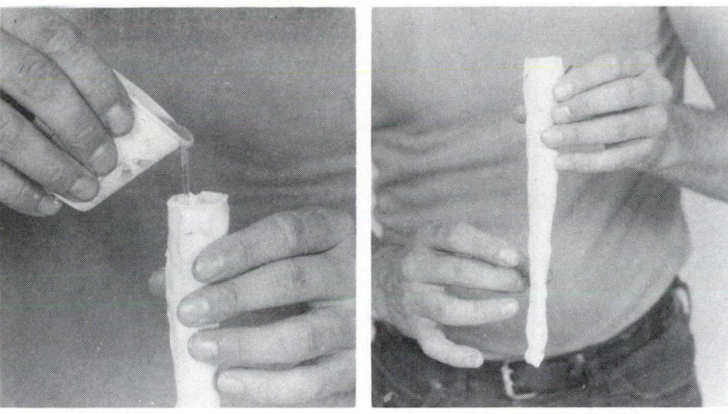

Artificial Water resin was catalyzed, mixed thoroughly, and then poured into the rubber mold. The mold was gently tapped to allow any air bubbles to escape, and then the mold was set aside to allow the resin to cure.

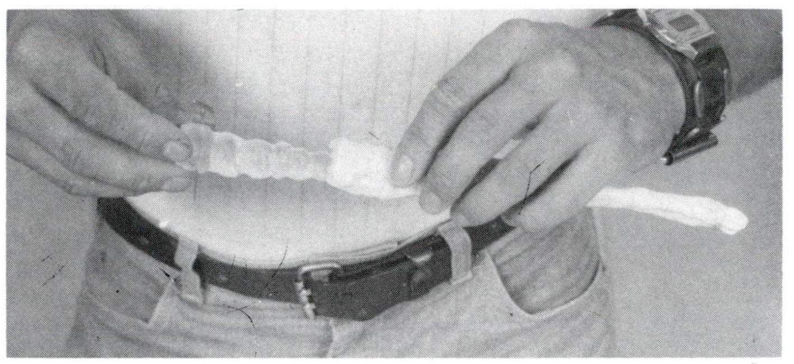

Once the icicle was fully cured, it was removed by slipping it out of the rubber mold. Additional resin can be poured over the icicles at anytime to improve their appearance. Simply place a cup underneath the icicle to catch excess resin and then pour the resin over the icicle to increase it's length, width, remove finger prints, etc., or to simply change it's character.

The icicles can be modified at this point by grinding with a Foredom tool to any particular shape preferred. Although grinding away any Artificial Water will scratch the surface, a smooth, transparent surface can be restored at any time by simply pouring additional catalyzed resin over the icicle in the same manner.

Next, a hole was drilled in the icicle to allow a T-pin to be inserted through the top.

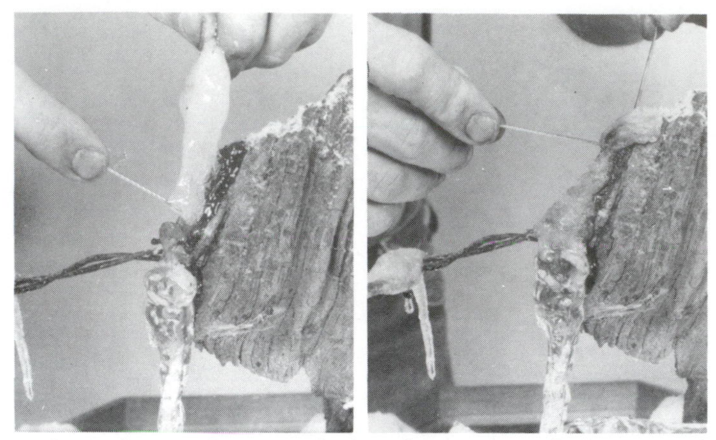

The icicle was then attached to the stump with the pin and covered with "cotton ice" (make this the same as for the small icicles as previously described) This completes the icicle portion and the exhibit was now ready for the snow application.

Now would be the time to apply "fine powdered" textured snow to the exhibit. First lightly mist water with a hand sprayer over the areas to be covered with this texture of snow.

Next, apply Snow Flocking over the areas using screen wire to sift it through or by using a flocking gun. (The Snow Flocking contains a self adhesive that is activated when it comes in contact with water.) Repeat as often as necessary to obtain the desired depth. Note: The illusion of drifts and depth can be created to a great extent when building the contours of the base. Carved styrofoam or urethane can be used to elevate certain areas, then a light covering of snow over these contours can actually be made to "appear" several inches thick.

Fine powdered snow (and fluffy snow) should show the effects of the wind. Use an air compressor to accomplish this. Set the regulator to approximately 10 PSI (or simply use a fan or blow on it) as it is sifted. Try to make it drift and gather to one side to appear wind blown.

The final texture to apply to the exhibit is the "light fluffy" texture snow. Use spray adhesive or a very thin coat of Fin Cream or Ultra Seal sprayed through a hand sprayer on the areas that you wish to apply it. Then, using a coffee can with holes punched in the plastic lid top, lightly sift the Artificial Snow HM600 over the adhesive. Try to actually simulate a light snowfall by gently sprinkling while blowing the snow in the direction that the wind is portrayed as coming from. Repeat this until the desired thickness has been obtained. Footprints can be added at this point if desired by simply using the foot of a specimen and placing it in the snow to make an impression.

Dirt can also be added to the snow to resemble dirt that has been kicked up by a running animal.

Now use diamond dust, iridescent glitter, and glass beads to add sparkle to the scene. Use very small amounts of these, as too much will ruin the effect. Misting a very light coating of adhesive will secure the sparkle to the exhibit. This completes the winterizing of the scene.

The final step would be to cover the scene with a plexiglass or glass cover. All snow scenes must be protected, if not the exhibit will be ruined by dust in a very short period of time.

Randy Nelson has developed a unique method of creating snow scenes. Here is his step-by-step procedure.

# Creating Snow The Randy Nelson Way

***by Ken Edwards***

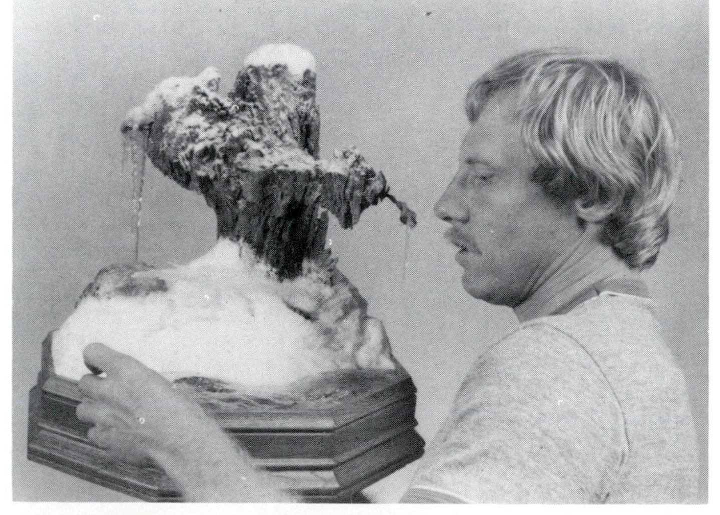

Randy Nelson of St. James, Minnesota, uses ordinary cotton as the base for his snow scenes and it is also the primary material for most of his snow scenes. The goal is to reproduce snow which is fluffy at the top and icy at the bottom—to recreate the natural appearance of actual snow. This is the procedure that Randy has developed over years of experimentation and invention.

The materials needed for this snow base are:

- Urethane Foam (2 to 4 lb. density)
- Oak base
- Ultra Seal (Elmer's white glue or similar product)
- Unsterilized cotton (available at any pharmacy)

- Resin Craft Surface Coat Spray
- Flocking
- Salt
- Diamond dust (glitter)

To begin building a snow-covered base, Randy first creates the basic shape of the base with a lightweight (2 to 4 lb. density) urethane foam. He carefully measures equal amounts of the A and B components (1 tablespoon of each), and mixes them together thoroughly in a small paper cup. He then pours the

liquid urethane foam onto his base in the shape desired, and allows it to rise. He repeats this procedure several times, mixing multiple small batches and pouring them over the previous

layers. This enables him to have some control over the design and composition of the general shape of the base. As each pour of the foam expands, the shapes (much like the shapes of billowy snowdrifts) begin to develop.

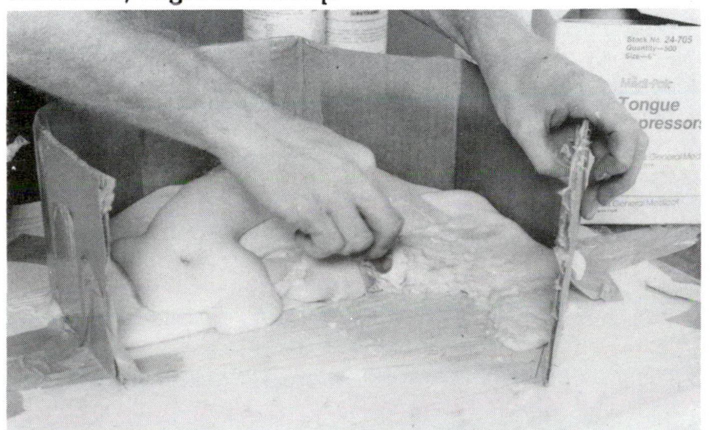

The foam can also be shaped or textured (if desired) while it is still slightly soft. When he is able to touch the surface of the foam without it sticking to his finger, it is at a good time for slight texturing and shaping. With snow scenes, however, manual alterations of the natural foam shapes will rarely be necessary. The multiple pours of small quantities of foam over each other will create a convincing shape for reproducing a fluffy snow effect.

Any stumps, rocks, fence posts, plants, or other habitat materials in the scene should be place into the foam at this stage while the multiple pours are being made. The extreme adhesive properties of the foam will securely lock these elements into position as the urethane expands to its full volume.

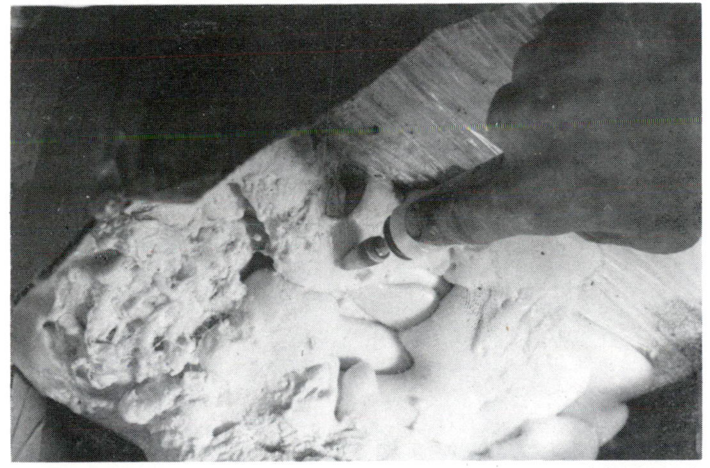

After the foam has cured it may be further modified with a Foredom tool, sanding screen, knife, or a small saw. This

may be necessary around the edges especially if the scene is to be placed in an existing base. (It is also very easy to simply pour the foam into the base from the start. Mask the wood with masking tape to avoid accidentally getting foam on it.)

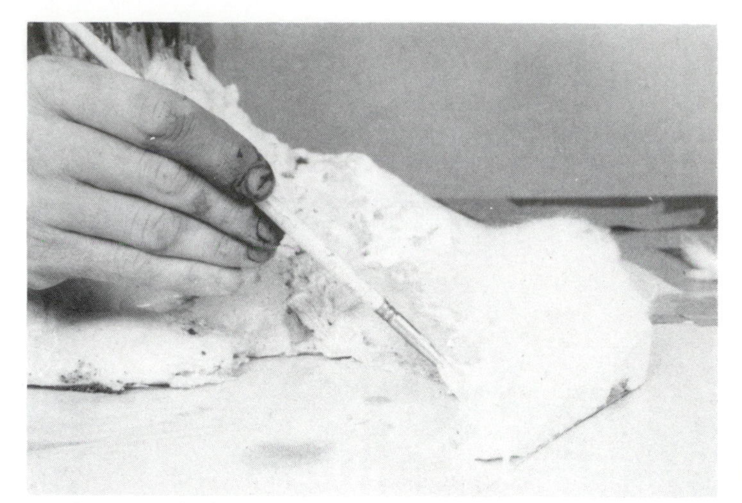

When satisfied with the final shape of the foam base, apply a thick coat of Fin Backing Cream, Ultra Seal, or Elmer's white glue to all areas to be covered by snow. Paint it on with a utility brush, and use "liberal" amounts.

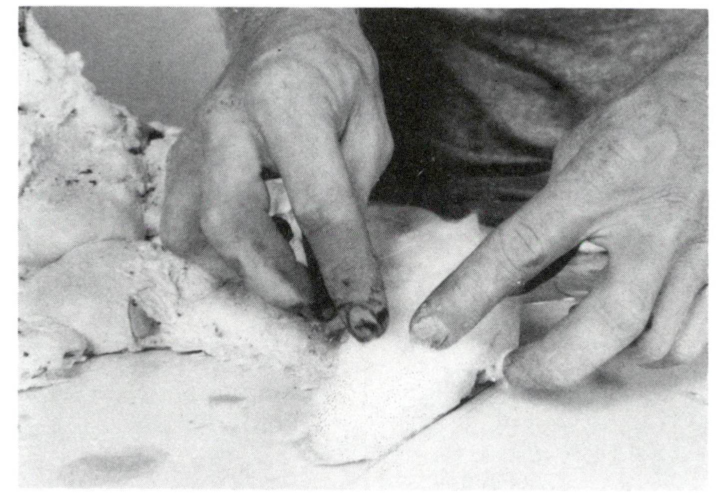

Next, tear ordinary cotton into small pieces and arrange them on the glue. By using different sized pieces of cotton, a varying surface texture will be created. Use larger pieces of cotton for snow "humps" or large clean areas to recreate drifts. Use smaller pieces of cotton to create small indentations, undercuts, or

lumpy "distressed" snow. Add more glue if necessary. Take adequate time during this step to allow for artistic decisions regarding the composition of the piece.

Next comes the step that changes the "cotton-looking" shapes into snow shapes. Using an aerosol can of acrylic glaze spray (such as Surface Coat Spray,) spray the cotton to texture and shape the surface. When spraying, think of the can of spray as the "wind" and use it to sculpt the snow much as the wind would. For larger rounded areas, hold the can farther away (12 to 18 inches). For small indentations and undercuts, hold the can much closer as it is sprayed. Change the direction that the spray is coming from to manipulate the surface as the wind

would. Actually "sculpt" the surface using both the acrylic spray from the can as well as the air pressure itself. Spray for longer and shorter times, varying both the distance and the amount of spray on the cotton.

In order for this method to work, a great deal of acrylic spray will be used. Even a small habitat will probably need an entire can of spray.

Remember that as snow melts (which it does, even on the coldest day), the bottom of the snow will be "icier," as the top layers of snow are fluffier. To create this icy look at the bottom of the snow, spray heavier on the bottom and lighter on the top. The heavier spray (the greater amount of acrylic saturating the cotton) will produce an excellent translucent icy snow effect. By varying the texture, distance, and amount of spray—and by remembering that snow becomes more icy towards the bottom, snow scenes created with this method will portray a stunning look of authenticity.

Once the snow shapes are textured with the acrylic spray, it is time to add the sparkle. For the crystalline look of snow, Randy likes to sprinkle a small amount of salt from above and/or throw it on from the side (while the acrylic spray is still wet). This gives the crystalline "granule-type" appearance of snow. One drawback to using salt, is that it will eventually yellow, spoiling the appearance of the habitat. Seal the salt by spraying again with acrylic sealer from far away.

To soften the snow and give it a dry, fluffy look, sprinkle a small amount of flocking onto the scene from above. (A screen works good to sift and sprinkle the flocking through.) Flocking also looks good blown into crevases on the sides of tree stumps, fence posts, etc.

For a final touch, sprinkle a fine grade of glitter (Diamond Dust) from above onto the snow surface (if the last coat of acrylic spray has dried, moisten the surface slightly with another light spray before applying the glitter). To adhere and seal the glitter, hold the can of acrylic spray far away from the scene and lightly mist over the top of the snow.

# Packed Snow

To create packed snow, the method is basically the same as for fluffy snow. However, the foam underneath should reflect the shapes of packed snow as opposed to billowy shapes. When spraying the cotton with acrylic spray, apply it very heavily. The more the cotton is saturated with the glaze, the more the snow will look packed. While the cotton is wet with the acrylic spray, the snow can be manually compressed for further compacting. Even footprints may be embedded into the surface at this point. Packed snow will not require any flocking, but do use salt and glitter, following each with an additional heavy spray of acrylic glaze. Also be aware that packed snow has generally been around longer than fluffy snow, so it should not look as clean. Some of the yellowing effects of salt could actually be beneficial in a packed snow scene. It may even be desirable to sprinkle (or blow) some dirt onto the snow between layers of acrylic spray.

# Snow Not On The Ground

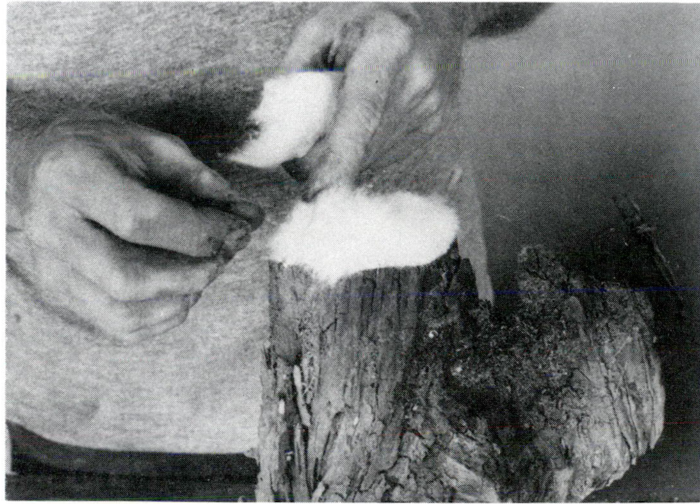

For snow on stumps, tree limbs, etc., Randy Nelson's procedure is similar to the fluffy snow method. If the area of the snow is large enough to require the use of cotton, simply paint the upper surface with a heavy coat of Ultra Seal, Fin Backing Cream, or Elmer's glue, followed by torn pieces of cotton.

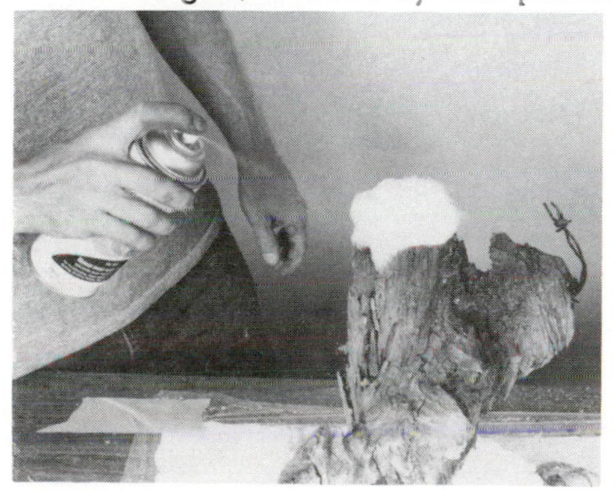

Shape and sculpt the cotton using acrylic spray as if it were the wind. Spray heavier coats on the bottom of the snow to give it a melted icy look. Between subsequent light coats of acrylic glaze, sprinkle salt, flocking, and glitter to complete the effect.

There is almost always ice under elevated snow on tree limbs, stumps, etc. The finishing touch is to add this ice (or even small icicles) by dipping small pieces of cotton in catalyzed Artificial

Water (casting resin) and placing the saturated cotton into position around and under the bottom of the snow. If the cotton above was sprayed heavily with acrylic spray, the transition will be invisible. At this point you can very easily add small icicles by merely pulling down a few strands of the saturated cotton with a couple of T-pins. The final effect should be a smooth transition from (starting at the top) fluffy dry snow, fluffy wet snow, packed snow, icy snow, wet icy snow, to icicles.

# Sheet Ice

This unique method of recreating partially melted sheet ice is the kind of ice that clings to the banks of a creek or tree after the water level has receded.

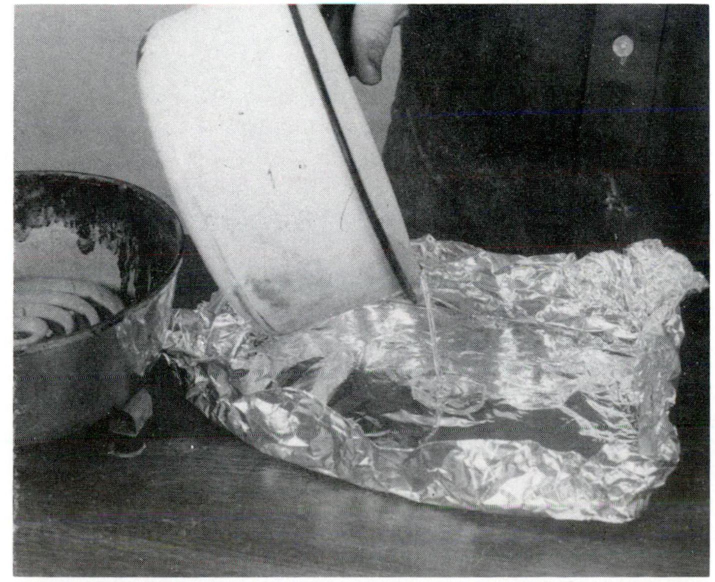

Randy begins by melting parrafin wax in a pan over a double boiler. (The double boiler assures that the wax won't become hot enough to ignite.) A shallow pan is constructed with aluminum foil and the melted wax is poured onto the foil, creating a thin sheet of wax in which to imprint the texture.

After the wax has hardened, it will need to be slightly warmed to become soft enough to texture. Randy places the thin wax sheet in a pan of hot water (about 105 degrees, or hot enough to hold your hand in without discomfort). The wax is ready for impression when it is approximately as soft as clay.

The wax is textured by pressing a granite-type rock into the surface of the wax. Randy chooses a rock with a texture similar to sheet ice which has a finely-textured look. When the surface of the wax is sufficiently detailed, allow the wax to cool, harden and then remove from the foil.

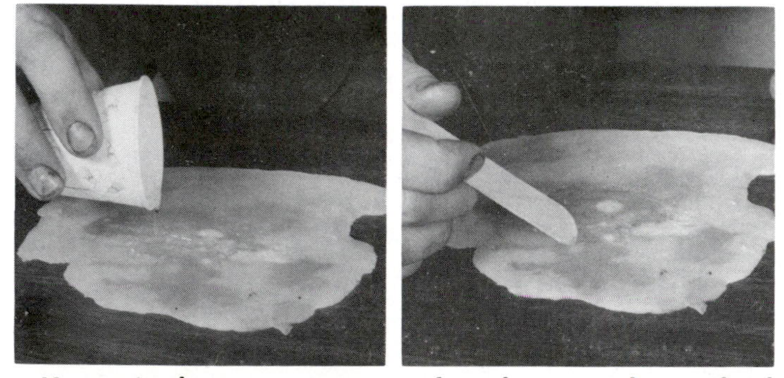

Now mix the proper amount of catalyst into the Artificial Water resin. Pour the resin onto the center of the textured surface of the wax. Using a T-pin, toothpick, or tongue depressor, drag strings of resin out of the center of the wax mold in random patterns. As the strings of resin settle into the random pits and depressions of the wax, intricate and delicate contours will form as "fingers" along the edges of the ice sheet.

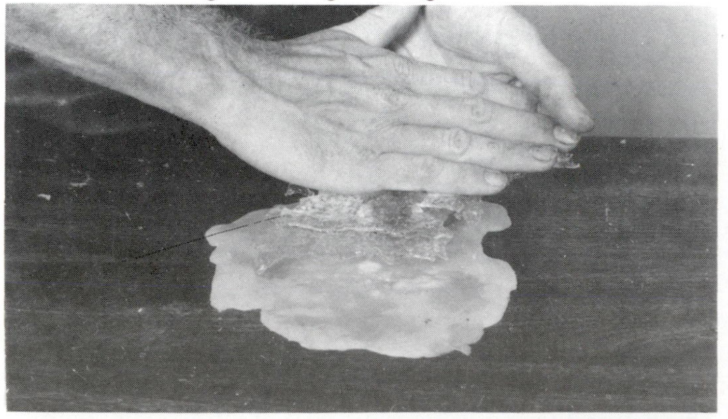

Allow the resin to cure overnight. After hardening, the resin can easily be removed from the parrafin. The natural color of the ice sheet is a cloudy white which will be picked up from the wax surface. Natural sheet ice is cloudy (not clear), so this ice is ready to use in a habitat.

If it is necessary for the ice to appear clear, this can be accomplished by a coat of acrylic spray, or by pouring a layer of catalyzed artificial water over the surface of the ice sheet.

To make the ice even whiter (cloudier), Randy mixes a small amount of silicone caulk into the resin before pouring the ice sheet.

For a hazy look, he brushes silicone caulk on the bottom surface of the ice sheet after it has hardened.

Ice sheets will appear more realistic if they are pieced together in layers. Use some cloudy sheets and some clear. Also, many ice sheets have small clumps of ice hanging down from the underside of the sheet. These are made by using catalyzed Artificial Water that has just begun to gel (when it will break off in a chip but is still sticky). Make these tiny ice chips by grating the semi-hard resin through a 1/8" screen (hardware cloth). The small ice chips can be arranged beneath the ice sheet(s) at this point. Since they are still not fully gelled, the chips will easily adhere to the sheet ice.

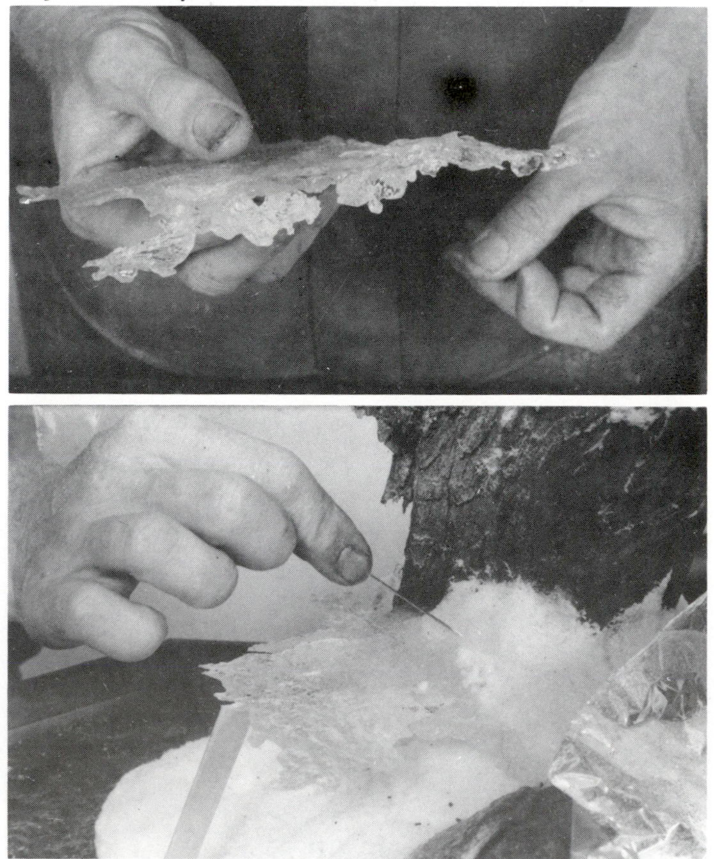

The ice sheets may now be placed within the habitat scene. Epoxy glue is used to attach the sheet to trees, fence posts, etc.

# CHAPTER 11

# Insects, Snails, Reptiles & Amphibians

*The collection, preservation, and use of various small "critters" in habitat scenes is outlined by wildlife artists Mark Belk of Provo, Utah, and Dan Blair of Peterson, Iowa.*

## How To Use Them Correctly

***by Mark Belk***

Insects—they're everywhere! On trees and grass, in the air, in the water, on animals, even in the house. Many animals, birds, and fish spend a large part of their time either eating or avoiding insects. It seems only natural then that taxidermists and carvers could use insects and other invertebrates to enhance and add realism to their mounts. Don't get me wrong, 'live" insects on any mount are bad news! But, incorporating preserved invertebrates into a mount adds greatly to the realism and uniqueness of the final presentation. As testimony to this realism, I have observed many spectators trying to "shoo" the flies off of a "fresh caught" fish mount, only later realizing it was actually a preserved part of the exhibit.

Like anything else, insects can detract from a mount if they are not used properly. I once observed a waterfowl mount depicted in a fall scene with five or six butterflies on the dried vegetation. As if this were not bad enough, the butterflies were made of feathers with paper bodies. Needless to say the overall display was a fiasco. As with anything else in wildlife art, some kind of background knowledge or, (you guessed it!), reference is required.

Reference depicting how insects fly, rest, stand, or swim can be quickly collected by visiting the nearest woodlot, field, stream, or pond. Once again, an example using butterflies: butterflies seldom hold their wings straight out from their body like they are displayed in many collections. Butterflies usually hold their wings in a folded vertical position above the body or opened to an angle of about 45 degrees from the level. A flashy butterfly displayed with wings laid flat can be compared to putting excessive frosting on an already sweet cake, however, displaying one with folded wings in a more natural pose could add realistic beauty to a mount without overdoing it. Another obvious fact often overlooked by careless artists is that insects have legs. Too often I see a beetle or fly on a mount with broken legs and the belly glued flat to the piece of wood or mounted object. This type of work looks like a mistake rather than a planned addition to the mount. With proper reference these and other mistakes can be easily avoided. The following article will explain my techniques for finding, preserving, and properly displaying insects to improve most any exhibit.

## Finding & Catching Insects

For most of us, insects are an unpleasant nuisance and we spend more time avoiding them than trying to find them. But, all too often, finding just the "right" specimen for a particular mount can be very difficult. Here are some easy methods for finding the more desirable types of insects.

For large fast-flying insects, such as dragonflies, butterflies, and wasps, search early in the morning when the air is still cool. Insects have to heat up each day so they are often in open areas "sunning" during the first hour of light. They also move slower when it's cool so they are much easier to capture without damaging them. Large beetles can often be found under loose tree bark, in rotting logs or under leaf litter. For nocturnal insects, hang a white bed sheet behind a florescent light at night. Moths and other night-flying insects will land on the sheet where they can be easily captured. A good method for getting cryptically-colored insects out of vegetation is to spread

a white sheet under the vegetation and then beat the branches and leaves with a small stick and watch for insects to drop onto the sheet. No matter how well-camouflaged they are, they will be easy to see against the white background.

## Killing And Care

Once the insects are located, the next step is to collect them and get them back to the shop without damaging the legs, wings, antennae or body. The first necessary piece of equip-

ment is a kill-jar. Insects can be rapidly killed in this jar without damaging them. A kill-jar should be made from a small glass jar with a tight fitting lid, such as a baby food jar. First, mix a small amount of Plaster of Paris and pour it in the jar to a depth of about one-half inch. Allow the plaster to dry and cure and then saturate it with ethyl acetate available from your local druggist, (isopropyl alcohol will also work but not as fast). The ethyl acetate will soak into the plaster and leave a dry surface. Be careful not to put too much liquid in as the surface of the plaster must not be wet. Insects placed in the jar should die in 2 to 3 minutes. The ethyl acetate will evaporate away eventually, so it needs to be renewed about every 2 to 3 weeks. Most insects can be placed directly into the kill-jar and then transferred to another jar when they are dead; however, special care must be taken with large-winged or brittle specimens, such as butterflies or dragonflies. Large-winged insects should first be placed in a small triangular paper envelope with their wings folded flat together and then place the envelope in the kill-jar. This precaution will keep specimens from fluttering around and damaging their wings in the jar. The envelopes need to be large enough to accommodate the whole insect without jamming the head or wings. Envelopes can be easily constructed by simply folding a small square of paper diagonally and then folding the open edges over to close the envelope.

If a kill-jar is not handy, insects can also be killed by putting them in the freezer; however, precautions should be taken to check with your spouse first as some do not like opening their freezer to find frozen insects inside!

## Drying And Attaching

It's best to position specimens in the desired pose soon after they die, while they are still soft and pliable. Insects that have already dried brittle can be rehydrated and positioned properly,

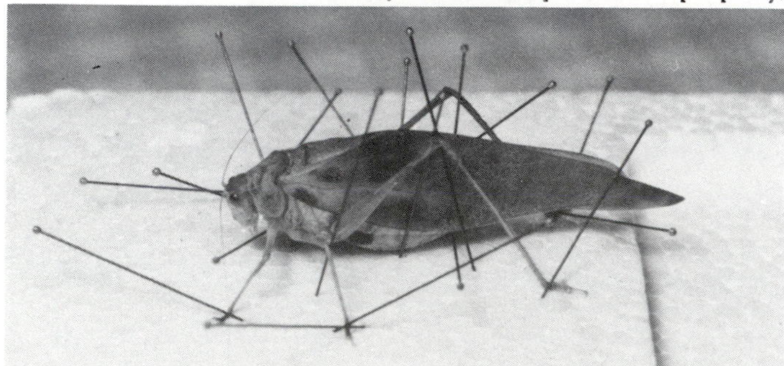

but this is extra work that can be avoided. Use a piece of styrofoam and insect pins to hold the insect in place. Pins should be used as props and need not be inserted into the body or head of the specimen. Insects generally have three pairs of legs, two pairs of wings, and one pair of antennae. Flies are an exception, and have only one pair of wings. Take care to prop each set of appendages in the right position. Also, be careful to support the abdomen as it will sag during the drying process. Set the insect in a box or tray where it won't get bumped and it should be dry in two days to two weeks depending upon the size of the specimen. Some types of insects will eat the dead and drying insect so make sure to place mothballs or napthalene crystals in the drying box. Very soft insects like caterpillars and spiders do not dry well using this method; however, they can easily be done in a freeze dryer if one is available. Injecting them with a 50/50 alcohol–glycerine solution and a very fine tipped syringe also preserves them fairly well.

When the specimen is fully dried it can be attached to the base using Instant Bonding Glue. Dry insects are very brittle so take it slow and easy. To attach an insect in a flying pose insert an insect pin into the body, snip off the pin head, and insert the other end into the supporting vegetation or object. Another method is to glue a very small black wire with Instant Bonding Glue to the inside of the insect's leg and then attach the other end to the base. This gives the impression of free flight since the black wire can only be seen at short distances.

## Rehydrating And Repairing

With a little extra work, insects that have been previously collected and dried or insects that are found dead and dried can both be utilized. Usually insects do not dry in a natural,

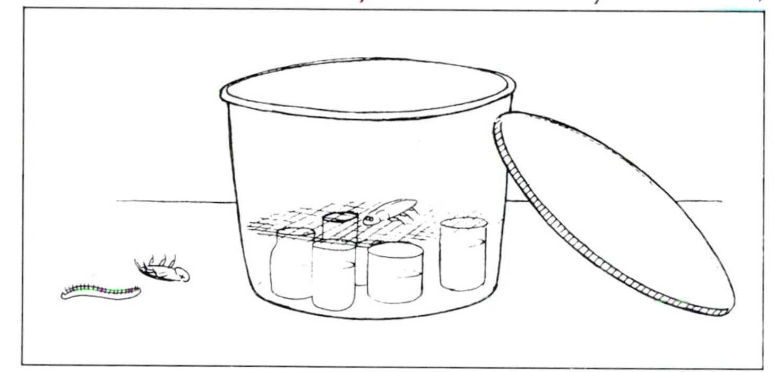

likelife pose so they must be rehydrated, positioned, and allowed to dry again. A rehydrating chamber is easy to make out of common household objects. To make one, obtain a plastic, cylinder-shaped container similar to a gallon ice cream bucket with a tight-fitting lid. Take six or seven small glass jars similar to baby food jars and place them right side up in the bottom of the plastic container. Cut a round piece of galvanized wire screen large enough to fit in the plastic container and lay it flat on top of the jars. Fill the jars with water, place the insect on top of the screen, and snap the lid on. This setup exposes the specimen to a high humidity environment without actually soaking it in water. The specimens should be pliable in 4 to 10 days, depending on size. They can then be repositioned and dried in the proper pose without damage. As an added bonus of using a rehydrator, several insects can be caught while they are abundant, dried, and later rehydrated and positioned for use in specific exhibits.

If legs, antennae, wings, or body parts are broken off during handling, they can easily be reattached with super glue. If a wing is broken or torn, it is usually better to start with a new specimen and throw the old one away. Repairing appendages on insects is tedious work, and requires a lot of patience.

I have had good success using these methods for collecting and preserving insects. Try a preserved insect on your next mount and make it come alive!

# Beetles, Bugs, and Butterflies (How to get'em—How to keep'em)

*by Dan Blair*

Anyone who's ever been faced with having to catch a supply of live fishing bait such as nightcrawlers and crickets knows the advantages of looking for them at night. Therein lies the simplest technique for collecting a *large variety* of insects in a short amount of time.

Particularly during the spring and summer months, just about any porch light, street lamp, or lighted business sign becomes a ready source for night time "collecting." The variety and quantity of specimens is almost endless, so don't limit the collecting to immediate needs only. The demand created from customers that view exhibits with insects incorporated into habitats may become difficult to fill during the cold months of fall and winter. Make it a point to take advantage of their availability during peak seasons.

Sometimes it is necessary to pursue a different sort of insect which rarely comes into evening yard lights. Dragonflies, grasshoppers, and butterflies, to name a few, are rarely seen after dark. But this is still the best time to search them out.

A flashlight is helpful in finding their nightly roosts, but the beam effect is such that it limits the range of your visibility. The better choice, if available, is a propane lantern which lights a broader area and helps your eyes catch movements of insects moving into the shadow and away from the light. (The brighter light may also *attract* other desirable insects.)

Conduct a search in obvious areas in which these specimens most commonly live. Fence rows, hedges, gardens, etc., are likely habitats in which to find grasshoppers. Butterflies can often be found in and around flowering shrubs, lilac bushes, and fruit trees, where they are seen in the daylight hours. Dragonflies are more apt to be found near ponds and streams in the willow thickets and cattails or weedy grasses. And, of course, water beetles are most likely to be found in shallow ponds and pools of quiet water with many small insects and fish life for them to prey upon.

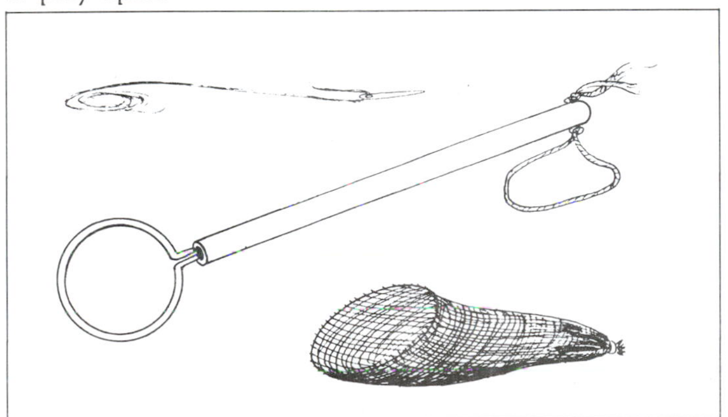

Actually "catching" these creepy critters needn't be difficult at all. If your local dime store or hobby shop doesn't stock a butterfly net, one can easily be made with a nylon net vegetable bag available from the local grocery store (usually free for the asking). Stretch the mesh over a wire loop on the end of a 3 foot wooden handle and secure it with a few wraps of string or even tape. This will provide a serviceable net the like of which I have used to catch crayfish, frogs, clams, and a "variety" of insects. For daytime pursuits, the nets are almost a must. But in the cooler hours of evening the metabloism of many insects slows down, making them as easy to catch as picking them up by hand. For capturing nightime beetles and bugs, I make a simple paper cone with a rolled sheet of paper. I cut the small end off which will allow me to scoop up the unsuspecting prey and funnel him right into an open container with a readily available snap-on lid.

An ideal "carrier" should be unbreakable, such as a plastic margarine or whipped topping bowl with the easy on and off lids. These do tend to be a bit shallow for larger insects like the big butterflies and moths, or for jumping insects like grasshoppers and crickets. It is better to use large plastic jars with the screw-on lids, or larger sized coffee cans with plastic lids for these insects.

Overcrowding the insects, or mixing incompatible species together can cause problems with keeping them in good shape. Try not to keep more than 5 or 6 to a container unless they are of the same species such as grasshoppers and/or crickets. Needless to say, some insects prey upon each other and it's not a good idea to mix them together. (A praying mantis, as an example, can't be put with "any" other insect.) Also, several moths or butterflies in the same container destroy each other's wings by their attempts to fly in a confined space. Therefore, it's a good idea to keep the large winged insects in envelopes or store them in a container such as a metal Band-aid can.

Usually an hour or so on an evening's quest will net enough insects to keep the artist busy for another hour or two the

next day. Once the insect safari has been conducted and a quantity of likely specimens has been collected, steps must be taken to preserve and display the insects naturally.

Unless ready to do the preservation and finish work immediately, the best method of reducing damage caused by their efforts to escape is simply to place their container in a freezer overnight (or utilize the "kill jar"). For some insects that means dying, but for others it will only mean dormancy. So don't be surprised if after thawing one gets up and walks away.

Some insects just don't make good habitat specimens unless they're used in glass cases where they cannot be handled or accumulate dust. Butterflies and large moths are absolutely beautiful in an exhibit until their wings tear and their brittle legs and antenna fall off. Dragonflies are equally exciting additions to a leaping largemouth bass mount but can be ruined if the transparent wings crack and the long slender abdomen disengages from the chest and legs. Be careful with the butterfly and moth family because of the fragile hair-like surface of their wings and body. Most other insects (including wooley caterpillars) can be collected and saved, and lend themselves better to exhibit work.

Observe the collection of beetles, bugs, and butterflies in the light of day, and sort out and separate the good, the bad, and the ugly. (Let's face it! Some bugs have a face not even a mother could love.) Some beetles and bugs are just not noticeably attractive enough to bother including in a habitat. On these, don't waste time; simply discard them. As for the "bad," these would include specimens not considered *prime*, with missing legs, torn wings, etc., and should also be discarded.

Once the collection has been "high-graded," the insects must be preserved.

Most insects used in a habitat will be doing one of two things; walking or flying. Obviously they must be positioned accordingly before being allowed to dry. This positioning can be done immediately after thawing out the insect if frozen. On the other hand, if dealing with a live insect, use a hypodermic syringe with alcohol or lacquer thinner and inject the insect in both the head and vent areas. This will quickly kill it.

At this time, if the insect is to be flying, insert a fine wire into the vent and run it all the way through the body, neck, and out of the forehead. Now pull the wire back slightly into the head until it can not be seen. This step will secure the *entire* length of the insect. Next, I very carefully card the wings with typing paper and staples. Then to finish a flying insect or to prepare a walking one, position the feet and legs on a sheet of cardboard or styrofoam using straight pins. Finally, set them aside in a fairly warm area for a few days to thoroughly dry.

Once the insects are stiff, it's a simple procedure to put a drop of Instant Bonding Glue to the *joints* of each leg, foot, and wing. I recommend a touch of glue also at the junction of the head and chest, and also the chest and tail. The glue helps to secure these points during the final step.

After completing the gluing step, prime the entire body, legs (and wings if applicable), and head of the insect with a clear coat of *Polytranspar*™ Fungicid..l Sealer FP/WA220. This sealer coat helps to "waterproof" the insect for the next step. After the insects have been primed with sealer allow them to dry for at least 1/2 hour.

Using Fin Backing Cream which has been thinned with water. Use a fine, soft artist brush to paint a fine coating to the legs, antenna, and all of the body and head. If the insect is flying or has exposed wings like a dragonfly, give *both* sides of the wings a coat. (This should likewise be applied with a very soft sable-type brush.) Be sure to cover all the body and leg

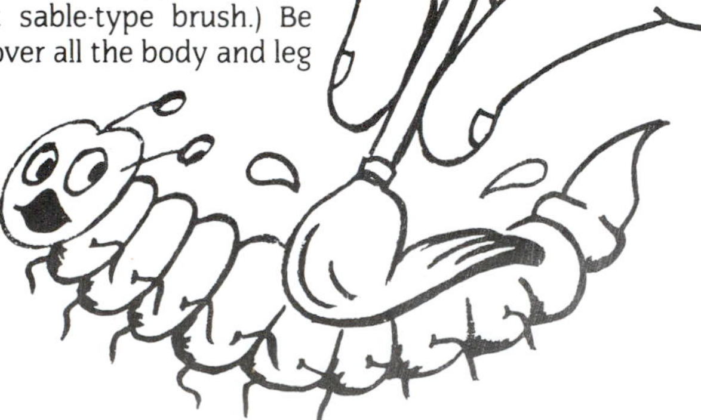

as joints as these are the weakest points on the insect. Once the Fin Backing Cream has dried, inspect for color loss and touch up wings or body with airbrush paints to the desired shades. Always add color sparingly to "tint" rather than "paint" missing colors, particularly on the transparent wings of dragonflies.

Butterflies and moths, being covered with small downy hairs on their wings and body shouldn't be handled any more than necessary, as these hairs make up the delicate colors of these beautiful insects. Fingerprints will become noticeable "bruises," which detract from the beauty and quality of the mount. Extra care is required to keep them in original life-like condition.When handling these type of insects hold the wire inserted through their body rather than touching the delicate insect. Additionally, the ends of the wing cards may be held but be careful not to hold it too close to the wing and crush it.

Large moths such as the Giant Cecropia may be best preserved by freeze drying but can also be dried much the same as butterflies. For the most part, these moths should *not* be used in open displays and gluing is out of the question because of the growth of fuzz covering their entire body, legs, and wings. Butterflies usually have naked legs and bodies which may have super glue and Fin Backing Cream applied to them without much difficulty. But avoid getting any of either product on the wing hairs as it is difficult to remove without damage to the wings and coloration.

After the insects are dry and have been touched up as needed with an airbrush, it is but a simple step to use them in a habitat. Walking insects are simply located where best suited and attached with a dab of Instand Bonding Glue. Flying insects such as the butterfly and moth can be wired to a stem of vegetation by simply gluing the wire into a pin hole. It may be preferable to use a wire which protrudes from the head than one from the abdomen. This could portray the insect being suspended with its tongue in a bud or flower.) Finished

dragonflies can very easily make exciting additions to a leaping bass mount. To attach the dragonfly, first make a pin hole in the roof of the fish's mouth near the natural separation in the top, front row of teeth. Then, insert the insect's tail wire (cut to the proper length) to allow the teeth and tail to make contact. Only a slight drop of glue on the end of the wire before insertion will be needed to hold the insect at the right depth. The wire may be bent somewhat at this point to further position the dragonfly at the proper angles.

The next time a customer or a competition judge compliments you on the outstanding and likelike details of your exhibit, you can honestly tell him, "There's nothing to it really. . .once you get all the bugs worked out."

# Reptiles & Amphibians

## Small Additions for Making Big Impressions!

*by Dan Blair*

All though classed as "lower vertebrates," the addition of any of these creatures to a habitat base may move a competition mount even "higher" in the point standings. But only if proper steps were first taken to ensure the quality of the addition and the probability of its being where you put it.

According to Audubon's Field Guide To North American Reptiles And Amphibians, there is a vast assortment to pick and choose from; 283 native and introduced species of reptiles, and 194 species of amphibians.

To make it a bit more simple to comprehend, there are snakes (115), turtles (49), lizards (115) and alligators, caimen, and crocodiles (one each) making up the group we know as reptiles.

Of the eight families of salamanders found in the northern hemisphere above the equator, 7 families are found in North America north of Mexico. 112 species of them, along with 82 species of frogs and toads, make up the group we know as amphibians (accustomed to life on land or in water).

With a good reference book, such as the one previously mentioned, it is easy to identify both the specie and the geographical location from which it originated. Habitat is usually included so the artist can be consistantly accurate in portraying the specimen in its correct environment. Exceptional color plates aid in painting the skin pigments to renew life-like detail.

Since a majority of the amphibians are nocturnal and "amphibious," their capture is best effected at night near ponds and sloughs. Search for them on roadways in the headlights of your auto, or with flashlights along the edges of waterways. (To be certain such activity is legal, consult your state conservation office before hand.)

Reptiles are a bit more dispersed in their habitat, making their capture more complicated. Whereas the turtles and aligators tend to be more aquatic even than some of the amphibians. The snakes and lizards on the other hand tend to be found more often in the arrid or more drier regions and environments. They can be found just about anywhere, from underground, in burrows, rock piles, wood piles, up trees, and in the thickest weed patches. Some prefer the early morning and evening hours for foraging while still others favor night hunting.

Most often, the specimens taken are an exception to the rule; discovered and taken by accident rather than intentionally. The alternative to premeditated "collecting" is the outright purchase of specimens from pet shops and exotic pet dealers. It might be wise to use disgression here to avoid dealing in *protected* species. It is advisable to consider the source and *know* the reach and ramifications of your state game laws regarding such practices. (Most states require special licenses for such dealers or game breeders.) Be safe!!! Know your source is a licensed professional *before* you begin any transactions.

Freeze dry wholesalers are usually good sources of preserved specimens and many offer a wide selection including frogs, turtles and toads, lizards and snakes (as well as crayfish, panfish, insects, chipmunks, and baby game birds). Consider having your own specimens dried by one of these companies, but compare prices and quality before selecting one. There *is* a difference and quality (particularly the lack of it) may cost you more than money.

Certain select specimens are also available in frozen form from suppliers who generally offer their selection and prices to customers in the classified ads of taxidermy magazines such as *Breakthrough*. Availability is usually on a limited basis and there may be a waiting list in some cases. However, the wait can be worth it if an artist can order exactly what he wants *and get it!*

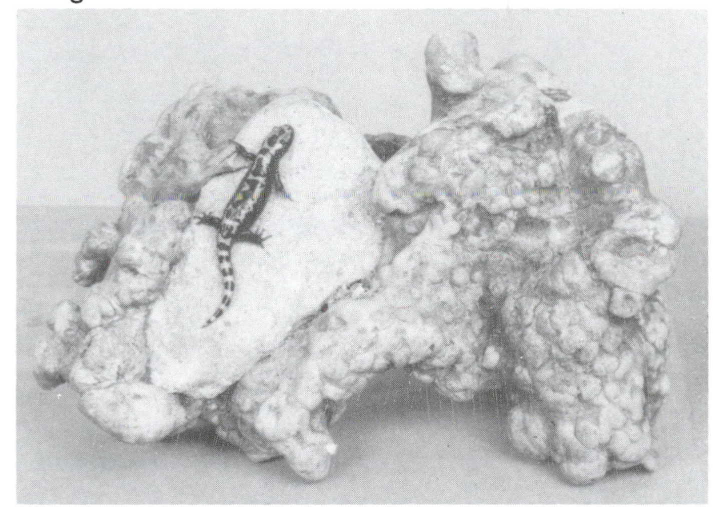

As a final reminder, be certain the selection of specimens is compatible with the habitat they'll be placed in. And just as important, the reptile or amphibian must be compatible (in a sense) with other birds, fish, or animals in the diorama. It stands to reason that an Egret (from the swamp) and a Horned Toad (from the desert) are two "non-compatibles," but that same Egret with an aquatic salamander or frog would be a more accurate portrayal.

As to how to use the finished reptile or amphibian, that remains entirely the decision of the artist. To make a suggestion would be like telling a wildlife painter where to put ducks on a canvas. It is sufficient to suggest that the addition of these often shy and reclusive creatures be kept in perspective; subtle and inconspicuous rather than blatant. Even in a scene where the salamander is the *only* animal life form, the probability is he would still have to be sought out and "discovered." After all, it's the *discovery* that makes an adventure of just about every walk, glance, or conversation we make.

The inclusion of the "little things" in a habitat are a very definite part of an artist's form and style; that extra step that stands him one step above and beyond the competiton as proof that "small additions can make BIG impressions."

# Molding Reptiles & Amphibians

A quick and simple method of preserving a live specimen for mold making is to freeze it. Freezing is also a relatively quick and efficient method of dispatching a specimen too small (such as a tadpole), or too dangerous to handle (such as a rattlesnake).

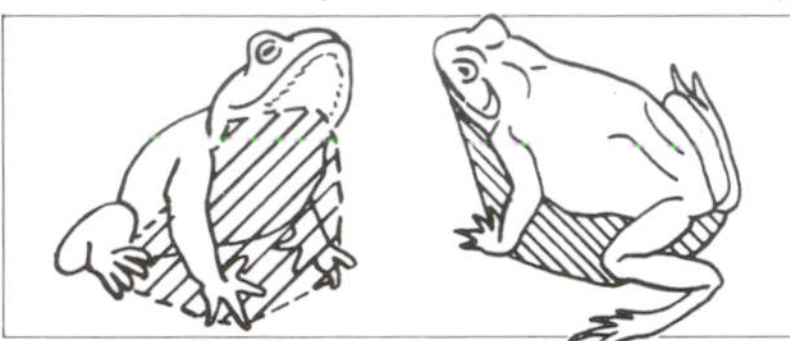

To mold an amphibian such as a frog, toad, or salamander, use modeling clay to build dams half way up the chest, under legs, and to build up below tail or haunches. Use a fine wire or T-pin pushed point first up from beneath the specimen to hold its chin up if need be. Later, after the first half of the mold had set up, the pin can be removed. At this same time, the damming material can be carefully removed, keeping the specimen secured in the first mold as much as possible. Or, if necessary, remove the specimen from the mold and clean it completely and then reinsert it into the mold.

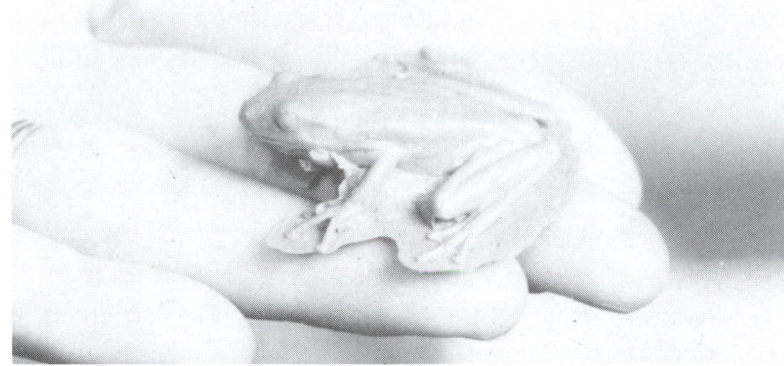

*This frog reproduction was made from a one-piece mold.*

Once the first mold has been cleaned, prepare it for the second pouring by putting vaseline on the outer edge as a separator or mold release, keeping the two halves from bonding together. At this point you are ready to wrap the first side in a carboard dam and pour the second side of your mold.

Smaller specimens may be cast in paper cups or plastic bowls, etc. For the larger forms, you may need to use a cake pan, or even a suitably sized cardboard box to hold in the pouring and to conserve on the amount required to actually encase your specimen.

After both sides of your mold have cured sufficiently to remove the specimen without risk of damage to the details of the mold, take the specimen from the mold. Unless you no longer need it for molding in different poses, rinse it in *cold* water, remold it, or refreeze it until ready for the next casting.

Depending on your efficiency at making molds, and the cure time of your mold making materials, you may be able to use some specimens for 2 or 3 seperate molds, but rarely more. The number of molds to be made can be directly related to how much heat is created by the mold as it cures. (Alginate is the exception to this in that it cures in 2 to 3 minutes without heat and stays cool and wet, even while setting up.)

# Vented Two-Piece Molds

On occasion, particularly when doing a hollow casting with latex of a larger sized specimen such as a lizard or snake, it will be necessary to "vent" the mold. This venting allows the escape of gasses formed by the catalytic actions of the components. Otherwise bubbles become trapped in the high spots and create unwanted voids which are difficult, if not totally impossible to perfectly repair.

If making a two piece mold of a lizard as an example, vent the mold with a small hole (1/4 inch or less) drilled or carved into the plaster at the end opposite of the filling access. This is usually sufficient to bleed off unwanted gasses, but care should be taken to place the vent "above" the normal fill line, and to keep it from becoming plugged.

When actually making a hollow latex pouring into the vented mold, pour enough latex into the mold to fill it just less than half full. When making this initial pour, keep a finger over the vent orifice to retard leakage. Then lay the mold flat and using a cotton swab stick or wire, clear the vent of latex. Allow the mold and latex to stand for about 3 to 4 hours to begin its cure.

At this point plug the vent hole with modeling clay, add another 25 percent latex to the original pouring and turn the mold upside down. Every 2 hours the mold should be reversed, causing the latex to spread evenly back and forth between the two halves, creating the hollow effect.

Latex castings have the distinct advantage of being somewhat flexible and can be positioned fairly easily before the latex cures completely. This is particularly valuable when casting snakes and lizards which can be molded straight and then readjusted to any attitude when taken out of the mold. Fresh out of the mold, a latex snake can be curled and coiled, but once the latex has become entirely cured, it will hold the final shape indefinitely.

# Snails: Quick & Simple

One small additive to an underwater habitat which never fails to catch a viewer's attention is the addition of one or more snails. And because snails come in both land and aquatic varieties, they can be accurately utilized in a number of habitats.

Because their body and antennae are so completely retractable at the slightest stimulation, snails are almost impossible to cast in *any* form. But there is an exceptionally quick and simple solution to recreating these common creatures in three easy steps.

First, a variety of either land or water (or both) snail shells should be found or purchased. The shells that work best are those already dead and "cleaned out." If the only ones found are alive, they can be cleaned by dropping them in boiling water for a few minutes and then removing them from the shell with tweezers. If that seems repulsive, supposedly ants will gladly clean them out for you. However both times I tried this, the ants in my locale showed no interest and the snails just dried up in the sun (and in the shell).

With empty shells at hand, use a hot-melt glue gun to fill the inside of the shell with glue. If the snail is bigger around than a dime, it might be advisable to stuff cotton or paper towel into the shell first and then fill the remaining 1/3 with glue. Do this to all the shells before the next step.

When the first injection of glue has cooled, apply more glue, working with the tip of the glue gun, to create a tail and a head. The tail tapers off like a pencil point, but the head is somewhat bulbous.

It may be necessary to blow on the glue to cool it quickly and to "set" a desireable shape before it droops out of the preferred position. Having ice water or cold tap water on hand to dip the snail in can also be helpful. Again, complete this (head & tail) step on all snail shells before going to the next procedure. Each of the three steps works best if allowed to cool between each one.

The final step is the addition of the antennae to the heads of the snails. This is accomplished by touching the head in the correct spot and pulling slowly away with the glue gun while blowing gently on the string of glue attached to the head. When cool, cut both to the same length. A nodule can be added to their ends by touching the tips lightly with hot glue.

CHAPTER 12

# Lighting and Dioramas

## Display Lighting

*by Jim Hall*

One subject often overlooked by wildlife artists when they wish to display their "wares" is the proper lighting of the display. This is especially true if color is of utmost importance in the display, such as the deep hues of a well-painted fish, or the subtle shimmering effects obtained by many of our leading woodcarvers. Many of the effects artists spend hours attempting to create and perfect can be lost completely if the specimen is placed in an area of improper or insufficient lighting. In many situations, particularly competition events, the artist really doesn't have much choice how his piece of work is illuminated, but this is not always true, and providing additional lighting may be permissible and definitely should be considered.

## Types of Lighting

There are two types of lighting commonly used today; namely *incandescent* and *florescent*, and each has its own characteristics, advantages, and disadvantages.

Incandescent, or tungsten filament, is the type used in common light bulbs, spot lights, etc. It is brighter in the portion of the light spectrum that encompasses the red/orange colors, and thereby yields warmer, richer colors. This type of light is preferred by many artists while they are working on a rendering, and is definitely preferred during the painting operations because it more closely resembles natural or outdoor light. These types of lights can also be easily located and positioned right at a work area by any one of a dozen different desk-type lamps or clamp on spot lamp holders.

Florescent lighting is stronger in more of the blue/green range of the light spectrum, and yields a much softer, pale light. It has the advantage of offering an inexpensive light for large areas, and is the type generally used in commercial buildings. It has the disadvantage of altering the apparant color hues and values of any object that was satisfactorily painted under natural or incandescent light. This is the reason why many art exhibits appear bland and lifeless, even though they appeared very natural and lifelike in the artist's studio, and is the reason why individual lighting should always at least be considered.

## Competitive Events

As we have stated, competition events (which are generally located in commercial buildings containing florescent lighting), are poor places to display color in works of art. At many shows, providing individual spot lighting is either impractical because of the location of electrical outlets, or is even denied by the management. Some competitors and management personnel feel that someone using or trying to use individual lighting is taking unfair advantage, even though the advantage is available to all. Some competitions are so crowded that space availability will dictate whether individual lighting would be permitted. A simple phone call to the show manager will quickly answer these questions.

Lighting a small display at a competition area should be kept *very* conservative. There are many very small lamps available that will provide additional incandescent light, and these are available at any large convenience store. In most instances, only enough light is needed to put a sparkle in an animal's eye, or to bring out that neat red color value on a fish, or to show the iridescence obtained on that special feather so that the judge can *also* see it.

## Commercial Advertising Shows

Since the financial value of the works of artists depends, by and large, on the whims of the public, sooner or later the artist must display his work. This can be done at any public location, such as store front windows, sport shows, private art displays, local businessmen's functions, and the like. Under these conditions, the artist usually has much more freedom concerning individual lighting. At some locations, such as sport shows, the sky is the limit because there is more room to provide lights. We have attended shows where artists utilized up to six or eight 150 watt spot lamps or flood lights to illuminate only one eight by ten foot booth. The displays were literally flooded with light, and the improved effect was dramatic. Many artists use the inexpensive clamp-on reflector flood lights while others go to the bother of making elaborate floor brackets to hold the special lighting. It all depends on what type of item they wish to illuminate, and the best angle from which to provide the light. Always make sure that the lighting does not reflect or shine in the viewer's eyes, and make the source of the light as inconspicuous as possible.

## Contained Exhibit Lighting

Many wildlife artists prefer to display their work in some type of case which may vary in size from a mini-diorama to a walk-in museum exhibit. The lighting problems are the same but a few other problems must be considered.

If the artist wishes to illuminate a small case exhibit, he must consider the heat given off by incandescent lighting. Since the case is small, only small light bulbs should be used, such as 25 watt or smaller, and adequate ventilation must be provided for the heat to escape. Another alternative is to switch to florescent fixtures, which generate much less heat, but try to obtain color-corrected (corrected to approach natural light) lamps for the fixtures. These are often very difficult to obtain in the smaller sizes so check with your local lighting dealer for availability.

Larger displays or dioramas may require the use of florescent lighting to provide enough light at a reasonable cost; however, in the larger sizes of figures, the artist will have a greater selection of color-corrected lamps. If there is a certain item within the diorama that requires a special highlight, incorporate one or more incandescent spotlights in conjunction with the florescent lights.

It will be well worth the time for the wildlife artist to visit a local lighting distributor to find out what the latest thing is in display lighting.

# The Museum Diorama

*by Dan Blair*

In a museum diorama, there are at least six basic factors which must be kept in mind from the very beginning. You can start at the back of the scene being constructed and follow down and around the area as if drawing a circle to picture these factors more clearly in mind. The greatest visual impact most often comes from a well designed and executed "background." It sets the scene for the entire habitat which will be built upon the "base." The foundation of the diorama is this base. Lack of attention to detail here can lose the entire effect and success of the finished presentation. "Foreground" is the next factor to consider. Foreground helps to convey the depth of view within the scene and emphasizes the three dimmensional effect. Most museum diorama's will also require "protection" from spectators, dust, etc. and to adequately provide this security, a barrier of some form must be installed. As we reach the top of our imaginary circle, we find "lighting" one of our very important factors. And the sixth factor is contained within the atmosphere of the enclosed scene; that being "climate." Each of these factors must be considered individually, and yet as a whole because each plays an integral part in the flow and continuity of the overall effect.

**BACKGROUND:** Several methods work well if artisticly applied with the *utmost* attention to detail. Some background scenes may be entirely habitat such as rocks, earth or vegetation and trees. That may be an easy way out, but not for the artist who intends depicting a mountain goat or sheep above the timberline with an emphasis on the far reaching "wide open spaces" of a mountain range. In such a case, the need is for *sky* and lots of it. It will most often be painted directly to a perfectly smooth background by hand. The same features in reverse apply to an underwater scene with fish; details hand painted with artist or airbrush paints, or both.

Budget may limit the number of man hours required to hand paint a large background scene, in which case an alternative may well be the use of photo wall murals. These murals are usually "photo quality" and come in a large variety of outdoor scenes taken from a range of environments; from tropical islands to seashores, streamsides, lakes and even mountains. Prices are reasonable, being from as low as $25 up to a couple of hundred dollars for murals 8 feet high by 12 feet long or more. Most wallpaper and paint stores can show samples of the murals and order those not already in stock. The murals come in easily applied sections which are pasted to the wall as simply as wall papering.

**BASE:** Virtually any combination of materials can constitute this portion of the diorama, depending on the environment and habitat found within it. But one important key to using it correctly is to remember to "blend" the background and the base together. Keep colors in the background scene compatible with vegetation and rocks, etc. in the base and foreground. Ideally, no one should be able to readily see (if at all) where the back and bottom of the scene merge.

**FOREGROUND:** This is the area usually "between" you and the specimen and should be constructed in such a way as to; balance the scene, draw attention to the main subjects, and as mentioned earlier, give a better definition to the three dimmensional effect and depth of view. Foreground should blend into the base as unnoticeably as the background did.

**PROTECTION:** Glass or plexiglass is used to seal off the scene from potential damage from a variety of sources which need not be described. The glass barrier is installed in the same manner as one would install windows and may, depending upon the fragility of the enclosed specimen and habitat, be caulked or sealed as well.

**LIGHTING:** This particular portion of the diorama can present difficult challenges. Lighting must be done inconspicuously. Foot lights or ceiling lights must be projected in such a way to prevent casting unnatural shadows, keeping in mind that in nature, there is but one light source and shadows naturally recede away from it. Sometimes it is necessary to remove natural shadows with lighting and then resuggest them with applications of painted shadows applied with an airbrush. Shine or reflections off the background scenery is also undesirable and may be prevented by using certain types or colors of bulbs. Heat can be damaging to certain mounts and materials, making cooler flourescent lighting a must. Flourescent bulbs can be tinted with transparent *Polytranspar*™ paints to create greens, blues, or any other hue for lighting effects throughout the diorama. Try to paint ingredients using the same lighting as used in the diorama as colors change when moved from incandescent to florescent lighting.

**CLIMATE:** Because temperature or humidity fluctuations can deteriorate many of the ingredients in a diorama, the atmoshere or climate of the environment must be controlled. Humidity is a definite threat as the host of a variety of pending problems. Bacteria, mildew, mold and even mushrooms (the real thing) might appear in a scene in which humidity is allowed to collect unchecked. A dehumidifier system may be needed in conjunction with ductwork vented into and concealed within the habitat to draw unwanted moisture away. Likewise, ductwork may be needed to regulate the temperature variations inside the enclosure.

Larger museums usually assume the life expectancy of a diorama to be about 20 years and design it to be accessible to maintenance and restoration as needed. This accessability is normally through the plexiglass barrier which should be kept removeable by installing its framework with screws rather than nails.

Another precaution is to rope off the viewing area to prevent fingerprinting or scratching of the window materials. Usually a distance of 3 feet is sufficient, but a few signs stategically placed suggesting PLEASE DO NOT TOUCH EXHIBITS might be added assurance that the display will remain untouched and undamaged.

*This beautiful diorama is by the Deaton Museum Studio, Newton, Iowa.*

# CHAPTER 13

# Miscellaneous Tips & Effects

## Imitation Broken Eggs

*(the Tom Ridge method)*

Broken eggs are a real attention getter in a habitat. They can be made exceptionally realistic with a minimum of time and effort.

The larger end of a broken egg shell makes a perfect mold in which to make the yolk. Rub some vaseline with your finger tip into the shell which will act as the mold release. Mix some yellow or yellow orange oil paint with casting resin to the desired egg yolk color and add catalyst. Pour the mix into the egg shell mold. At the point in which it begins to gel, turn the mold upside down and drop the "yolk" in place on your habitat.

The final step is simply adding Cabosil to thicken the resin slightly before mixing hardner into it and pouring the "egg white" over the "yolk." The Cabosil helps to take away a "slight" amount of unwanted transparency.

Before the egg white begins to cure, add bits and pieces of broken shell to it including the two larger halves.

Practice this one on a smooth, vaseline covered base and then take the results home to the kitchen floor. That's the quickest and most effective place to test your success. You'll find out *real quick* just how real it looks and if it will fool the experts.

## Fish Flesh

*(the Tom Ridge method)*

Tom ridge uses a very realistic method of creating a trout or other fish that has a sizeable bite from it exposing raw flesh, rib bones, and egg clusters. Similar techniques can be used to simulate a bite or tear in the flesh of birds and mammals as well.

After completely mounting and finishing the fish on a foam body, Tom cuts out the shape of the cavity he wishes to create in the dried mount. Using Sculpall,™ he rebuilds flesh, intestines, etc. While the Sculpall is fresh, Tom inserts actual rib bones from a comparable sized fish into the appropriate locations and positions and allows the Sculpall to cure around them.

Sculpall is also used to create the shapes of the egg clusters by rolling it between your hands to the right size. While still fresh, the Sculpall is rolled in turnip seeds to completely cover it. After drying the replicas, they are painted to the proper shade of yellow or cream color.

Now comes a unique step which adds ultra realism to the overall effect. The egg clumps are fitted into a latex finger from a surgical glove which, when stretched tightly, duplicates the outer membrane of a fish's egg sack. Small tears can be created at various points with more seeds (eggs) glued to the glove. Add blood vessels with a red ink pen. The clusters are then fitted into the stomach cavity and glued into place.

The only thing missing from this ultra realistic looking mount is flies, but when done accurately, it might even fool them. If not, they can always be added as a final touch.

## Hollow Log Construction

*(the Irving Comeaux method)*

According to Irving Comeaux, the "best" glue he's found for creating this realistic recreation is a 50/50 combination of dextrine glue and *Polytranspar*™ Hide Paste. Its heavy duty consistency, superior holding power, and light weight (once dried) makes it the ideal choice to duplicate the Comeaux hollow log technique.

Begin with three comparable sized 1/2 to 3/4 inch plywood circles the diameter of the log desired. Since rotting hollow logs usually have sunk into the earth, these circles should be flat on the "bottom" side. All around the top edge of one circle, staple 1/8 inch hardware cloth. Then repeat this step on the remaining two plywood circles, keeping the flat bottoms aligned. Over this wire mesh, brush an even layer of the dextrine/Hide Paste mix.

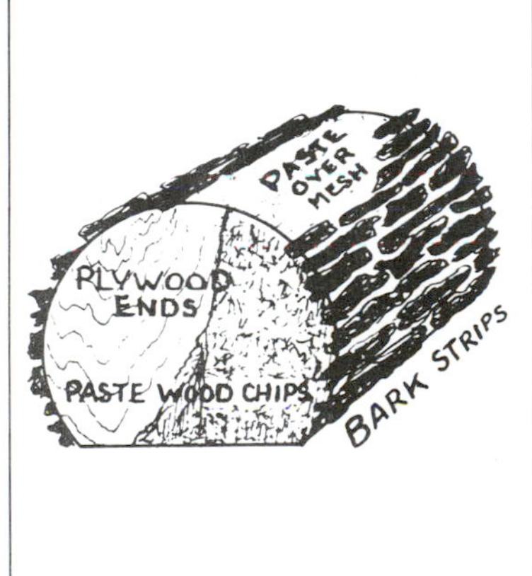


Using narrow strips of coarse tree bark (not wider than 2 or 3 inches) and using the paste mixture brushed onto the bark strips, glue them individually and lengthwise on the framework. With practice, the pieces will fit together like a jigsaw puzzle. Be careful to keep pieces narrow, as wider ones give the impression of the log having flat sides. The roundness of the "log" can be exaggerated by using thinner pieces toward the bottom. Allow the ends of the log to remain jagged, giving the impression the log broke off.

After all the bark strips have been glued in place and while the glue is still wet, sprinkle shredded wood from a dry-rotted tree over the seams between the bark strips. Try to use shredded wood close to the color of the bark. (Rotting wood usually gets darker in color the closer it is taken from the ground. It should also be dried, sifted and fumigated before being used.) If necessary, blend the bark and shredded wood together with airbrush paints.

The same type of dry-rotted wood can be pasted to the plywood ends, or even over sculpture detailed in mache to reproduce the inner tree if it shows.

For large specimens like bears, cougars, etc. mounting boards or blocks can be glued inside the log where needed to securely hold mannikin leg bolts.

## Mellon Slices

Carved from a wedge of foam and primed with *Polytranspar* Fin Backing Cream or shellac, mellon slices are easily imitated and made more realistic by painting to lifelike colors using a variety of irridescent and transparent airbrush paints.

Real melon seeds are glued into slots and onto the mellon after the final painting and the whole slice is wetted down with Wet Look Gloss (WA/FP 240)

Applied to the outside or rind of a muskmellon slice is a thin coat of Sculpall™ textured as required with impressions from a real mellon rind.

# Picture Frame Bases

One simple but effective method of finishing off a square or rectangular habitat base is by simply hot gluing it from the back side (bottom) onto the *face* of an appropriate sized picture frame.

Picture frames can be somewhat expensive if purchased new, but can be found very low priced at flea markets, garage sales, and second hand stores.

For a competition piece, it is better to measure the base and have a frame custom fitted. A large selection of examples are usually on hand in a custom frame shop to test the effect of a style, size and color.

# Fish Eggs—Overly Easy

***by Dan Blair***

The addition of fish eggs to an underwater scene can enhance the realism. Many kinds of fish and insect eggs are formed in gelatin clusters attached to the undersides of rocks, driftwood, plants, or embankments of earth. One way of duplicating these lucid gatherings is by building a pod of *Polytranspar*™ Fin Backing Cream in which Permafrost glass beads (available from WASCO) have been premixed. Shape and smooth the clusters with a soft, damp artist's brush. A similar method with a more transparent appearance is accomplished by pouring the beads *onto* the wet Fin Cream and then tamping them in lightly with a fingertip or sculpting tool.

For larger eggs like those referred to as "salmon eggs" and used for trout fishing, synthetic eggs made from the same material as "rubber" fishing worms can be purchased from most bait & tackle shops in a small variety of colors and one or two sizes. Attach these to the gravel bottom with Fin Cream or hot glue.

Craft and hobby supply stores sometimes carry clear acrylic beads in about four sizes; all appropriate for creating imitation fish eggs. Tinting to match the colors of real fish eggs can be easily accomplished using the transparents and pearls in the *Polytranspar* line of airbrush paints available from Wildlife Artist Supply Company.

# Saliva

Some artists use resin epoxies or casting resins (*Polytranspar* Artificial Water) to make saliva in the mouths of drooling mammals, etc. But there is a less expensive and more easily applied method to use than resin.

For open mouth mounts, the teeth, gums, and tongues can be made to appear slobbery wet by applying coats of Fin Backing Cream selectively to the tips of teeth and tongue, or heavily in the valleys along the gums and the tongue.

The advantages of Fin Cream over the resins are; cost is minimal, no health or fire hazards involved, easy clean up with soapy water instead of hazardous chemicals, considerably less risk of defacing painted surfaces or damaging nylon teeth or jaw sets.

Removal of a mistake in resin is next to impossible without major effort and risk. Once set, it's too late to make changes. Fin Cream, with a slower drying time and being water soluable, is a more reassuring and forgiving product for the amateur (and even professional) artist to work with. Finished results are very comparable.

For a bubbly effect, use a fine needled hypodermic syringe to inject bubbles into the wet Cream. Do this before it begins to "skin over" as the bubbles will leak back out if the hole doesn't heal shut right away.

# Blood Without Sweat & Tears

One simple solution to creating realistic blood is the addition of Winsor & Newton Acrylic Paint or dry powdered tempera paints from to Fin Backing Cream. Variations in the amount of paint or combinations of colors can create the impression of fresh, red blood, or browner coagulated blood. Experimentation is the key word here, much the same as if you were "painting" the addition to a flat canvas rendering.

Be certain to premeditate the application of fake blood before actually applying it, because cleaning up mistakes may leave tell-tale stains in feathers or fur.

An "easy applicator" for laying down the mixed Fin Cream and paint can be made from a plastic squeeze-type catsup or mustard bottle. Just spatulate the mixed "blood" into the bottle, invert, and squeeze out a thin line or puddle any where desired. Touch up the details with a soft haired artist's brush dampened with clear water.

"Freshening" or "aging" the blood even more can be done by carefully spraying the appropriate *Polytranspar* color directly onto the imitation blood. Be careful not to overspray because, again, cleaning up mistakes may leave unwanted stains.

# Duck Weed Painting Tube

Small round pieces of paper punched by a Telex machine or printing company can be used as artificial duckweed in many scenes. The trick is in painting this miniature confetti without making a mess. One easy way is to tape a small section of screen to the ends of a bathroom tissue roll core. If the paper is placed inside the tube, an airbrush can be used to paint the paper directly through the screen (see diagram below).

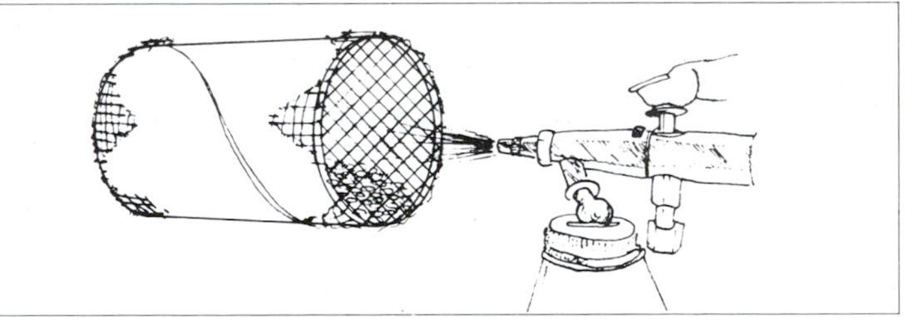

# Shadow Box Dioramas

On occasion a small, simple and yet effective diorama is called for. One of the easiest, least expensive and yet eye catching displays is one I call the "Picture Frame" or "Shadow Box" diorama because it's almost as simple and easy as framing a picture.

Knowing the size of the subject determines the size of the frame which can be as small as 5 × 7 inches, on up. But be careful not to dominate the display area with an overstated and menacing monstrosity that threatens to fall the first time someone slams a door.

Keep in mind that this type of diorama is a wall hanging and depth perception is limited. The "box" won't be much deeper than the specimen is wide. For lizards, butterflies and moths, small birds, etc., the box housing the entire habitat and specimen need only be two or three inches deep. For large fish such as bass or members of the pike family, and for medium sized birds like quail, teal, and partridge, a deeper box is needed; perhaps as deep as 8 inches or more. Going deeper means coming out from the wall farther and possibly creating that "menacing monster" effect. Try to avoid it if possible!

Ideally, especially in a museum display, it is more desireable when depth is required, to build "into" rather than away from the wall.

The box can be built to fit into a frame or, if one is good at cutting accurate joints on a miter box, the frame can be custom made to fit any size box. (Custom framing shops will gladly frame the box for you, but the cost may be a bit more prohibitive.) A large selection of frame moldings offer color, style, and dimension to choose from, making compatibility with the subject easier to achieve.

Lighting can be a problem, especially if the box is a large one. The larger the box, the more light is required to illuminate it satisfactorily. The alternative to actually installing electricity into a large display box is to leave the top portion of the box transparent. Instead of 1/4 inch plywood or chipboard, use plexiglass. Depending on how and where the habitat is installed in the display, it may be advisable to fabricate both ends with plexiglass as well.

If the display is susceptible to damage from dust, etc. such as a snow scene, plexiglass or glass should also be used to cover the front of the diorama.

Hangers on the back side should be securely installed with screws into blocks of wood hidden in the interior of the display. Remember to place the hangers at the right width to engage wall studs which in most homes are on 16 inch centers.

Once the box is built, securely glued together, and dry enough to be safely handled, build the interior habitat the same as you would a pedestal base except that in this case you have only one angle of view. All attention to details should be from that perspective to make your shadow box diorama "picture perfect."

A "background" tip when using the shadow box method for fish mounts: check the local pet shop for underwater habitat photographs sold by the yard (or foot) from bulk spools. They make a good foundation to work up (or out) from. Aquarium plants, rocks, and gravel from the same shop can often be matched to those in the background photo.

Once you have completed the interior of your diorama and installed your mount, attach the box to the picture frame. If the box is to be completely enclosed, the fram should be glassed in first and then added to the box. Clear silicone tile caulk from Dow Corning, the type used for sealing or repairing aquariums, works well for this step. The adhesive isn't as obvious as if hot glue were used.

# Free Mixing Cups

Handy little mixing cups come in all shapes and sizes and free for the asking, if you know who to ask. Ask the wife, kids, folks or whomever to please save all those little "plastic bubble packs" which house so many things we purchase now days (batteries, screws, small toys, cosmetics). These plastic packs can be used for mixing epoxy glues, paints, resins, and just about everthing else in. Some are uniquely shaped and are well suited for molding smaller items like mushroom caps, bird tongues, etc. Cleanup may be as easy as turning the container upside down and "popping out" the casting material. If it doesn't come out cleanly, just throw it away and use a new one. An easy way to store them close at hand is in a plastic bag taped to the end of your work table or bench.

# Shells and Coral

Because mankind (and womankind) has such a fascination for shells and coral, almost every vacation to the seashore brings home a new variety. Eventually, these end up at a rummage sale in the 5 cent grab box. Pet shops offer a larger variety and higher quality shells than any sandy beach you'll ever discover, but the prices are considerably higher. Particularly when used in a saltwater habitat, an artist's arrangement of shells can be very eyecatching. Coral, when used carefully, can become the netire base to which the finished fish mount or carving is attached. For extra support, drill a hole into the stem of coral and into the back of the fish in which to epoxy or hot glue a wire or screw shank with the head removed. Epoxy the coral and fish together and while the resin is still wet, sprinkle it with crushed coral (or salt).

*Black-tipped shark miniature carving by Dan Blair on natural coral base.*

# Droppings

It's a simple matter to use white acylic paints to "white wash" a streamside rock or fence post upon which to stand a heron or a hawk, but bring a buffalo into the scene and the "matter" can get out of hand real fast.

Like any other part of a competition quality rendering, one shouldn't just let droppings fall where they may. Remember that members of the canine family have the habit of leaving droppings upon rocks or logs. Felines of all sizes, including cougars, tend to cover or "scrape" debri over thiers.

Imitations can be formed with Fish Fill, Sculpall, mache, etc. and made authentic looking by adding bits of the food particles the animal would have ingested; bone fragments, animal hair, feathers, etc. in the meat eater's stool; and grasses, twigs, etc. in that of the herbivores like the buffalo.

Perfect (?) reproductions can be made of the well known "cow pie" by allowing urethane foam to catylize from it's container up and expand through a hole in a cardboard or plywood lid. The cured foam can be sprayed with adhesive and then sprinkled with dried grass, sand or soil. Final detailing is done with an airbrush and locked in place with a clear gloss or matte finish, depending on the age of the dropping being imitated.

# CHAPTER 14

# Financial Considerations

*Habitats can easily generate substantial additional income for the wildlife artist. In this chapter, Bob Williamson explains how to calculate direct costs, indirect costs, and profit.*

Habitat and exhibit building can be a very lucrative sideline. In fact several artists are now using these as their "primary earning" source. Galleries, gift shops, tourist shops, sportsmen shops, your studio showroom, and the like are all prime candidates for potential sales.

Be sure and abide by all game and fish regulations before embarking on this cause. Migratory species are governed by federal regulations while others are largely controlled by state regulations which tend to vary widely. "Before selling anything" it is highly advisable to check with federal, state, and local officials to verify the laws in your area. Game farm species provide a good source and in most cases meet legal requirements for resale.

In order to sell exhibits, customers must "see" them. Photos and brochures are one method of allowing customers to see your work, however, there is no substitute for actually viewing an exhibit in person in order to make a "sale" (the name of the game). A good showroom in your studio is a necessity. Placing exhibits in galleries, gift shops, sporting goods stores, etc., is another way to allow the customer a first hand look. Remember this is 3 dimensional art and 2 dimensional photos just cannot always do these renderings justice.

It's a fact that many customers are "impulsive" and often buy luxury items on a whim. Be prepared to "close" the deal if the opportunity presents itself. Don't be an "order taker", be a "salesman." "Sell your work!" As with just about any merchandising selection it is advisable to have at least three choices of different priced items—Inexpensive, medium, and expensive. Customers of all three categories will view your work and buy according to their preference and income, so why not accommodate all of them. Often artists make a mistake in just producing work of the very expensive category. More money could be made by offering a wider selection. Just make sure that a minimum "solid" margin of profit exists on every item that you sell whether it is priced as an inexpensive item or expensive.

In order to make a profit, the "cost" of the project must first be determined. No amount of adjectives can describe how important it is to accurately "cost" each project that is intended to be sold. After all, how else can an item be priced if the cost isn't known?

To determine cost, first figure direct costs, then indirect costs, and then add in the desired profit margin.

Calculate these as follows:

**DIRECT COSTS:**

**A:** Total Supplies & Materials Used For The Project

**B:** Total Labor (Number of Hours × Hourly Wages Spent Building The Project)

**INDIRECT COSTS:**

The artist must then calculate the overhead per the number of Hours Expended On Project. Any knowledgeable business owner would realize that an indirect or "overhead" figure would need to be calculated and added to the direct costs. Calculate the total overhead, such as utilities, rent, equipment, auto, commissions, etc. (see The *Breakthrough* Business Management Manual for detailed method of calculating overhead), and divide it by the number of working hours in the month. For example, if there are 20 working days at 10 hours per day in a month, you would have 200 working hours in the month. If your overhead is $1000 per month, then divide $1000 by 200 hours, 200 ÷ $1000 = $5.00, or $5.00 per hour shop time. Now multiply this times the total number of hours required to do the project, i.e., 2.25 hrs. X $5.00 = $11.25. This then would be your indirect costs for the project. Now, add this to your direct costs:

| | |
|---|---|
| Total Supplies | $25.00 |
| Total Labor | $22.50 |
| **Total Direct Costs** | $47.50 |
| **Indirect Costs** | $11.25 |
| **Total** | **$58.75** |

## Profit

Now, any business must add in a profit margin (if they are to stay in business). The artist should first calculate the total costs of the project then add in the desired profit percentage.

**HOW TO CALCULATE PROFIT PERCENTAGE ("MARK-UP"):**

As has been pointed out, as a business owner you are entitled to make a "mark-up" (profit) on your work. Unless you know *how* to calculate this, you could well be in for a very rude awakening! Percentages can play tricks on all of us!

Most people would assume that by adding 30% to their costs that they would end up *getting* a 30% profit. This is not true! To explain this apparent contradiction, we'll use $10.00 as a supply cost figure. By taking 130% (i.e. 1.30 × $10.00) you would *assume* that you are making 30% profit on this sale. Now look at example A to see what really happens.

| | | | | |
|---|---|---|---|---|
| Example A: | Selling price 1.3 × $10.00 = | $13.00 | = | 100.0% |
| | Cost | $10.00 | = | 76.9% |
| | Profit | $ 3.00 | = | 23.1% |

Had you marked the cost up 42.85% ($10.00 × 1.4285) the results would have been per Example B:

| Example B: | Selling price | $14.29 | = | 100.0% |
|---|---|---|---|---|
| | Cost | $10.00 | = | 70.0% |
| | Profit | $ 4.29 | = | 30.0% |

Such a miscalculation could prove disastrous if for example you wanted to give some sort of discount, say 30%. Please note what would happen in example C:

| Example C: | Selling price | $13.00 | = | 100.0% |
|---|---|---|---|---|
| | Cost | $10.00 | = | 76.9% |
| | Profit | $ 3.00 | = | 23.1% |
| | Discount (30% = 130.00) | ($ 3.90) | | (30.0%) |
| | Loss | ($ .90) | | ( 6.9%) |

Instead you should have used calculations per example D.

| Example D: | Selling Price | $14.29 | = | 100.0% |
|---|---|---|---|---|
| | Cost | $10.00 | = | 70.% |
| | Profit | $ 4.29 | = | 30.0% |
| | Discount (30% = 14.29) | ($ 4.29) | | (30.0%) |
| | (Profit/loss) | $ 0.00 | | 0.0% |

It should be obvious from these examples that making the apparently *"logical"* mathematical computations may result in a totally *illogical* and *incorrect* answer! Here are some examples of mark-ups to cost to achieve the desired profit margins.

| Desired Profit Margin: | Multiply by Factor of: | = | $ (Based on $10.00) |
|---|---|---|---|
| 20% | 1.2500 | | $12.50 |
| 25% | 1.3333 | | $13.33 |
| 30% | 1.4285 | | $14.29 |
| 35% | 1.5385 | | $15.39 |
| 40% | 1.6667 | | $16.67 |
| 45% | 1.8182 | | $18.18 |
| 50% | 2.0000 | | $20.00 |

For factors not listed: subtract the desired profit percentage from 100. Then divide the remainder into 100 to get the factor, i.e.,

If the desired profit margin is 16%
100 – 16 = 84 100 ÷ 84 = 1.1904 Factor: 1.1904

If the desired profit margin is 60%
100 – 60 = 40 100 ÷ 40 = 2.5000 Factor 2.5000

A 40 percent profit added to your total direct and indirect costs would bring what you "should" be charging to $97.94. Total Costs $58.75 X 1.667 (profit factor for 40 percent) = $97.94.

For this example, a taxidermist should charge $97.94 for this project in order to make a profit of 40%. If customers in your area would not pay that much, then you have a decision to make. First, we would assume that you are a good business owner and have carefully analyzed your costs and have cut them and they are rock bottom, (and without lowering quality—something you never should do). I also assume that you have analyzed your operation and that it is 100 percent efficient and your overhead is likewise rock bottom. If so, then you are: (1) faced with lowering your price and settling for less profit (30 percent is bare minimum for survival), (2) convince your customers that your mounts are worth that much, or (3) remove that item from your price list.

If you will take the time to set your prices as outlined you will assure yourself a fair profit (as other businesses make) and you will have a better chance of survival. Know your exact costs, price right, advertise heavily, keep your quality of work high, give good service, work hard, and success should be yours. *Offer "quality" products, give good service at a fair price,* and you can't lose. Also, it doesn't hurt to be friendly and courteous to your customers.

*Bob Elzner of Apache Junction, Arizona, won Best in World Mammal at the 1986 World Taxidermy Championships with his rendering entitled "Hauling Buns."*

# ACKNOWLEDGEMENTS

The editors gratefully acknowledge the valuable assistance of the following people for their contributions to the success of this manual.

Curtis Badger, Salisbury, Maryland
Larry Barth, Stahlstown, Pennsylvania
Mark Belk, Provo, Utah
Bob Berry, El Cajon, California
Tan Brunet, Galliano, Louisiana
Wendy Christensen-Senk, Hales Corners, Wisconsin
Irving Comeaux, Morgan City, Louisiana
Jeff Compton, Crosby, Minnesota
Sallie Dahmes, Loganville, Georgia
Bob Elzner, Apache Junction, Arizona
Mozelle Funderburk, Stone Mountain, Georgia
Alan Gaston, Eagan, Minnesota
Vic Heincher, St. Louis, Missouri
John Lager, Littleton, Colorado
Rick Laurienti, Denver, Colorado
Ralph Lehrman, Hoagland, Indiana
Tom Lenort, Clarks Grove, Minnesota
Clarice Mechling, Idaho Falls, Idaho
Randy Nelson, St. James, Minnesota
Frank Newmyer, Union Lake, Michigan
Tom Ridge, Blaine, Tennessee
Rich Rossiter, St. James, Minnesota
John Scheeler, Mays Landing, New Jersey
Kelly Seibels, Huntsville, Alabama
Garry Senk, Hales Corners, Wisconsin
Tom Sexton, Monroe, Georgia
Roger Smith, Leslie, Michigan
Scott Souders, Columbus, Nebraska
Jim Stanley, Tucker, Georgia
Ed Thompson, Lilburn, Georgia

(Our apologies to anyone inadvertently omitted).